POULTRY PRODUCTION IN HOT CLIMATES

Second Edition

POULTRY PRODUCTION IN HOT CLIMATES

Second Edition

Edited by

Nuhad J. Daghir

*Faculty of Agricultural and Food Sciences,
American University of Beirut, Lebanon*

CABI is a trading name of CAB International

CABI Head Office	CABI North American Office
Nosworthy Way	875 Massachusetts Avenue
Wallingford	7th Floor
Oxfordshire OX10 8DE	Cambridge, MA 02139
UK	USA
Tel: +44 (0)1491 832111	Tel: +1 617 395 4056
Fax: +44 (0)1491 833508	Fax: +1 617 354 6875
E-mail: cabi@cabi.org	E-mail: cabi-nao@cabi.org
Website: www.cabi.org	

© CAB International 2008. All rights reserved. No part of this publication may be reproduced in any form or by any means, electronically, mechanically, by photocopying, recording or otherwise, without the prior permission of the copyright owners.

A catalogue record for this book is available from the British Library, London, UK.

Library of Congress Cataloging-in-Publication Data

Poultry production in hot climates / edited by N.J. Daghir. -- 2nd ed.
     p. cm.
  Includes bibliographical references and index.
  ISBN 978-1-84593-258-9 (alk. paper)
 1. Poultry. 2. Poultry industry. 3. Poultry--Climatic factors. I. Daghir, N. J. (Nuhad Joseph), 1935-

  SF487.P8645 2008
  636.5--dc22
                                                                                  2007042039

ISBN-13: 978 1 84593 258 9

Typeset by AMA DataSet Ltd, UK.
Printed and bound in the UK by Cromwell Press, Trowbridge.

The paper used for the text pages in this book is FSC certified. The FSC (Forest Stewardship Council) is an international network to promote responsible management of the world's forests.

Contents

Contributors		vii
Preface to the first edition		x
Preface to the second edition		xii
1.	Present Status and Future of the Poultry Industry in Hot Regions N.J. Daghir	1
2.	Breeding for Resistance to Heat Stress R.S. Gowe and R.W. Fairfull	13
3.	Breeding Fast-growing, High-yield Broilers for Hot Conditions A. Cahaner	30
4.	Behavioural, Physiological, Neuroendocrine and Molecular Responses to Heat Stress R.J. Etches, T.M. John and A.M. Verrinder Gibbins	48
5.	Poultry Housing for Hot Climates M. Czarick III and B.D. Fairchild	80
6.	Nutrient Requirements of Poultry at High Temperatures N.J. Daghir	132
7.	Feedstuffs Used in Hot Regions N.J. Daghir	160

Contents

8. **Mycotoxins in Poultry Feeds** — 197
 N.J. Daghir

9. **Broiler Feeding and Management in Hot Climates** — 227
 N.J. Daghir

10. **Replacement Pullet and Layer Feeding and Management in Hot Climates** — 261
 N.J. Daghir

11. **Breeder and Hatchery Management in Hot Climates** — 294
 N.J. Daghir and R. Jones

12. **Waterfowl Production in Hot Climates** — 330
 J.F. Huang, Y.H. Hu and J.C. Hsu

Index — 377

Contributors

About the Editor

Dr N.J. Daghir was born and raised in Lebanon, where he received his primary and secondary education at the International College in Beirut. In 1954, he lived and worked on poultry farms in the states of Indiana and Arkansas as the first Lebanese participant in the International Farm Youth Exchange Programme, sponsored by the Ford Foundation. He received his BSc from the American University of Beirut (AUB) in 1957 and was immediately appointed by AUB to provide agricultural extension services to the central and northern Beqa'a region in Lebanon, to where he introduced commercial poultry production. In 1958, he left for the USA for graduate work, where he earned both his MSc and PhD degrees from the Iowa State University in 1959 and 1962, respectively. In 1962, he helped establish a Lebanese branch of the World Poultry Science Association and became president of that branch until 1984. During the same year, he started his teaching and research career at the AUB as Assistant Professor of Poultry Science and Nutrition. In 1967, he was promoted to associate professor and in 1975 to full professor. His teaching covered regular undergraduate and graduate courses to students from all over the Middle East area. He has served as adviser for over 40 MSc graduate students, many of whom have later received PhD degrees from US universities and are now occupying key positions all over the world.

Dr Daghir spent two sabbatical years in the USA, one at the University of California, Davis, in 1969, as visiting associate professor and one at the Iowa State University in 1979, as visiting professor. He is a member of several professional and honorary organizations, such as the American Institute of Nutrition, the American Poultry Science Association and the World Poultry Science Association. He has travelled widely in over 60 different countries of Asia, Europe, Africa and America. He has served as a consultant to poultry companies in Lebanon, Jordan, Syria, Iraq, Iran, Egypt, Kuwait, Tunisia, Saudi Arabia and Yemen, and participated in lecture tours on poultry production in these countries, sponsored by organizations such as ASA, USFGC, WPSA, etc. Dr Daghir has also served on special assignments for the Food and Agriculture Organization of the United Nations (FAO) and Aramco, and participated in preparing feasibility and pre-tender studies for poultry projects in several Middle East countries. Dr Daghir has had over 100 articles published in scientific journals and the proceedings of international meetings as well as several chapters in books and compendia. His research has covered a wide range of subjects, such as factors affecting vitamin requirements of poultry, utilization of agricultural by-products in poultry feeds, nutrient requirements of poultry at high-temperature conditions, seeds of desert plants as potential sources of feed and food, single-cell protein for poultry, and plant protein supplements of importance to hot regions. His research has received funding from the US National Institutes of Health, the International Development Research Centre and LNCSR, as well as from AUB. He has served in many administrative positions at the university, such as Chairman of the Animal Science Department, Associate Dean and Acting Dean of his Faculty and, for 2 years (1984–1986), served as team leader of the American University of Beirut technical mission to Saudi Arabia. From September 1986 to June 1992, he served as Director of Technical Services at the Shaver Poultry Company in Cambridge, Ontario, Canada, and nutrition consultant to ISA breeders in North America. He served as Dean of the Faculty of Agricultural Sciences and Professor of Poultry Science at the United Arab Emirates University, Al-Ain, UAE, from 1992 to 1996, and as Dean of the Faculty of Agricultural and Food Sciences at the American University of Beirut, from 1996 to 2006. He is at present Professor and Dean Emeritus at the same University.

Other Contributors

A. Cahaner, *Faculty of Agriculture, The Hebrew University, Rehovot, Israel.*
M. Czarick III, *The University of Georgia, Athens, Georgia, 30622, USA.*
R.J. Etches, *Origen Therapeutics, 1450 Rollins Road, Burlingame, California, 94010, USA.*
B.D. Fairchild, *The University of Georgia, Athens, Georgia, 30622, USA.*
R.W. Fairfull, *Centre for Food and Animal Research, Research Branch, Agriculture Canada, Ottawa, Ontario, Canada K1A 0C6.*
R.S. Gowe, *Centre for Genetic Improvement of Livestock, Department of Animal and Poultry Science, University of Guelph, Guelph, Ontario, Canada N1G 2W1.*
J.C. Hsu, *Department of Animal Science, National Chung Hsing University, Taiwan.*
Y.H. Hu, *Department of Livestock Management, Ilan Branch, Livestock Research Institute, Council of Agriculture, Executive Yuan, Taiwan.*
J.F. Huang, *Department of Livestock Management, Ilan Branch, Livestock Research Institute, Council of Agriculture, Executive Yuan, Taiwan.*

T.M. John, *Department of Zoology, University of Guelph, Guelph, Ontario, Canada N1G 2W1.*

R. Jones, *9 Alison Avenue, Cambridge, Ontario, Canada.*

A.M. Verrinder Gibbins, *Department of Animal and Poultry Science, University of Guelph, Guelph, Ontario, Canada N1G 2W1.*

Preface to the First Edition

This book was first envisioned in the early 1970s when the editor was teaching at the American University of Beirut, Lebanon. It was felt at the time that a reference was needed in the teaching of a senior-level course on Poultry Production in Hot Climates. When the literature was screened at that time it was found that the references on the subject did not exceed 100 and covered very few areas. Today, more than 20 years later, there has been extensive work in the areas of breeding for resistance to heat stress, heat-stress physiology, housing for improved performance in hot climates, nutrient requirements at high temperatures, feedstuff composition and nutritional value, and management of broilers, layers and breeders in hot climates, and a book covering information on these various subjects that includes over a thousand references became possible.

Chapter 1 gives an overview of the poultry industry in the warm regions of the world (Africa, Latin America, Middle East, South Asia and South-east Asia) and covers some of the constraints to future development of the industry in those regions. Chapter 2 covers research on breeding for heat resistance and concludes that it is feasible to select for resistance to heat stress, but the challenge of the breeder is to introduce heat-stress tolerance while retaining and improving the economic traits needed in commercial chickens. Chapter 3 discusses several aspects of heat-stress physiology and the behavioural, physiological, neuroendocrine and molecular responses to heat stress. It also includes a section on heat-shock proteins. Chapter 4 addresses several issues in housing for improved performance in hot climates. It deals with the principles related to housing design as well as factors affecting poultry-house design for hot climates. Poultry-house maintenance and monitoring of house performance are also covered. Chapter 5 highlights some of the findings on nutrient requirements of chickens and turkeys at high environmental temperatures and emphasizes the fact that nutritional manipulations can reduce the detrimental effects of high temperatures but cannot fully correct them, for only part of the impairment in performance is due to poor nutrition. Because of the shortage and high cost of feed ingredients in many hot regions of the world, Chapter 6 was included to focus on the importance of research on the composition and nutritional value of available ingredients that can be used and those that are being used for poultry feeding in hot regions. It is hoped that data presented in this chapter will help countries in these regions to take full advantage of the knowledge available on these feedstuffs. Since mycotoxins in poultry feeds are a serious problem in hot climates where there are prolonged periods of high temperature accompanied at times by high humidity, Chapter 7 has been included to highlight this subject. This chapter focuses mainly on aflatoxins since these are the most widely spread in those regions. Mycotoxins such as citrinin, fumonisins, ochratoxins, oosporein, T-2 toxins, vomitoxin and zearalenone are also covered. Chapters 8 and 9 deal with selected aspects of feeding and management of broilers, replacement pullets and layers in hot climates. The reason that these two chapters present a combination of nutrition and management strategies is because the author believes that this is the most adequate approach to overcome problems of heat stress in poultry meat and egg stocks. Finally, Chapter 10 describes some of the important features of breeder management in hot climates as well as hatchery management and operation problems in such a climate.

This book is aimed at advanced undergraduate and graduate students studying animal production and specifically interested in production problems of hot and tropical countries. It will also serve as a reference for scientists working in the poultry industry in those regions. The book presents not only a review of published research on this subject during the past two decades, but also specific recommendations for poultry producers and farm supervisors.

It was not the intention of the authors to include in this book standard material that is usually present in a poultry production textbook. Only those areas that have actually been investigated in hot climates or studied at high temperatures have been covered.

The reference lists at the end of each chapter are not exhaustive, particularly in the case of certain chapters. However, they have been chosen to provide the reader with a suitable entry to the literature covering research reports on each facet of poultry science research in hot climates.

The authors of this book are indebted to research workers throughout the world who have conducted the research reported in this book. The editor is indebted to Dr Michel Picard for his review and criticism of Chapter 5, to Dr J.L. Sell for his valuable review of Chapter 6, to Dr M.L. Lukic for his review of Chapter 7, and to Dr J.H. Douglas for his review of Chapter 10. The editor would like to express his deepest thanks for the help, assistance and understanding of members of his family all through the preparation of this book.

N.J. Daghir

Preface to the Second Edition

The first edition of this book was published in 1995. Since that date a great deal has been written on various sectors of the poultry industry in hot climates. It was felt, therefore, that a new revised edition should be prepared to include as much of the new published information as possible. The new edition includes 12 chapters, three of which are new and were not included in the first edition. Chapter 3 is on breeding fast-growing, high-yield broilers for hot climates, which covers recent work on broiler breeding and serves as an update of Chapter 2 from the first edition. Chapter 5 is on housing for hot climates, which covers new developments in both naturally ventilated as well as power-ventilated houses. Chapter 12, which is also new, is on 'Waterfowl Production in Hot Climates'. It was added because it was felt that this was one of the deficiencies of the first edition.

Chapter 1, which covers an overview of the poultry industry in the warm regions, has been completely revised with new tables of production and consumption figures. Chapters 2 and 4 have not been changed or revised in any way since they are very basic and cover valuable information that should be part of this edition. Chapters 6, 7, 8, 9, 10 and 11 have been updated.

As mentioned in the preface to the first edition, this book is aimed at advanced undergraduate and graduate students in the animal and poultry science field and those who are specifically interested in problems of production in hot and tropical climates. It should serve as a reference for scientists working in those regions. Although most chapters cover reviews of published work on the subject, they also include specific recommendations for producers and industry people in general. FAO classifies poultry enterprises all over the world into four production systems. These are: (i) industrial; (ii) large-scale commercial; (iii) small-scale commercial; and (iv) backyard production. This book is addressing mainly the first two systems and to a limited extent the third. It does not provide any information on the fourth system, which we realize is important and comprises a large percent of the poultry raised in many developing countries besides being a rich pool of poultry genetic resources. Several articles and books have been published by FAO and others on the backyard system, which the interested reader can refer to.

We have tried to include at the end of each chapter as many references as possible so that the readers can later pursue their specific research interest. This second edition has over 300 new references in addition to the 950 that were included in the first edition. We continue to be indebted to research workers throughout the world who have worked on the research reported in this book. Besides those recognized in the preface to the first edition of this book, the editor would like to thank Dr M. Sidahmed for reviewing and correcting Chapter 6, Mr Musa Freiji for reviewing Chapters 1 and 6, and Dr M. Farran for reviewing Chapter 12. The editor would also like to express his thanks to Ms Haifa Hamzeh and Mrs Yamamah Abdel-Walli for their help in typing the manuscript and last, but not least, to his wife Mona for her understanding and patience during the preparation of this book.

N.J. Daghir

1 Present Status and Future of the Poultry Industry in Hot Regions

N.J. DAGHIR

Faculty of Agricultural and Food Sciences, American University of Beirut, Lebanon

Introduction	1
Present status	2
Africa	6
Latin America	7
Asia	7
Middle East	7
South Asia (India, Pakistan, Bangladesh, Nepal, Sri Lanka and Bhutan)	8
South-east Asia (Brunei, Cambodia, Hong Kong, Indonesia, Laos, Malaysia, Myanmar, Philippines, Singapore and Thailand)	9
Future development	9
References	11

Introduction

Domestication of poultry is said to have started in Asia and there is evidence of domesticated chickens in China that goes back to 3000 BC. It is believed that today's breeds originated in India, since the earliest record of poultry dates back to about 3200 BC in that country. Green-Armytage (2003) reports, however, that archaeological evidence of chickens in India dates back only to 2000 BC. Ketelaars and Saxena (1992) indicate that the first domestication of the fowl took place in China and not in South-east Asia. Chickens appear in writings and artwork of ancient Greece. They may have been brought to Greece from Persia, and Persian soldiers brought them from India (Green-Armytage, 2003).

Chickens have been bred in captivity in Egypt since about 1400 BC. The red jungle fowl, an Asian breed, is assumed to be the ancestor of our modern poultry breeds (West and Zhou, 1989). The warm regions of the world were the areas from which all modern breeds of chicken have evolved. Poultry were kept by farmers in China, India and East Asia long before they were known to the Europeans and Americans (van Wulfeten Palthe, 1992). Poultry as a business, however, was not known before the twentieth century. It was not until R.T. Maitland wrote his *Manual and Standards Book for the Poultry Amateur*, in which he describes the husbandry, care and breeding of poultry with a short description of all poultry strains present at the time.

The World's Poultry Congresses have helped, since their inception in 1921, to spread knowledge about poultry production. The International Association of Poultry Instructors and Investigators was founded in 1912. This organization was later transformed to become the World's Poultry Science Association, and the first congress was held in The Hague in 1921. Out of 22 congresses held between 1921 and 2004, only five were held in the warm regions of the world (Barcelona, 1924; Madrid, 1970; Rio de Janeiro, 1978; India, 1996; and Turkey, 2004). This may be one of the reasons why knowledge of modern poultry husbandry has not reached the warm regions of the world as rapidly as it has the temperate areas.

Present Status

The poultry industry has in recent years occupied a leading role among agricultural industries in many parts of the world. Table 1.1 shows the remarkable growth in poultry meat and eggs during the last 35 years. These products have increased much faster than beef and veal or pig meat. Since the publication of the first edition of this book in 1995, poultry meat has surpassed beef and veal production, and 2005 poultry meat production was 20.6 t higher. Hen egg production has tripled during the past 35 years, reaching nearly the volume of that of beef and veal.

Table 1.1. Development of global meat and hen egg production between 1970 and 2005, data in 1000 t (Windhorst, 2006).

Year	Beef and veal	Pig meat	Poultry meat	Hen eggs
1970	38,349	35,799	15,101	19,538
1975	43,724	41,674	18,684	22,232
1980	45,551	52,683	25,965	26,215
1985	49,285	59,973	31,206	30,764
1990	53,363	69,873	41,041	35,232
1995	54,207	80,091	54,771	42,857
2000	56,951	90,095	69,191	51,690
2005	60,437	102,523	81,014	59,233
Increase (%)	57.6	186.4	436.5	203.2

The potential for further growth is obvious in view of the value of eggs and poultry meat as basic protective foods in the human diet. Table 1.2 shows the world production of eggs and chicken meat and the increase in the production of these products in various continents. Africa, Asia and South America show the greatest increases in egg production, with decreases in both Europe and Oceania. Chicken meat production continued to increase in all continents, with the highest increase in Asia and South America. The worldwide rate of increase in egg production has averaged 5.3% per year, while for chicken meat it was slightly higher at 5.7% per year.

Tables 1.3 and 1.4 present production figures in the developing versus the developed regions of the world. During the past 35 years, the production of eggs continued to increase rapidly in the developing regions, which include most of the hot regions of the world. On the other hand, egg production came practically to a standstill in the developed regions. As for poultry

Table 1.2. World production of eggs and chicken meat in 1000 t (FAOSTAT, 2005a).

Continent	Egg			Chicken meat		
	1990	2005	% Change	1990	2005	% Change
Africa	1,420	2,230	+ 57	1,790	3,189	+ 78
N. & C. America	5,790	8,052	+ 39	12,830	22,653	+ 77
S. America	2,310	3,518	+ 52	3,850	13,697	+ 256
Asia	14,270	40,055	+ 181	9,390	21,989	+ 134
Europe (includes former USSR)	11,710	10,035	− 14	11,520	11,802	+ 2
Oceania	250	227	− 9	480	942	+ 96
World	35,750	64,117	+ 79	39,860	74,272	+ 86

Table 1.3. Development of hen egg production in developed and developing countries between 1970 and 2005, data in 1000 t (Windhorst, 2006).

Year	World	Developed countries	Developing countries	Developing countries % share of the world
1970	19,538	14,866	4,672	23.9
1975	22,232	16,204	6,028	27.1
1980	26,215	17,950	8,265	31.5
1985	30,764	18,667	12,097	39.3
1990	35,232	18,977	16,255	46.1
1995	42,857	17,490	25,367	59.2
2000	51,690	18,263	33,427	64.7
2005	59,233	19,170	40,063	67.6
Increase (%)	203.2	29.0	757.5	–

Table 1.4. Development of poultry meat production in developed and developing countries between 1970 and 2005, data in 1000 t (Windhorst, 2006).

Year	World	Developed countries	Developing countries	Developing countries % share of the world
1970	15,101	11,219	3,882	25.7
1975	18,684	13,409	5,275	28.2
1980	25,265	17,986	7,279	28.8
1985	31,206	20,775	10,431	33.4
1990	41,041	25,827	15,214	37.1
1995	54,771	28,392	26,379	48.2
2000	69,191	32,708	36,483	52.7
2005	81,014	36,663	44,351	54.7
Increase (%)	436.5	226.8	1,042.5	–

Table 1.5. Per capita consumption of eggs (number) and broiler meat (kg) in selected countries during 2005.

Country	Eggs*	Broiler meat**
Argentina	174	18.54
Brazil	130	37.27
South Africa	107	24.70
India	46	1.72
Iran	133	4.88
Thailand	105	10.54
Average	116	16.23
Canada	188	31.82
France	251	14.24
Hungary	295	27.81
Russian Federation	259	16.45
USA	255	44.24
UK	172	26.76
Average	237	26.78

* International Egg Commission (2006).
** FAOSTAT (2005b).

meat production, the increase during the past 35 years in the developing regions has been phenomenal, exceeding that of the developed countries since 2000.

The hot regions of the world have probably the greatest potential for further growth since the level of consumption is still very low. Table 1.5 shows the per capita consumption of eggs and broiler meat in selected countries from both the hot and the temperate regions of the world. Per capita egg consumption in countries located in the hot regions is still considerably lower than in the temperate areas, and per capita poultry meat consumption is on average lower than in countries from the temperate regions.

Table 1.6 shows that the contribution of the continents to global meat production has varied considerably since 1970. Asia now leads the world in poultry meat production, followed by North and Central America, which had the lead up until 1990. In 2005 Asia and South America contributed almost 50% to global meat production. Table 1.7 shows also that the contribution of continents to global hen egg production has undergone a drastic change since 1970. Europe was the leading continent in egg production in 1970, while in 2005 Asia was in the lead, with over 60% of the world's production.

Most governments in the hot regions are aware of the relative ease and rapidity with which the industry can be developed in those areas. They are also aware of the contribution that the poultry industry can make towards

Table 1.6. Changing contributions to global poultry meat production between 1970 and 2005, data in % (Windhorst, 2006).

Continent	1970	1990	2005
Africa	4.0	5.0	4.2
Asia	17.9	24.4	34.0
Europe	28.1	20.6	16.4
USSR	7.1	8.0	–
N. & C. America	36.2	31.3	28.4
S. America	5.8	9.5	15.7
Oceania	0.9	1.2	1.2
World	100.0	100.0	100.0

Table 1.7. Changing contribution of the continents to global hen egg production between 1970 and 2005, data in % (Windhorst, 2006).

Continent	1970	1990	2005
Africa	3.0	4.4	3.7
Asia	23.7	39.2	60.4
Europe	30.9	20.1	16.9
USSR	11.5	13.0	–
N. & C. America	25.3	16.4	13.6
S. America	4.3	6.3	5.1
Oceania	1.2	0.7	0.4
World	*100.0	*100.0	100.0

*Sum does not add because of rounding

improving the quality of human diets in their countries. There is therefore a very rapid expansion of the industry in many countries in these regions. This is very evident, for example, in Brazil in South America, Morocco and Nigeria in Africa, and Saudi Arabia in the Middle East.

Poultry meat and eggs are among the highest-quality human foods; they can serve as important sources of animal protein in those areas of the world that have protein insufficiency. Most countries in the hot regions of the world have daily per capita animal protein consumption below that recommended by the Food and Agriculture Organization (FAO) and the World Health Organization (WHO). In the Middle East and North Africa, the average daily per capita intake of animal protein does not exceed 25 g, compared with over 45 g in most of the developed countries.

In the following sections, a brief description of the known and reported status of the poultry industry is presented for Africa, Latin America and several parts of Asia.

Africa

The poultry industry is highly developed in South Africa and has seen a great deal of development in other African countries during the past two decades. Eggs and poultry meat are beginning to make a substantial contribution to relieving the protein insufficiency in many African countries. Sonaiya (1997) and Gueye (1998), however, reported that almost 80% of poultry production in Africa is found in the rural and peri-urban areas, where birds are raised in small numbers by the traditional extensive or semi-intensive, low-input–low-output systems. According to a study by Mcainsh et al. (2004), chickens are the most commonly kept livestock species in Zimbabwe. Chicken production is divided into large-scale and smallholder chicken production. The smallholder production is mainly free-range systems. Women are the care takers and decision makers of most chicken flocks, and chicken meat is eaten more than any other meat. Adegbola (1988) reported that only 44 eggs were produced in the African continent per person per year.

Per capita consumption of eggs and chicken meat has increased significantly in certain African countries, particularly where production has been rising, as one would expect. For example, per capita egg consumption in South Africa went up from 89 eggs in 1990 to 107 eggs in 2005, and chicken meat consumption from 15.5 kg in 1990 to 24.7 kg in 2005 (Viljoen, 1991; FAOSTAT, 2005b).

Chicken meat production in Africa as a whole went up from 1790 thousand tonnes in 1990 to 3189 thousand tonnes in 2005. The leading countries in chicken meat production in Africa are Algeria, Egypt, Morocco, Nigeria and South Africa (FAOSTAT, 2005a).

There is very little statistical information on the industry from other African countries, with few exceptions. In spite of the expansion that Nigeria has seen in its poultry meat industry, per capita consumption is still below 2.0 kg (FAOSTAT, 2005b), as was reported by Ikpi and Akinwumi (1981). The Moroccan industry had undergone tremendous growth since the early 1980s, when per capita consumption was estimated at 50 eggs and 7.6 kg of poultry meat (Benabdeljalil, 1983). Today, per capita consumption in Morocco is 102 eggs and 8.6 kg of poultry meat (FAOSTAT, 2005b). The top egg-producing countries of Africa (Nigeria, South Africa, Egypt and Algeria) have increased their production from 957,000 t in 1990 to 1,220,000 t in 2005 (FAOSTAT, 2005a).

Latin America

Commercial poultry development has been occurring in a number of Latin American countries, particularly Brazil, Chile, Colombia, Mexico and Venezuela. Per capita consumption of broiler meat in Brazil was reported to be 12.7 kg in 1990, while egg consumption was 93.4 eggs during the same year (USDA, 1990). Today per capita consumption is 109 eggs and 34.3 kg broiler meat. Per capita egg consumption in Mexico today is 282, which is one of the highest in the world. Broiler meat production in Brazil went up from 2,250,000 t in 1990 to 8,692,110 t in 2005 (FAOSTAT, 2005a), particularly due to increased exports.

Asia

China is by far the largest producer of eggs and poultry meat in the Asian continent, with a production of 28,674 thousand tonnes of eggs and 10,233 thousand tonnes of chicken meat (FAOSTAT, 2005a). India is the second largest producer, followed by Japan. Asia now produces over 60% of the total world production of eggs, an increase from only 24% in 1970 (Windhorst, 2006).

Middle East

The Middle East has gone through very rapid growth in the poultry sector during the last two decades of the 20th century. Massive investments were made in the development of environmentally controlled poultry houses equipped with evaporative cooling systems. In the Arab world alone, over 22 billion table eggs are produced annually, which account for over 2.5% of the total world production of table eggs (Morocco, Algeria, Egypt and Syria being the highest producers). Poultry meat production in these countries amounted to over 2500 thousand tonnes in 2005, with Saudi Arabia in the lead followed by Egypt and Morocco (Freiji, 2005). In Saudi Arabia per capita poultry meat consumption rose from 29.8 kg in 1990 to 38.7 kg in 2005, while in Lebanon it increased from 21.6 kg to 28.2 kg during the same period.

It is noteworthy that this tremendous development of the poultry industry in the Arab Middle East all started in Lebanon in the late 1950s and early 1960s. In 1954, income from poultry amounted to less than 15% of the total income from animal production in that country, whereas in 1967 it rose to 60% of the total (Taylor and Daghir, 1968). This growth had a significant effect on animal protein consumption, rising from 14.6 g to 28.3 g per person per day (McLaren and Pellett, 1970). Per capita protein consumption in Lebanon now is 76 g per day, 39% of which is from animal protein sources (Iskandar, 2004).

Potential for expansion in this area is still very great. The annual average per capita consumption of poultry meat and eggs in the Arab world is 8 kg and 70 eggs respectively, which is below world averages of 11.2 kg and 145 eggs. The potential for increased needs for poultry products in those countries is very obvious. Even in countries where production has increased significantly, the demand is still higher than the supply. Kuwait, for example, still imports over 50% of its needs for poultry meat and 45% of its table eggs (Al-Nasser, 2006). This increased consumption is going to be hastened by population growth, increased income from oil and increased rural-to-urban migration.

South Asia (India, Pakistan, Bangladesh, Nepal, Sri Lanka and Bhutan)

According to Panda (1988), the process of planned poultry development on scientific lines was initiated between the early 1960s and mid-1970s in South Asia. Despite sizeable growth of the poultry industry in those countries, per capita consumption varied from five to 35 eggs and from 0.25 kg to 1.25 kg of poultry meat.

Pakistan, with a population of 162 million people, has also gone through a sizeable growth in the production of poultry meat and eggs. Per capita availability went up from 23 in 1991 to 43 eggs in 2005 and poultry meat availability increased from 1.48 kg to 2.38 kg during the same period. Barua and Yoshimura (1997) reported that, in spite of the tremendous growth in the poultry sector in Bangladesh, the per capita availability of eggs is still about 20 and poultry meat 4 kg, which is far below world averages.

The development of the poultry sector in the Indian subcontinent (India, Pakistan, Bangladesh, Sri Lanka, Nepal and Bhutan) has been more rapid than that of any other animal sector. The rate of increase has been more than 10%. During 2005, this area produced more than two billion broilers, and the number of layers on farms exceeded 50 million. This is expected to continue increasing since per capita meat consumption is about 3 kg and per capita egg consumption is about 30 eggs (Botwalla, 2005). This is a tremendous improvement in 15 years compared with consumption figures of 150 g of poultry meat and 20 eggs per person per year reported by Reddy (1991).

The potential for growth of the industry is high when we consider that 20 million tonnes of poultry feed is produced in this area, out of the world production of 240 million.

India has gone through significant improvements in both egg and chicken meat production, reaching 2492 thousand tonnes of eggs and 1900 thousand tonnes of chicken meat in 2005 (FAOSTAT, 2005a), compared with 1160 thousand tonnes of eggs and 250 thousand tonnes of chicken meat in 1990, as reported by Saxena (1992).

South-east Asia (Brunei, Cambodia, Hong Kong, Indonesia, Laos, Malaysia, Myanmar, Philippines, Singapore and Thailand)

Poultry-raising in most South-east Asian countries continued to be characterized by traditional and small-scale systems of farming and operated predominantly by small and low-income farmers. The bulk of the national supplies of poultry meat and eggs in this region are derived from the small and subsistence-type farms rather than from large commercial operations.

With the exception of Thailand and the Philippines, it is estimated that 80% of poultry are located in backyard operations and only 20% on commercial farms. Countries in this region have therefore a tremendous potential for increasing poultry production through the development of small- and medium-scale commercial operations. Success of these operations depends on the adoption of modern practices of poultry production in an integrated approach. This development is going to be closely linked to availability of feed sources that do not compete with human food. Indonesia, Thailand, Malaysia and the Philippines are the leading countries in egg production in this region, estimated to be about 2748 thousand tonnes (FAOSTAT, 2005a). They are also leading in chicken meat production, estimated to be about 3970 thousand tonnes. Leong and Jalaludin (1982), more than two decades ago, reported that the total poultry numbers in those countries did not exceed 500 million. This shows the tremendous development in poultry production that has occurred in these regions.

Future Development

Based on the poultry industry's development during the last two decades and the need for increased animal protein sources in the hot regions of the world, there is general agreement that these areas are going to witness further expansion in the current decade. Richardson (1988) estimated that world production of eggs will reach 51 million tonnes by the year 2000, the greatest increase being in the developing countries. As for poultry meat, he estimated that world production will reach 47 million tonnes, with demand in the developing countries reaching 16 million tonnes. Production has already exceeded the projections made by Richardson, as figures in Table 1.2 show.

Table 1.8. Poultry meat consumption by region (Shane, 2006).

Region	Population (m)	Per capita poultry meat consumption (kg)	Estimated change in consumption (2000–2010) (%)
Asia	3917	6.7	+ 12.8
Africa	888	4.1	+ 23.7
Europe	725	17.8	−1.1
N. & C. America	518	37.4	+ 11.9
S. America	372	24.3	+ 13.9
Oceania	33	26.3	+ 12.3
World	6453	11.2	+ 12.5

Shane (2006) presented data which shows that there will be an increase of 12.5% in consumption of poultry meat during the present decade (2000–2010). The highest increase will be in Asia, Africa and South America, the main warm regions of the world (see Table 1.8). There is no doubt that this increase in the availability of eggs and poultry meat will contribute significantly to the improvement of the nutritional status of the people in the developing countries.

Although the need for more eggs and poultry meat is obvious and the availability of these products can go a long way to meet the protein needs of several populations in hot regions, there are several constraints to the future development of the poultry industry.

The first and foremost is the availability of capital. With the exception of the oil-rich countries, these regions are in general poor with low per capita incomes. If their governments are not able to provide loans, they will not be able to go into commercial poultry production activities. These investments have to be made not only for initiating commercial poultry production enterprises but also for reclaiming and cultivation of irrigated lands to produce the needed feedstuffs for the industry.

The availability of adequate supplies of grain and protein supplements necessary for the formulation of poultry feeds is a major constraint for development. Production of feed grains and oilseeds to support a feed industry is a prerequisite for further growth in the poultry industry. It is true that many countries in those regions have plentiful supplies of agricultural and industrial by-products that can be used in poultry feed formulations but, before these can be used at relatively high levels, they need to be evaluated both chemically and biologically to determine proper levels of inclusion in poultry feeds.

A third constraint on future poultry industry development in these areas is the need to develop the various supporting industries necessary for commercial poultry production. Production of poultry equipment, pharmaceuticals, packaging materials, housing materials, etc. is practically non-existent and needs to be developed alongside the development of commercial poultry production.

The lack of poultry-skilled people for middle management positions in these areas is a real hindrance to further growth in the industry. Training programmes in poultry management for secondary school graduates can go a long way in supplying these badly needed skills for the various integrated poultry operations. The creation of a stable political environment in many of these regions is a must for further development of the industry. Local governments need to establish laws that encourage investment, guarantee gains and control abusive practices.

Disease diagnosis and control are of primary importance in the development and continued growth of any animal production enterprise. It is of utmost importance for modern poultry enterprises because of the intensive nature of these enterprises. There is need in those areas for the establishment of poultry disease diagnostic laboratories as well as for the training of veterinarians and technicians in poultry pathology and poultry disease diagnosis. It is also important that these veterinarians receive some training in poultry management, since disease control in commercial poultry production needs to be looked at as part of a total management package. Disease control practices have been in the limelight in recent years because of the spread of the deadly Asian flu virus (H_5N_1), which has expanded its range among domestic birds from South-east and central Asia to the Middle East, Africa, Europe, North and Central America.

The most obvious constraint on poultry production in these regions is the climate. High temperature, especially when coupled with high humidity, imposes severe stress on birds and leads to reduced performance. Fortunately, during the past three decades, there has been a great deal of development in housing and housing practices for hot climates, and most modern poultry houses have been properly insulated. Poultry equipment companies have come up with various devices that contribute to lowering house temperatures and reducing heat stress. These are discussed in Chapter 5 in this book. These innovations in housing have probably been one of the most significant developments in poultry production practices for hot climates in recent years. Finally, protection of the local industries from imported poultry products, especially poultry meat from low-cost producing countries or from countries that subsidize their production, might help these regions in developing their local poultry industry much faster.

References

Adegbola, A.A. (1988) The structure and problems of the poultry industry in Africa. *Proceedings of the 18th World's Poultry Congress*, pp. 31–38.

Al-Nasser, A. (2006) Poultry industry in Kuwait. *World's Poultry Science Journal* 62, 702–706.

Barua, A. and Yoshimura, Y. (1997) Rural poultry keeping in Bangladesh. *World's Poultry Science Journal* 53, 387–394.

Benabdeljalil, K. (1983) Poultry production in Morocco. *World's Poultry Science Journal* 39, 52–60.

Botwalla, S. (2005) Poultry development in the Indian Subcontinent. *Poultry Middle East and South Africa* 184, 70–74.

FAOSTAT (2005a) http://faostat.fao.org/site/340/default.aspx, viewed on 19–22/03/2007.

FAOSTAT (2005b) http://faostat.fao.org/site/346/default.aspx, viewed on 12–14/03/2007.

Freiji, M. (2005) Poultry industry in the Arab World: situation and skylines. *Poultry Middle East and South Africa* 184, 108–111.

Green-Armytage, S. (2003) *Extraordinary Chickens*. Harry, N. Abrams, Inc. Publishers., New York, pp. 320.

Gueye, E.F. (1998) Village egg and fowl meat production in Africa. *World's Poultry Science Journal* 54, 73–86.

Ikpi, A. and Akinwumi, J. (1981) The future of the poultry industry in Nigeria. *World's Poultry Science Journal* 37, 39–43.

International Egg Commission (2006) Egg market review on per capita egg consumption for selected countries in 2005.

Iskandar, M. (2004) Diet and physical activity as determinants of non-communicable disease risk factors in Lebanon, MSc. thesis, American University of Beirut, Lebanon.

Ketelaars, E.H. and Saxena, H.C. (1992) *Management of Poultry Production in the Tropics*. Benekon, Holland, 11 pp.

Leong, E. and Jalaludin, S. (1982) The poultry industries of South East Asia – the need for an integrated farming system for small poultry producers. *World's Poultry Science Journal* 38, 213–219.

Mcainsh, C.V., Kusina, T., Madsen, J. and Nyoni, O. (2004) Traditional chicken production in Zimbabwe. *World's Poultry Science Journal* 60, 233–246.

McLaren, D.S. and Pellett, P.L. (1970) Nutrition in the Middle East. *World Review of Nutrition and Dietetics* 12, 43–127.

Panda, B. (1988) The structure and problems of the poultry industry in South Asia. *Proceedings of the 18th World's Poultry Congress*, pp. 39–44.

Reddy, C.V. (1991) Poultry production in developing versus developed countries. *World Poultry* 7 (1), 8–11.

Richardson, D.I.S. (1988) Trends and prospects for the world's poultry industry to year 2000. *Proceedings of the 18th World's Poultry Congress*, pp. 251–256.

Saxena, H.C. (1992) Evaluation of poultry development projects in India, Middle East and Africa. *Proceedings of the 19th World's Poultry Congress*, Vol. 2, pp. 647–651.

Shane, S.M. (2006) The future of the world's broiler industry. Zootechnica International, June, pp. 12–19.

Sonaiya, E.B. (1997) Sustainable rural poultry production in Africa. *African Network for Rural Poultry Development*. Addis Ababa, Ethiopia.

Taylor, D.C. and Daghir, N.J. (1968) *Determining Least-cost Poultry Rations in Lebanon*. Faculty of Agricultural Sciences, Publ. 32, American University of Beirut, Lebanon, pp. 11.

USDA (1990) World poultry situation. *Foreign Agricultural Service Circular Series*, Florida and P 1–90, January, pp. 1–30.

van Wulfeten Palthe, A.W. (1992) *G.S. TH. Van Gink's Poultry Paintings*. Dutch branch of the World's Poultry Science Association, Beekgergen, the Netherlands. pp. 10–13.

Viljoen, W.C.J. (1991) The poultry industry in the Republic of South Africa. *World's Poultry Science Journal* 47, 250–255.

West, B. and Zhou, B. (1989) Did chickens go north? New evidence for domestication. *World's Poultry Science Journal* 45, 205–218.

Windhorst, H.W. (2006) Changes in poultry production and trade world wide. *World's Poultry Science Journal* 62, 588–602.

2 Breeding for Resistance to Heat Stress

R.S. GOWE[1] AND R.W. FAIRFULL[2]

[1]Centre for Genetic Improvement of Livestock, Department of Animal and Poultry Science, University of Guelph, Guelph, Ontario, Canada N1G 2W1; [2]Centre for Food and Animal Research, Research Branch, Agriculture Canada, Ottawa, Ontario, Canada K1A 0C6

Introduction	13
Population differences in resistance to heat stress	15
Major genes that affect heat tolerance	16
Naked neck (*Na*) – an incompletely dominant autosomal gene	16
Frizzle (*F*) – an incompletely dominant autosomal gene	17
Dwarf (*dw*) – a sex-linked recessive gene	18
Slow feathering (*K*) – a sex-linked dominant gene	19
Other genes	19
Interactions among major genes	19
Use of major genes in developing heat-resistant strains	20
Experiments on selection for heat tolerance	21
Use of quantitative genes in developing heat-resistant strains	23
Feasibility of developing commercial poultry stocks with heat resistance	23
Genotype × location (tropical versus temperate) interaction	23
Selecting under a controlled environment or a tropical environment	24
Summary	25
References	26

Introduction

Most of the major international poultry breeders are located in temperate countries (Canada, France, Germany, the Netherlands, the UK and the USA), although the southern part of the USA has a warmer climate than some of the other temperate countries. Nevertheless, much of the world's poultry production takes place under more extreme temperature conditions than even the southern USA. Often humidity conditions are very extreme, in addition to the prolonged periods when the temperature is over 30°C. The question then arises whether commercial stocks developed in more moderate climates are optimal for the high heat and humidity conditions of a very large segment of the poultry-producing areas in the world. Stocks developed in temperate climates are now being sold and used throughout the world but are they the best genetic material for all conditions? Would it be possible to select strains that are resistant to heat stress and that have all the other economic characteristics and are therefore more profitable under these conditions?

This chapter will examine the evidence available on selecting for heat resistance. However, before looking at this subject in detail, it is important to point out that selection for resistance to heat stress alone will not lead to a profitable commercial bird. In laying stocks, all the other traits, such as egg production, sexual maturity, egg size, eggshell quality, interior egg quality, disease resistance, fertility, hatchability, body size and feed efficiency, must be improved or maintained. For meat birds, traits such as growth rate, meat yield, conformation and leanness must also be considered. Otherwise, the stocks will not be economically competitive with the widely sold commercial stocks because these traits are under continuous improvement by the major breeders. Exactly how resistance to heat stress, if genetically based and of sufficient economic importance, should be introduced to a comprehensive breeding programme will not be dealt with in this chapter. The subject of multi-trait selection is a very complex one, which has been partially dealt with in a recent book (Crawford, 1990).

For a more general discussion of the value of the new methods for estimating breeding values and some of the advantages and disadvantages, the reader is referred to the papers by Hill and Meyer (1988) and Hartmann (1992). The recent papers by Fairfull *et al.* (1991) and McAllister *et al.* (1990) outline one procedure to incorporate best linear unbiased prediction (BLUP) animal model estimates of breeding values for seven key egg-stock traits. These authors use an economic function to weight the seven different traits. That still leaves several traits that these authors suggest are best incorporated into the selection programme by using separate independent culling levels. The use of independent culling levels for some traits and an index of some kind for others has been further elaborated by Gowe (1993) and Gowe *et al.* (1993). Even with such complex plans that use both indices and independent culling levels, selection in meat stocks occurs at more than one age. Individual selection at broiler age for growth, conformation, feed efficiency, viability and other traits, such as the level of abdominal fatness, is usually practised before selection based on adult characteristics is employed.

If the heat tolerance trait is inherited in a classical quantitative way, with many genes with small effects influencing the trait, then it may be necessary to incorporate the trait into several grandparent lines since most commercial stocks are two-, three- or four-way crosses to make use of heterosis. As a quantitative trait, it would be easier and perhaps more effective to test for heat tolerance at a young age, especially in broilers, while costs per bird are low and population size is higher and before adult characters are assessed.

Although this chapter is not designed to answer the many complex problems of how to incorporate each trait into a breeding programme, it is important to emphasize that demonstrating a genetic basis for a useful economic trait is only the beginning. It is very important to evaluate the cost of introducing each trait to an already complex programme. After evaluating the information available on the genetic basis for heat stress, the final section of this chapter will comment on the development of heat-resistant stocks for hot climates.

Population Differences in Resistance to Heat Stress

There is a large body of literature on breed and strain differences in resistance to heat stress dating back to the 1930s. Some of the results are based on field observations and some on critical laboratory experiments. Many of these experiments have been reviewed before (Hutt, 1949; Horst, 1985; Washburn, 1985) and will not be reviewed in detail here. In general, the early literature showed that there were significant breed and strain within breed differences. The White Leghorn (WL) has been shown to have a greater tolerance for high temperatures than heavier breeds such as Rhode Island Reds (RIR), Barred Plymouth Rocks (BPR), White Plymouth Rocks (WPR) and also Australorps. However, this generalization only held true where there was water available during the stress period (Fox, 1951). WL strains also differed significantly in their ability to withstand heat stress (Clark and Amin, 1965). The heritability of survival of WL under heat stress was high, varying from 0.30 to 0.45 depending on the method of calculation used (Wilson *et al.*, 1966).

To test whether a native breed (a desert Bedouin bird of the Sinai), developed under hot desert conditions, could tolerate very high temperatures better than WL, an experiment was run under controlled conditions by Arad *et al.* (1981). Although the numbers of birds used were not large, the Sinai bird withstood the higher temperatures used (up to 44°C) better than the WL. However, because of its much higher initial production level, the WL was still laying at a higher rate at the highest temperatures than the acclimatized Sinai bird. A series of later papers (Arad and Marder, 1982a,b,c) presented further evidence that the Sinai fowl is more tolerant of high temperatures than the Leghorn, and that crosses of the Sinai and Leghorn came close to the Leghorn in egg production under severe heat conditions, but they were much inferior in egg size. They suggested that the Bedouin (Sinai) bird could be used in breeding programmes to develop strains tolerant of extreme heat.

Others have tested indigenous strains of birds under general tropical conditions and have compared them with crosses utilizing indigenous and improved stocks (Horst, 1988, 1989; Zarate *et al.*, 1988; Abd-El-Gawad *et al.*, 1992; Mukherjee, 1992; Nwosu, 1992). Although the indigenous breeds performed better under higher levels of management than under village conditions, they still do not perform competitively under commercial conditions.

Most workers expressed the opinion that the acclimatization traits of indigenous breeds can be best made use of in synthetic or composite strains and in cross-breeding programmes, since most indigenous stocks lack the productive capacity of the highly improved commercial stocks. However, as Horst (1989) has pointed out, a major contribution of native fowl may be in their contribution of major genes that have an effect on adaptability, which is to be discussed next.

Major Genes that Affect Heat Tolerance

There are several genes that affect heat tolerance. Some, such as the dominant gene for naked neck (*Na*), affect the trait directly by reducing feather cover, while others, such as the sex-linked recessive gene for dwarfism (*dw*), reduce body size and thereby reduce metabolic heat output. Each of the genes important in conferring heat tolerance will be discussed separately, and other potential genes will be mentioned briefly as a group.

Naked neck (*Na*) – an incompletely dominant autosomal gene

Although the naked-neck gene (*Na*) is incompletely dominant, the heterozygote (*Na na*) can be identified by a tuft of feathers on the ventral side of the neck (Scott and Crawford, 1977). The homozygote (*Na Na*) reduces feather covering by about 40% and the heterozygote reduces feathers by about 30%. As described by Somes (1988), the *Na* gene reduces all feather tracts and some are absent. Feather follicles are absent from the head and neck, except around the comb, the anterior spinal tract and two small patches on each side above the crop. The extensive literature describing the characteristics of both broiler and egg-type birds carrying the *Na* gene, when compared with normals (*na na*), has been summarized by Merat (1986, 1990), and the reader is referred to these excellent reviews for more details. In summary, meat-type birds carrying the *Na* gene and grown in a warm environment (usually over 30°C) have larger body weight, better feed efficiency, a lower percentage of feathers, slightly more fleshing, higher viability (when temperatures are very high) and sometimes a lower rate of cannibalism. The homozygote is slightly superior in most tests to the heterozygote for body weight and feed efficiency.

In egg-type birds tested at higher temperatures, the *Na* gene improves heat tolerance, as indicated by higher egg production, better feed efficiency, earlier sexual maturity, larger eggs with possibly fewer cracks, and lower mortality when compared with *na na* birds with similar genetic backgrounds. The *Na* gene may have more positive effects in medium-weight layers than in light birds (Merat, 1990).

Recently, Cahaner *et al.* (1992, 1993) reported on experiments to determine the effect of the *Na* gene in a high-performing broiler sire line and crosses of this line. The *Na* gene was introduced through six backcross generations to a commercial sire line so the stock carried the *Na* gene in a homozygous state but most of the genome was made up of the genes of the high-performance commercial stock. It was then crossed to another commercial sire line (*na na*) to produce *Na na* stock. Several crosses were made to produce comparable *Na na* and *na na* progeny. These progeny were tested under Israel spring and summer conditions, where the temperature ranged between 31 and 22°C. All five flocks tested showed that the 7-week broiler weight of the *Na na* birds was higher than that of the *na na* birds. Samples that were tested under a controlled, constant high temperature (32°C) showed the same results. These studies are important as they show that the wide range of benefits of the *Na* gene demonstrated previously for various relatively slow-growing stocks (see Merat, 1986, 1990) are applicable to stocks that grow in the range of the modern commercial broiler. The three genotypes, *Na Na*, *Na na* and *na na*, were also compared under a controlled, constant high-temperature experiment (32°C) and a controlled lower temperature of 23°C (after 3 weeks of age) to broiler market age. Since the birds grew to market size slower at the higher temperature, they were slaughtered at 8 weeks while those raised at 23°C were slaughtered at 6 weeks of age. The *Na na* and *Na Na* birds grew faster with better feed efficiency (although not significant at 23°C), lower feather production, higher breast meat yield and lower skin percentage at both temperatures, demonstrating the advantage of the *Na* gene even at the lower temperatures. Previous reports suggested there was either no advantage or a small disadvantage to *Na* at lower temperatures. The authors attribute this to the fact that their results were obtained with birds that grow close to the rate of modern commercial broiler stocks, while earlier studies used slower-growing birds.

The studies by Washburn and colleagues (Washburn *et al.*, 1992, 1993a,b) also showed that the effect of the *Na* gene in reducing heat stress as indicated by growth is greatest in more rapid-growing stocks than in slow-growing stocks. In addition, they demonstrated that small-body-weight birds had a higher basal body temperature at 32°C and a smaller change in body temperature when exposed to acute heat (40.5°C) for 45 min than larger-body-weight birds.

The many advantages of the *Na* gene are associated with an increase in embryonic mortality (lower hatchability), particularly in its homozygous state (see Merat, 1986, 1990; Deeb *et al.*, 1993). In any commercial use of this gene, this loss will have to be balanced against the positive effects of the gene under hot conditions. Selection for hatchability in *Na* stocks may reduce this disadvantage.

Frizzle (*F*) – an incompletely dominant autosomal gene

The frizzle (*F*) gene causes the contour feathers to curve outward away from the body. In homozygotes, the curving is extreme and the barbs are extremely curled so that no feather has a flat vane. Heterozygotes have less extreme effects (Somes, 1988).

There is less information available on the effect of the *F* gene on heat tolerance than the *Na*, and most of it is recent. Nevertheless, there is some evidence now to indicate that this gene may be useful in stocks that have to perform under hot, humid conditions. This gene will reduce the insulating properties of the feather cover (reduces feather weight) and make it easier for the bird to radiate heat from the body.

Haaren-Kiso *et al.* (1988) introduced the *F* gene into a dual-purpose brown egg layer sire line by repeated backcrossing for several generations. Heterozygous males (*Ff*) of this sire line were crossed with a high-yielding female line (*ff*). The *Ff* and *ff* progeny were compared under two temperatures, 18–20°C or 32°C, and egg production data obtained. In the hot (32°C) environment, the birds carrying the *F* gene laid 24 more eggs over a 364-day laying period, while in the cooler environment the *F* gene birds laid only three fewer eggs on average. They also reported that the *F* gene favourably affected egg weight, feed efficiency and viability under the 32°C environment, but gave no supporting data. Later, Haaren-Kiso *et al.* (1992) reported that the *F* gene (as a heterozygote) caused a 40% reduction of feather weight at slaughter and an increase in comb weight.

Deeb *et al.* (1993) reported that the *F* gene (as *Ff*) reduced the feather weight of broilers in addition to the reduction caused by the *Na* gene.

Dwarf (*dw*) – a sex-linked recessive gene

The main effect of the dwarf gene (*dw*) is to reduce the body weight of the homozygous males by about 43% and that of homozygous females by 26–32%. There are many other associated physiological and biochemical effects of the gene. The reader is referred to the reviews by Merat (1990) and Somes (1990a) for details.

Some reports show an advantage to the small-body-weight dwarf (*dw w*) hen at high temperatures over comparable normal (*Dw w*) hens. There is less depression of egg size and egg production. Other reports show no advantage (Merat, 1990). Industrially, the *dw* gene has been used with some success (particularly in Europe) in the female parent of the commercial broiler, since there are substantial savings in feed and housing for this smaller broiler mother. The slight loss in growth of the heterozygote male broiler (*Dw dw*) that results from matings of *Dw Dw* male × *dw w* female may be offset by the lower cost of producing broiler hatching eggs. This seems to hold true in situations where feed costs for parents are high or the broilers are slaughtered at a small size and the cost of the hatching egg is relatively more important to the overall costs.

Horst (1988, 1989), however, in a test carried out in Malaysia, reported that Dahlem Red breeding birds (an egg-type synthetic cross) with the *dw* gene were substantially poorer in egg production and egg size than the normals (*Dw*) of the same stock. There is a body of evidence to show that under heat stress conditions the optimum body size of egg-type birds is found in the smaller breeds or strains and that within strain the optimum size is intermediate (Horst, 1982, 1985).

Slow feathering (*K*) – a sex-linked dominant gene

The slow feathering (*K*) gene has been widely used to 'auto-sex' strain and breed crosses. At hatching, the primary and secondary feathers of the recessive birds (*kw* or *kk*) project well beyond the wing coverts while those of slow-feathered chicks (*Kk* or *Kw*) do not. There are two other alleles in the series, both dominant to the wild-type or *k* gene, which are not used commercially (Somes, 1990b). Horst (1988) also credited the *K* gene with the indirect effects: (i) reduced protein requirement; (ii) reduced fat deposit during juvenile life; and (iii) increased heat loss during early growth, all of which may assist the bird in resisting heat stress. Merat (1990) reviewed other effects of this gene.

Other genes

It has been suggested by Horst (1988, 1989) that several other genes may be useful in making fowl tolerant of tropical conditions. The recessive gene for silky (*h*), which affects the barbules on the feathers, may improve the ability to dissipate heat. The dominant gene for peacomb (*P*) reduced feather tract width, reduced comb size and changed skin structure. These may improve the ability to dissipate heat. The recessive, sex-linked, multiple-allelic locus for dermal melanin (*id*) may improve radiation from the skin. However, to date no serious investigations have been reported on the use of these genes to develop heat tolerance in commercial-type birds.

Interactions among major genes

There is evidence that the *Na* and *F* genes can interact to improve the performance of egg stocks under heat stress (Horst, 1988, 1989). In a brief report, Mathur and Horst (1992) claimed that the three genes *Na*, *F* and *dw* interact so that the combined effects of one or two genes are lower than the sum of their individual and additive effects but still higher than the individual gene effects. The cross of the Dahlem Red naked-neck strain with the Dahlem White frizzle strain (both developed at Berlin University) has competed successfully in the Singapore random sample test (Mukherjee, 1992), suggesting the interaction of the genes *Na* and *F* had a positive effect on performance.

In broiler stocks, there is one report that 6-week-old broilers with the *Ff Na na* genotype had fewer feathers than the *ff Na na* comparable stocks, which had fewer feathers than the *ff na na* stock. The effects, which are not fully additive, indicate that it might be advantageous to introduce both genes (*F* and *Na*) into a broiler sire line (Deeb *et al.*, 1993). Birds from strains selected for slow feathering (*S*) within the *K* genotype and carrying the *Na* gene had lower feather cover than *S* birds not carrying *Na*, indicating that both *Na* and selecting for slow feathering within the *K* genotype were improving performance in a warm environment (Lou *et al.*, 1992).

Selecting for quantitative genes for slower feather growth in a broiler line breeding true for the *K* gene (Edriss *et al.*, 1988; Ajang *et al.*, 1993) reduced feathers at 48 days, reduced carcass fat, increased carcass protein and carcass meat, and increased growth and feed efficiency. Although these results were obtained at a moderate test temperature of 20°C, selection for slow feather growth in birds with the *K* gene should increase their heat tolerance by enhancing their ability to dissipate heat.

Use of major genes in developing heat-resistant strains

There is little doubt now that the advantages of using the naked-neck gene for birds to be grown under high temperature (30°C +) conditions outweigh the disadvantage of the slightly lowered hatchability associated with this gene (Merat, 1986, 1990). Besides the advantage of fewer feathers on the neck and other parts of the body, there are other positive effects of this gene for broiler stocks, such as increased carcass weight and meat yield, higher body weights, lower fat content and better feed efficiency (Merat, 1986; Cahaner *et al.*, 1993). Also, the positive effects of this gene have recently been shown to be present when the birds were grown under more temperate conditions (i.e. 23 ± 2°C from 3 weeks to slaughter), even though the benefits of the gene are greater at higher ambient temperatures (Cahaner *et al.*, 1993). The positive effects of the *Na* gene were clearly demonstrated for high growth rate in modern-type broilers when raised as commercial broilers under the spring and summer conditions of Israel, where temperatures ranged between 31°C in the day and 22°C at night (Cahaner *et al.*, 1992). The *Na* gene in this study was introduced by backcrossing for six generations to a commercial sire line.

Although there is less evidence that the *Na* gene, incorporated into high-producing egg stocks, will improve performance under high temperatures, the report of Horst (1988) on the Dahlem Red stock with this gene shows it to be superior in egg production, egg weight and body size to the comparable stock not carrying the gene. Production levels of this test were comparable to commercial performance in this region.

The *Na* gene can be successfully incorporated by backcrossing into high-performing meat or egg stocks. One procedure was outlined in detail by Horst (1989). If the gene can be obtained in a stock already improved, fewer backcross generations will be required. If it is from a relatively unimproved stock, from five to eight generations would be needed.

Although there is less conclusive evidence for the use of the frizzle gene, the limited evidence available suggests that breeders could seriously consider using this gene along with the *Na* gene to develop stocks specifically for the hot humid tropics (Horst, 1988; Haaren-Kiso et al., 1992; Mathur and Horst, 1992; Deeb et al., 1993). Also, there appear to be advantages to the use of the double heterozygote *(Na na Ff)* for stocks to be reared in the hot humid tropics. Both *Na* and *F* could be backcrossed into sire lines at the same time. It would take larger populations and a few more generations if the two genes came from different sources.

Whether the sex-linked recessive dwarf gene (*dw*) will be useful in female parent broiler stocks for the tropics, beyond its well-known characteristics of reduced body size and bird space requirements and improved feed efficiency, depends on the economics of broiler production of the region. Perhaps it may be even more useful in combination with the *Na* gene, the *F* gene or both, since the *Na* gene tends to increase egg size of stocks carrying that gene (Merat, 1986, 1990). Critical experiments are required before recommending the use of this gene in combination with *Na* and *F*.

Experiments on Selection for Heat Tolerance

The early research on the genetics of heat tolerance was mainly concerned with preventing losses when the chickens were exposed to high temperatures, often accompanied by high humidity during a heat wave in temperate climates (Hutt, 1949).

In an early experiment, Wilson *et al.* (1966, 1975) placed 4-week-old Leghorn chicks in a chamber at 41°C with a relative humidity of 75%. Survival time of families during a 2-h exposure was recorded and used to select parents for breeders to form two lines, one with a long survival period and another with the shortest survival period. Although the heritability of survival time of the 4-week-old chick was quite high (about 0.4) and the two lines diverged significantly, there was no evidence presented that the line that survived longer in this short-term test was able to perform better under high ambient temperatures normally found where chickens are grown. Even if this procedure did result in chickens being able to perform better under high ambient temperatures, the high levels of mortality that resulted from this kind of test would be unacceptable now in most countries of the world.

Despite the legitimate welfare objections that might be raised when chickens are exposed to temperature levels that result in over 50% mortality, a seven-generation selection experiment, to develop heat tolerance in a Leghorn strain, was recently reported by Yamada and Tanaka (1992). They exposed adult Leghorn hens to 37°C and 60% relative humidity for 10 days. Survival rate varied from 16% for generation zero to 69% for generation seven, a very positive genetic increase in tolerance to the 10-day heat treatment. For each generation, the survivors were reproduced after a recovery phase of 60 days. The survivors of generation seven recovered faster, as indicated by the reduced days to the start of lay after the 10-day period at 37°C.

Although the authors claimed superior egg production for the thermotolerant strain, all the survivors apparently regained their normal egg production and egg quality when returned to a 21°C environment, as was shown to be the case when laying birds were exposed to 32°C and returned to 21°C (de Andrade, 1976). When the thermotolerant strain birds of Yamada and Tanaka (1992) were exposed to 38°C for 38 days, the authors claimed these birds were also superior physiologically, but the exact mechanisms were not made clear.

Selection under 38°C for heat tolerance (HR) and susceptibility to heat (HS) in the low-growth-rate, Athens-Canadian, random-bred (AC) strain and in a faster-growing broiler (BR) strain has been briefly reported by El-Gendy and Washburn (1989, 1992). In the AC population, after five generations of selection in both directions for 6-week body weight in the 38°C environment, body weight of the HR line was significantly heavier (39%), while the HS line was significantly smaller (46%) than the genetic controls. The realized heritability of 6-week body weight in the HR line was 0.4, compared with 0.9 in the HS line. It is important to note the authors reported that mortality over the 6-week heat test was not increased in either line. The HR line had lower body temperatures than the HS line under 38°C. After only two generations of selection for 4-week body weight and heat tolerance (HR) or heat susceptibility (HS) in the fast-growing broiler lines (BR), the HR line was 7% heavier and the HS line 6% smaller than controls. The realized heritability of body weight for the HR line was 0.4 and 0.3 in the HS line. Both the selected lines had lower mortality than the control population. Although the detailed data are not yet available, this study showed that it is possible to successfully select for growth rate in broiler-type stocks under heat stress (38°C).

Bohren et al. (1981) selected Leghorn lines for fast and slow growth in a hot (32°C) environment and a normal environment (21°C) and then compared the lines under both environments. They also tested the lines under a higher heat stress (40°C) and found there was no significant difference in survival between lines selected in the two different environments. They postulated that the same growth genes were being selected for under both environments. This may be true for short-term selection for growth genes in Leghorns, but it may not be true for selection over many generations in large broiler-type birds (Washburn et al., 1992).

Selection for genetically lean broilers reduces fatness and also gives these leaner birds the ability to perform better under higher temperatures than birds selected for fatness (MacLeod and Hocking, 1993). Another two populations of meat birds successfully selected in two directions for leanness or fatness were tested under hot conditions (32°C) and a moderate environment (22°C). The lean birds grew to a greater weight than the fat birds at 32°C, also demonstrating a greater resistance to hot conditions (Geraert et al., 1993). Since selection for feed efficiency reduces body fat (Cahaner and Leenstra, 1992), the physiological mechanisms are probably the same. Selecting for feed efficiency and/or leanness would give broiler stock the additional advantage of greater heat tolerance.

Use of quantitative genes in developing heat-resistant strains

The research of Yamada and Tanaka (1992) demonstrates that it is feasible to successfully select for heat tolerance in a White Leghorn strain. Similarly, Washburn and colleagues (El-Gendy and Washburn, 1989, 1992) have shown that it is possible to select for heat tolerance in broiler stocks by selecting for body weight under heat stress. These studies have shown there is a high heritability for selection for body weight or survival under heat stress. Evidence is not yet available as to whether selection for better performance under heat stress can be effectively combined with selection for all the other necessary economic traits of egg and meat stocks.

Recent research discussed previously shows that selection for feed efficiency and/or leanness conveys an increased ability to withstand heat stress. Since these traits are independently valuable, it would seem important to emphasize these traits for stocks to be used in hot climates, more than might be indicated by their economic value in temperate climates.

Feasibility of Developing Commercial Poultry Stocks with Heat Resistance

Genotype × location (tropical versus temperate) interaction

There are no scientific reports that these authors are aware of on the comparison of commercial stocks that were developed by the major poultry breeders and tested in both temperate and tropical environments to evaluate the magnitude of the interaction. Similarly, there are no published studies on the comparison of different commercial genotypes tested under different controlled temperature conditions. However, some comparisons of strains developed by major breeding companies with indigenous stocks made in tropical environments have shown that the indigenous stocks generally cannot compete, even in the climate to which they are acclimatized. There is a need for much more definitive research in this area (see Arad *et al.*, 1981; Arad and Marder, 1982b; Horst, 1988; Mukherjee, 1992).

A comparison of sire progeny groups of a brown egg cross that were tested in both a temperate climate (Berlin) and a tropical climate (Kuala Lumpur) showed that there were highly significant sire × location interactions for sexual maturity, egg production, egg weight and feed consumption (Mukherjee *et al.*, 1980). There was a negative correlation (−0.39) between the breeding value for body weight and egg production in Kuala Lumpur and a small positive correlation (0.10) for these traits at Berlin. This led Mukherjee *et al.* (1980) to suggest that smaller-bodied birds might improve egg output in the tropics, presumably because of their ability to withstand heat better than larger birds, and supports the thesis of Horst (1985) that optimal body weight is critical for heat tolerance of egg stocks.

In a later experiment, Mathur and Horst (1988) tested pedigree laying birds in three environments: a controlled warm temperature at Berlin (32°C), a partially controlled temperate climate (20 ± 2°C) at Berlin and a natural tropical environment in Malaysia (28 ± 6°C). The genetic correlations of breeding values of sires whose progeny were tested in all three locations was low for egg production, egg mass, laying intensity and persistency but relatively high for egg weight and body weight. These authors concluded that if the aim is to improve productivity in the tropics then selection should take place under tropical conditions; however, there was no direct evidence to support this conclusion.

After testing half-sib groups of broilers to 8 weeks of age in two locations that differed in temperature and other conditions (Federal German Republic and Spain), Hartmann (1990) found the genetic correlation of progeny performance for body weight was 0.68 and 0.56 for conformation and grade. He concluded that the correlation was high enough to justify selection at one location for use in both locations since simple errors in recording would account to some degree for the correlation not being 1.0; therefore, most of the selection would be directed at the same families in both locations.

Selecting under a controlled environment or a tropical environment

Most of the studies on the genetics of heat tolerance have made use of controlled temperature chambers, with temperatures maintained constantly at about 32°C (e.g. see Cahaner et al., 1993; Eberhart and Washburn, 1993a,b). Although high temperatures of 40°C or higher have been used, particularly in early experiments (Wilson et al., 1966), they are not generally used now because of the high levels of mortality. Nevertheless, recently Yamada and Tanaka (1992) used a controlled temperature of 37°C for 10 days to successfully select for heat tolerance in adult hens, but their losses were very high.

There is no doubt that selection for heat tolerance under controlled heat conditions is feasible. The heritability of heat tolerance is quite high and rapid progress can be expected (Wilson et al., 1966; El-Gendy and Washburn, 1992; Yamada and Tanaka, 1992). However, there is no evidence that strains selected for multiple performance traits as well as for heat tolerance in a constant temperature chamber do better under the variable conditions found in most tropical environments than the strains that were selected under variable temperate conditions. Experiments are needed to clarify whether the heat tolerance selected for under short-term controlled temperature conditions is the kind of heat stress tolerance that is needed in tropical and subtropical countries.

If the physiological mechanisms for resisting heat stress were better understood, it might be possible to more directly select for the trait required. It is possible as the physiological role of the heat-shock proteins are better understood that breeders will be able to select directly for the specific protein(s) that enables the bird to withstand the high temperatures (see Chapter 4 for a discussion of heat-shock proteins).

Summary

There is no doubt that it is feasible to select for resistance to heat stress, although all the practical problems have not yet been resolved. Nevertheless, the evidence is rapidly accumulating that those breeders, whether in tropical or temperate countries, that add heat tolerance or heat resistance to the many traits already needed for successful commercial egg- or meat-stock breeding will have an advantage in the tropical and semi-tropical regions.

It is particularly important for meat-stock breeders to add heat tolerance to the breeding programme, since there is an antagonism between rapid growth, large-bodied birds and heat resistance. The rapid-growing bird needs a high feed-consumption rate and, in turn, this generates metabolic heat, which these large, rapidly growing birds find difficult to dissipate rapidly enough following meals. This leads to reduced feed intake, to reduced growth and, if the temperatures are high enough, to poorer feed-conversion ratios (Washburn and Eberhart, 1988; Cahaner and Leenstra, 1992). In egg stocks, it is also essential for high-production hens to maintain feed consumption under heat stress, or egg production will fall drastically. The large-bodied female parent of the broiler must also be able to dissipate heat.

The breeder has many approaches possible. Selection can be directed at the quantitative genes responsible for heat resistance by challenging under controlled conditions or natural conditions. The controlled-temperature approach was demonstrated as feasible for egg stocks (Yamada and Tanaka, 1992) and for meat stocks (El-Gendy and Washburn, 1989, 1992). It is not yet clear from the research to date whether most progress would be made by selecting under controlled, high-temperature conditions, or by selecting under the variable ambient temperatures of a tropical or semi-tropical climate. Mathur and Horst (1988) thought it would be best to select under the prevailing climatic conditions where the birds are to be used. Also, if a controlled-temperature environment is used for testing birds, studies are needed to determine the optimal temperature, humidity and length of exposure.

Reports indicate that the heritability of the trait is quite high (El-Gendy and Washburn, 1992; Yamada and Tanaka, 1992), and rapid progress has been made in uni-trait selection studies using constant temperature chambers. There is no published evidence of selecting for heat tolerance in a multiple-trait selection programme as yet. Although there is no reason why the trait couldn't be added to a selection programme, it would probably reduce selection pressure on other economic traits.

It is now clear that selection for weight gain under variable heat stress (as in Israel), when compared with stocks developed in the Netherlands, results in stocks that have better adaptability to conditions that are 'normal' for the Netherlands (from 15 to 47 days of age, the temperature varied from 28 to 20°C) than to lower temperature conditions (i.e. from 15 to 47 days of age, the temperature varied from 22 to 15°C). Stocks developed in the Netherlands did better than the Israel stocks at the above low temperatures (Leenstra and Cahaner, 1991). It is apparent that some general adaptability to the specific climates of the two countries was occurring in the broiler stocks. This research supports the concept of selection under the variable conditions in hot climates for performance in hot climates.

As discussed earlier, there would be a positive effect on heat tolerance for strains selected for either feed efficiency or leanness, or both. These traits are valuable in any climate.

Although the quantitative genetics approach discussed above may prove useful in the long run, the introduction of major genes such as *Na* and *F* into high-producing lines (both egg and meat), or perhaps only into the sire or the female parent of the commercial product, would appear to provide a quicker way to introduce heat tolerance. By backcrossing six to eight generations into high-performing sire lines, the benefit of the heat tolerance can be achieved with little or no loss of other essential performance traits (Horst, 1988, 1989; Cahaner *et al.*, 1993).

Native or indigenous chickens produce more than 50% of the poultry meat and eggs of the tropical countries (Horst, 1989; Mukherjee, 1992). Many of these breeds or strains must possess some tolerance to heat stress and adaptability to tropical conditions. These indigenous village chickens could be upgraded by mating them to males from improved strains of indigenous stock, or by crossing them to exotic stock that have higher productivity and possibly one or two major genes for heat tolerance (Horst, 1989). A small increase in the performance of these scavenging birds will bring large returns to many people. Some native stocks might be used in composites or synthetics that include high-performance strains. The adaptability genes of the indigenous stock would be selected for along with the performance genes of the commercial stocks in the composite if selection is carried out under tropical conditions. The addition of major genes such as *Na* or *F* into these composites (if not already there) might help ensure there is greater heat tolerance as well as general adaptability. Such new strains developed from synthetic strains might then be useful in crosses to be used under tropical conditions. However, this would be a long-range project with no guarantee that a useful strain would emerge.

The challenge for the poultry breeder is to introduce heat-stress tolerance while retaining and improving the wide array of other economic traits needed in commercial chickens.

References

Abd-El-Gawad, E.M., Khalifah, M. and Merat, P. (1992) Egg production of a dwarf (*dw*) Fl cross between an experimental line and local lines in Egypt, especially in small scale production. *Proceedings of the 19th World's Poultry Congress*, Amsterdam, the Netherlands, Vol. 2, pp. 48–52.

Ajang, O.A., Prijono, S. and Smith, W.K. (1993) Effect of dietary protein content on growth and body composition of fast and slow feathering broiler chickens. *British Poultry Science* 54, 73–91.

Arad, Z. and Marder, J. (1982a) Differences in egg shell quality among the Sinai Bedouin fowl, the commercial White Leghorn and their crossbreds. *British Poultry Science* 23, 107–112.

Arad, Z. and Marder, J. (1982b) Comparison of the productive performance of the Sinai Bedouin fowl, the White Leghorn and their crossbreds: study under laboratory conditions. *British Poultry Science* 23, 329–332.

Arad, Z. and Marder, J. (1982c) Comparison of the productive performances of the Sinai Bedouin fowl, the White Leghorn and their crossbreds: study under natural desert condition. *British Poultry Science* 23, 333–338.

Arad, Z., Marder, J. and Soller, M. (1981) Effect of gradual acclimatization to temperatures up to 44°C on productive performance of the desert Bedouin fowl, the commercial White Leghorn and the two reciprocal crossbreds. *British Poultry Science* 22, 511–520.

Bohren, B., Carsen, J.R. and Rogler, J.C. (1981) Response to selection in two temperatures for fast and slow growth to nine weeks of age. *Genetics* 97, 443–456.

Cahaner, A. and Leenstra, F. (1992) Effects of high temperature on growth and efficiency of male and female broilers from lines selected from high weight grain, favourable feed conversion and high or low fat content. *Poultry Science* 71, 1237–1250.

Cahaner, A., Deeb, N. and Gutman, M. (1992) Improving broiler growth at high temperatures by the naked neck gene. *Proceedings of the 19th World's Poultry Congress*, Amsterdam, the Netherlands, Vol. 2, pp. 57–60.

Cahaner, A., Deeb, N. and Gutman, M. (1993) Effects of the plumage reducing Naked Neck *(Na)* gene on the performance of fast growing broilers at normal and high ambient temperatures. *Poultry Science* 72, 767–775.

Clark, C.E. and Amin, M. (1965) The adaptability of chickens to various temperatures. *Poultry Science* 44, 1003–1009.

Crawford, R.D. (ed.) (1990) *Poultry Breeding and Genetics.* Elsevier, Amsterdam.

de Andrade, A.N. (1976) Influence of constant elevated temperature and diet on egg production and shell quality. *Poultry Science* 55, 685–693.

Deeb, N., Yunis, R. and Cahaner, A. (1993) Genetic manipulation of feather coverage and its contribution to heat tolerance of commercial broilers. In: Gavora, J.S. and Boumgartner, J. (eds) *Proceedings of the 10th International Symposium on Current Problems in Avian Genetics*, Nitra, Slovakia, p. 36.

Eberhart, D.E. and Washburn, K.W. (1993a) Variation in body temperature response of naked neck and normally feathered chickens to heat stress. *Poultry Science* 72, 1385–1390.

Eberhart, D.E. and Washburn, K.W. (1993b) Assessing the effects of the naked neck gene on chronic heat stress resistance in two genetic populations. *Poultry Science* 72, 1391–1399.

Edriss, M., Smith, K. and Dun, P. (1988) Divergent selection for feather growth in broiler chickens. *Proceedings of the 18th World's Poultry Congress*, Nagoya, Japan, pp. 561–563.

El-Gendy, E.A. and Washburn, K.W. (1989) Selection for extreme heat stress in young chickens. 3. Response of the Sj generation. *Poultry Science* 68 (Suppl. 1), 49 (Abstract).

El-Gendy, E.A. and Washburn, K.W. (1992) Selection for heat tolerance in young chickens. *Proceedings of the 19th World's Poultry Congress*, Amsterdam, the Netherlands, Vol. 2, p. 65.

Fairfull, R.W., McAllister, A.J. and Gowe, R.S. (1991) A profit function for White Leghorn layer selection. In: Zelenka, D. (ed.) *Proceedings of the Fortieth Annual National Breeders Roundtable.* St Louis Poultry Breeders of America, Tucker, Georgia, pp. 36–49.

Fox, T.W. (1951) Studies on heat tolerance in the domestic fowl. *Poultry Science* 30, 477–483.

Geraert, P.A., Guillaumin, S. and LeClerq, B. (1993) Are genetically lean broilers more resistant to hot climate? *British Poultry Science* 34, 643–653.

Gowe, R.S. (1993) Egg genetics: conventional approaches: should all economic traits be included in an index? In: Gavora, J.S. and Boumgartner, J. (eds) *Proceedings of the 10th International Symposium on Current Problems in Avian Genetics*, Nitra, Slovakia, pp. 33–40.

Gowe, R.S., Fairfull, R.W., McMillan, I. and Schmidt, G.S. (1993) A strategy for maintaining high fertility and hatchability in a multiple trait egg stock selection program. *Poultry Science* 72, 1433–1448.

Haaren-Kiso, A.V., Horst, P. and Zarate, A.V. (1988) The effect of the Frizzle gene (F) for the productive adaptability of laying hens under warm and temperate environmental conditions. *Proceedings of the 18th World's Poultry Congress*, Nagoya, Japan, pp. 386–388.

Haaren-Kiso, A.V., Horst, P. and Zarate, A.V. (1992) Genetic and economic relevance of the autosomal incompletely dominant Frizzle gene (F). *Proceedings of the 19th World's Poultry Congress*, Amsterdam, the Netherlands, Vol. 2, p. 66.

Hartmann, W. (1990) Implications of genotype–environment interactions in animal breeding: genotype–location interactions in poultry. *World's Poultry Science Journal* 46, 197–210.

Hartmann, W. (1992) Evaluations of the potentials of new scientific developments for commercial poultry breeding. *World's Poultry Science Journal* 48, 16–27.

Hill, W.G. and Meyer, K. (1988) Developments in methods for breeding value and parameter estimation on livestock. In: Land, R.B., Bulfield, G. and Hill, W.G. (eds) *Animal Breeding Opportunities.* British Society of Animal Production Occasional Publication No. 12, Edinburgh.

Horst, P. (1982) General perspectives for poultry breeding on improved productive ability to tropical conditions. *2nd World Congress on Genetics Applied to Livestock Production*, Madrid, Spain, Vol. 8, pp. 887–892.

Horst, P. (1985) Effects of genotype x environment interactions on efficiency of improvement of egg production. In: Hill, W.G., Manson, J.M. and Hewitt, D. (eds) *Poultry Genetics and Breeding.* British Poultry Science Ltd, Longman Group, Harlow, UK, pp. 147–156.

Horst, P. (1988) Native fowl for a reservoir of genomes and major genes with direct and indirect effects on productive adaptability. *Proceedings of the 18th World's Poultry Congress*, Nagoya, Japan, pp. 99–105.

Horst, P. (1989) Native fowl as a reservoir for genomes and major genes with direct and indirect effects on the inadaptability and their potential for tropically oriented breeding plans. *Archiv fur Geflugelkunde* 53, 93–101.

Hutt, F.B. (1949) *Genetics of the Fowl.* McGraw-Hill, New York, Toronto and London.

Leenstra, F. and Cahaner, A. (1991) Genotype by environment interactions using fast growing lean or fat broiler chickens, originating from the Netherlands or Israel, raised at normal or low temperatures. *Poultry Science* 70, 2028–2039.

Lou, M.L., Quio, O.K. and Smith, W.K. (1992) Effects of naked neck gene and feather growth rate on broilers in two temperatures. *Proceedings of the 19th World's Poultry Congress*, Amsterdam, the Netherlands, Vol. 2, p. 62.

McAllister, A.J., Fairfull, R.W. and Gowe, R.S. (1990) A preliminary comparison of selection by multiple trait culling levels and best linear unbiased prediction. In: Hill, W.G., Thompson, R. and Woolliams, J.A. (eds) *Proceedings of the 4th World Congress on Genetics Applied to Livestock Production*, Vol. 16, University of Edinburgh, Edinburgh.

MacLeod, M.G. and Hocking, P.M. (1993) Thermoregulation at high ambient temperature in genetically fat and lean broiler hens fed *ad libitum* or on a controlled feeding regime. *British Poultry Science* 34, 589–596.

Mathur, P.K. and Horst, P. (1988) Efficiency of warm stall tests for selection on tropical productivity in layers. *Proceedings of the 18th World's Poultry Congress*, Nagoya, Japan, pp. 383–385.

Mathur, P.K. and Horst, P. (1992) Improving the productivity of layers in the tropics through additive and non-additive effects of major genes. *Proceedings of the 19th World's Poultry Congress*, Amsterdam, the Netherlands, Vol. 2, p. 67.

Merat, P. (1986) Potential usefulness of the *Na* (Naked Neck) gene in poultry production. *World's Poultry Science Journal* 42, 124–142.

Merat, P. (1990) Pleiotropic and associated effects of major genes. In: Crawford, R.D. (ed.) *Poultry Breeding and Genetics*. Elsevier, Amsterdam, pp. 429–467.

Mukherjee, T.K. (1992) Usefulness of indigenous breeds and imported stocks for poultry production in hot climates. *Proceedings of the 19th World's Poultry Congress*, Amsterdam, the Netherlands, Vol. 2, pp. 31–37.

Mukherjee, T.K., Horst, P., Flock, D.K. and Peterson, J. (1980) Sire x location interactions from progeny tests in different regions. *British Poultry Science* 21, 123–129.

Nwosu, C.C. (1992) Genetics of local chickens and its implications for poultry breeding. *Proceedings of the 19th World Poultry Congress*, Amsterdam, the Netherlands, Vol. 2, pp. 38–42.

Scott, T. and Crawford, R.D. (1977) Feather number and distribution in the throat tuft of naked neck chicks. *Poultry Science* 56, 686–688.

Somes, R.G. (1988) International registry of poultry genetic stocks. Bulletin 476, Storrs Agricultural Experiment Station, University of Connecticut, Storrs.

Somes, R.G., Jr. (1990a) Mutations and major variants of muscles and skeleton in chickens. In: Crawford, R.D. (ed.) *Poultry Breeding and Genetics*. Elsevier, Amsterdam, the Netherlands, pp. 209–237.

Somes, R.G., Jr. (1990b) Mutations and major variants of plumage and skin in chickens. In: Crawford, R.D. (ed.) *Poultry Breeding and Genetics*. Elsevier, Amsterdam, pp. 169–208.

Washburn, K.W. (1985) Breeding of poultry in hot and cold environments. In: Yousef, M.K. (ed.) *Livestock Physiology*, Vol. 3, *Poultry*. C.R.C. Publications, Boca Raton, Florida, pp. 111–122.

Washburn, K.W. and Eberhart, D. (1988) The effect of environmental temperature on fatness and efficiency of feed utilization. *Proceedings of the 8th World's Poultry Congress*, Nagoya, Japan, pp. 1166–1167.

Washburn, K.W., El-Gendy, E. and Eberhart, D.E. (1992) Influence of body weight and response to a heat stress environment. *Proceedings of the 19th World's Poultry Congress*, Amsterdam, Vol.2, pp. 53–56.

Wilson, H.R., Armas, A.E., Ross, J.J., Dominey, R.W. and Wilcox, C.J. (1966) Familial differences of Single Comb White Leghorn chickens in tolerance to high ambient temperatures. *Poultry Science* 45, 784–788.

Wilson, H.R., Wilson, C.S., Voitle, R.A., Baird, C.S. and Dominey, R.W. (1975) Characteristics of White Leghorn chickens selected for heat tolerance. *Poultry Science* 54, 126–130.

Yamada, M. and Tanaka, M. (1992) Selection and physiological properties of thermotolerant White Leghorn hen. *Proceedings of the 19th World's Poultry Congress*, Amsterdam, the Netherlands, Vol. 2, pp. 43–47.

Zarate, A.V., Horst, P., Harren-Kiso, A.V. and Rahman, A. (1988) Comparing performance of Egyptian local breeds and high yielding German medium heavy layers under controlled temperature and warm environmental conditions. *Proceedings of the 18th World's Poultry Congress*, Nagoya, Japan, pp. 389–391.

3 Breeding Fast-growing, High-yield Broilers for Hot Conditions

A. CAHANER

Faculty of Agriculture, The Hebrew University, Rehovot, Israel

Introduction	30
Heat stress and its effects on broiler performance	31
Genetic association between heat tolerance and GR	32
Slow-growing versus fast-growing lines	32
Slow-growing versus fast-growing families within lines	34
Selection on GR under hot conditions	34
The effect of AT on feed intake in broilers differing in GR	35
The effects of feather coverage on heat tolerance in broilers	36
The effects of the naked-neck gene (*Na*) on feather coverage and heat tolerance in broilers	36
Discovery of the scaleless gene (*Sc*), which eliminates feather development	37
Comparing featherless and naked-neck broilers	37
Heat tolerance of featherless broilers versus their feathered counterparts	38
Superior performance under high AT of featherless broilers compared with their feathered counterparts	39
Featherless broilers produce more high-quality meat in hot conditions and high stocking densities	40
Summary	41
Featherless broilers in hot conditions	42
Breeding and production	43
References	43

Introduction

Tremendous genetic progress has been observed in broiler growth rate (GR) and meat yield since the 1950s (Havenstein *et al.*, 1994a,b, 2003a,b). However, this

© CAB International 2008. *Poultry Production in Hot Climates* 2nd edition (ed. N.J. Daghir)

dramatic increase in GR, and the consequent reduction in time needed to achieve marketing weight under optimal conditions, have not been accompanied by similar improvements under suboptimal conditions, e.g. high ambient temperatures (AT). The modern broiler industry continues to expand to regions where climatic conditions are suboptimal, mainly warm ones, and climatic control of broiler houses is lagging, due to high installation and operational costs, and unreliable supply of electricity. Nevertheless, commercial breeding is conducted mainly in temperate climates under optimal conditions. Broiler breeding companies continue to improve the productivity (mainly GR and meat yield) of their broiler lines through intensive selection under optimal conditions. However, results from several studies, to be presented below, suggest that due to the increase in genetic potential for rapid GR and meat yield, broilers are becoming more sensitive to rather small increases over optimal AT. Furthermore, also in developed countries, many broilers are reared under nearly hot conditions.

Heat Stress and its Effects on Broiler Performance

Chickens, like all homeothermic animals, maintain a constant body temperature (BT) over a wide range of AT. In birds, heat loss is limited by feathering and by the lack of sweat glands. The ability of an animal to maintain its BT within the normal range depends on a balance between internally produced heat and the rate of heat dissipation. The amount of internal heat produced by broilers depends on their BW and feed intake, and the rate of heat dissipation depends on environmental factors, mainly AT, and on the broilers' feather coverage. When the physiological and behavioural responses to high AT are inadequate, an elevation in BT occurs, causing a decrease in appetite and in actual GR. Consequently, the time needed to reach marketing weight is increased, leading to lower efficiency and profitability of poultry meat production (Cahaner and Leenstra, 1992; Leenstra and Cahaner, 1992; Ain Baziz *et al.*, 1996; Yalcin *et al.*, 1997a; Settar *et al.*, 1999). Moreover, hot conditions also negatively affect the yield and quality of broiler meat (Leenstra and Cahaner, 1992; Mitchell and Sandercock, 1995a,b, 1997; Ain Baziz *et al.*, 1996; Sandercock *et al.*, 2001), often leading to PSE (pale, soft, exudative) meat (Barbut, 1997; Woelfel *et al.*, 2002). Thus, high AT has been the major factor hindering broiler meat production in hot climates, especially in developing countries where farmers cannot afford costly artificial control of AT in broiler houses.

Modern broilers exhibit maximal GR at AT levels of about 18 to 20°C, with a decline as AT increases (Yahav *et al.*, 1998). In addition to reduced GR in hot conditions, increased mortality can occur during heatwaves, when AT sharply increases above the prevailing levels. Studies on the physiological effects of heat stress have yielded several approaches to increasing the thermotolerance of broilers, in order to minimize mortality and heat-related reduction in productivity. These approaches include acclimation to heat (e.g.

De Basilio *et al.*, 2003), feed intake manipulation and feed deprivation (Ait Boulahsen *et al.*, 1993, 1995) and air velocity (May *et al.*, 2000). However, as commercial breeding continues to increase the potential GR of contemporary broilers, the rate of their internal heat production will also increase because they consume more feed per time unit. Therefore the management and nutritional practices used for alleviation of heat stress are limited as they provide only short-term or costly solutions. This chapter presents findings obtained during the last two decades, which are relevant to various possible approaches to utilize breeding for genetic improvement of broiler production under hot conditions. Successful breeding programmes are expected to provide sustainable adaptation of broilers to high AT, thus allowing to maintain and further improve of broiler performances in warm and hot conditions.

Genetic Association between Heat Tolerance and GR

Slow-growing versus fast-growing lines

The heat-stress effect has been found to be more pronounced in fast-growing commercial broiler stocks than in non-selected broiler lines (Washburn *et al.*, 1992; Eberhart and Washburn, 1993; Emmans and Kyriazakis, 2000; Yalcin *et al.*, 2001) or broiler lines selected for traits other than GR (Leenstra and Cahaner, 1991, 1992; Cahaner and Leenstra, 1992). In the latter study, males and females from five lines were reared under three AT: high AT (constant 32°C), normal AT (gradually reduced, averaging 22°C between 4 and 7 weeks of age), and low AT (more rapidly reduced, averaging 15°C between 4 and 7 weeks of age). Two lines were characterized by rapid GR, similar to contemporary commercial broilers, following selection for BW in the Netherlands and Israel. The three other lines were GR-relaxed due to several generations of selection only for better feed efficiency (in the Netherlands) or divergently on abdominal fat (in Israel). Weight gain (WG) was significantly affected by interactions between the Israeli versus Dutch lines and low versus normal AT environments (Leenstra and Cahaner, 1991). The increase in feed intake and GR due to lower AT was more pronounced in the Dutch lines than in the Israeli ones, suggesting specific adaptation to the typical AT during selection, which is lower in the Netherlands than in Israel. A significant genotype-by-environment (G × E) interaction was found between normal versus high AT environments and the genotypes (lines) differing in their potential GR. The magnitude of GR reduction due to high AT was larger in the high-GR lines than in those with lower GR. Under high AT, BW of the GR-selected lines was similar to that of the GR-relaxed lines at 6 weeks of age (about 1400 g) and lower at 8 weeks of age (about 1650 versus 1750 g, respectively) (Cahaner and Leenstra, 1992). The AT also interacted significantly with sex: within each line, males exhibited the expected 15% higher BW than females under normal AT, whereas under high AT, males were heavier than females only in the GR-relaxed lines and at 6 weeks of age. At 8 weeks of age, both sexes had similar BW in the GR-relaxed lines, and females were heavier than males in the high-GR lines (Cahaner and Leenstra, 1992). This unique interaction, with females exhibiting higher GR and BW than males, also affected breast meat yield and carcass fat content. It appears to be the only report on females reaching higher BW than males.

The three Israeli-developed lines from the previous study, one high-GR and two GR-relaxed lines (selected divergently for high fat and low fat), were tested for their response to dietary protein under high AT (constant 32°C). A significant diet-by-genotype interaction was found: relative to the low-protein diet, a high-protein diet increased GR of the GR-relaxed lines and reduced GR of the high-GR line (Cahaner et al., 1995). Apparently, the high-protein diet increased the sensitivity of the fast-growing broilers to high AT. This could explain the substantial interactions found between normal versus high AT and genotypes differing in GR (due to selection or sex) in the previous study that was conducted in the Netherlands. On the research farm there, the same diet was used for the entire growth period, containing 21.5% crude protein (Leenstra and Cahaner, 1991), almost like the high-protein grower diet (22.7%) in the present study (Cahaner et al., 1995). The exceptionally higher BW of females versus males under high AT (Cahaner and Leenstra, 1992) may have resulted from the high-protein grower diet used in that study. Thus, since high dietary protein amplifies the effect of high AT on fast-growing broilers, it may lead to overestimated effects of heat stress. On the other hand, it can be used to increase the sensitivity of testing (or selection) for heat tolerance.

Broilers from three stocks with different breeding and selection histories were reared under controlled normal and high AT (Deeb and Cahaner, 2001a). There was a commercial sire line (G1), a commercial dam line (G2), and a non-commercial line derived from the dam line but kept for five generations without selection (G3). High AT reduced growth and meat yield in the progeny of all three stocks, but it was highest in the sire-line bred most intensively for high GR and large breast. This case of G × E interaction can be interpreted as an indication of higher sensitivity to high AT in broiler genotypes with a higher potential for rapid GR and meat yield. It was suggested that the Cornish breed, the base population of today's commercial sire-lines (such as G1 in this study) is more adversely affected by heat than the White Plymouth Rock breed, from which the dam-lines (G2 and G3) were derived.

A study with commercial broiler hybrids took advantage of climatic differences between summer (hot) and autumn (temperate) seasons in Izmir (Turkey). The broilers for both seasons were obtained from the same breeder flocks of three broiler stocks that have been bred in the UK, Germany, and the USA (Yalcin et al., 1997a). The three stocks exhibited similar GR in the temperate season, but differed in the magnitude of heat-induced GR depression in the hot season, leading to a significant stock-by-season interaction. Interestingly, the highest summer performance was exhibited by the USA-bred parent stock, which was imported to Turkey from a grandparent flock in Israel, whereas day-old chicks of the UK- and German-bred parent stock flocks were directly imported to Turkey. This difference, and the similar difference between the Dutch-bred and Israeli-bred lines in their response to AT (Leenstra and Cahaner, 1991), support the concept of 'localized breeding' as a means of developing genetic adaptation to local suboptimal conditions that characterize broiler production in a certain region (Cahaner, 1990, 1996). However, the feasibility of breeding heat-tolerant, fast-growing broilers by selection under hot conditions depends on the within-line genetic correlation between these traits (Cahaner et al., 1998, 1999).

Slow-growing versus fast-growing families within lines

The association between the genetic variation in GR and susceptibility to heat was experimentally investigated in a commercial sire-line. A single cycle of selection for high BW was conducted, similar to commercial broiler breeding (Deeb and Cahaner, 2002). Three males and 15 females were selected from the birds with highest BW at 35 days of age in a large flock (n ≅ 10,000), and three males and 15 females with average BW were sampled from the same flock. The males were mated with the similarly selected/sampled females, producing two groups of progeny: GR-selected and non-selected. After brooding, each group was equally divided into controlled normal AT (constant 22°C) and high AT (constant 32°C). At 42 days of age, the selected group was heavier than the non-selected group by 283 g (2386 versus 2103 g) under normal AT, and by only 46 g (1452 versus 1406 g) under high AT, leading to a significant GR-by-AT interaction. This G × E interaction was more pronounced for the GR during the last 2 weeks: under normal AT, it averaged 66 g/day in the non-selected group versus 78 g/day in the selected group, whereas under high AT, both groups exhibited the same mean GR, 25 g/day, during these weeks (Deeb and Cahaner, 2002). These results indicate that the genetic GR enhancement in response to selection on GR under normal AT, is hardly expressed under high AT.

The G × E interaction with regards to GR and AT was also evaluated under natural temperate (spring) and hot (summer) environments in western Turkey. The experimental population was produced by a full-pedigree, random-mating scheme where 29 commercial broiler males served as sires, each mated to five females from the same stock and produced progeny in a spring hatch and a summer hatch (Settar et al., 1999). The G × E interaction was evaluated by correlating sires' means of their spring and summer progeny. The correlation between the two seasons for GR from 4 to 7 weeks and BW at 7 weeks was negative (not significantly lower than $P = 0$). Sire-by-season ANOVA also revealed highly significant G × E interaction effects. These results suggest that genotypes (i.e. sires) with a potential for higher GR under normal AT exhibit lower GR under the hot conditions.

Selection on GR under Hot Conditions

The studies reported above indicate that the breeding of broilers for rapid GR under optimal conditions, as practised by the major broiler breeding companies, cannot improve their actual GR under hot conditions. Breeding for improved adaptation to a particular stressful environment should be the strategy of choice when G × E interaction significantly affects economically important traits (Cahaner, 1990; Mathur and Horst, 1994). Such a breeding may take place in a particular stressful location ('localized breeding') or under artificially induced stress. Experimental selection of broilers under controlled hot conditions was reported only once, about 15 years ago (El-Gendy et al., 1992), probably indicating that this approach was not successful. Commercial localized breeding of broilers under suboptimal conditions, including heat, has been applied successfully in India (Jain, 2000, 2004). When compared under local conditions, the locally selected broilers out-performed the imported broiler stocks. However, the performance under suboptimal conditions of the broilers selected commercially for suboptimal conditions is inferior to the potential of the broiler stocks that had been selected under optimal conditions, thus confirming the reported experimental results indicating that broilers cannot be bred to exhibit rapid GR and high body weight under hot conditions. So far, the latter fact has not been an important limitation in most hot-climate countries where customers prefer to buy small-body (about 1.5 kg) live broilers. However, large body weight at marketing is essential for the economic efficiency of mechanical slaughtering and processing of broilers. Thus, with the expected increase in the consumption of carcass parts and deboned meat, it will be impossible to avoid the negative effects of heat by marketing small-body broilers.

The Effect of AT on Feed Intake in Broilers Differing in GR

A better understanding of the limiting factors that inhibit broiler growth under high temperature may facilitate a genetic adaptation and higher GR and body weight under hot conditions. In the study which consisted of one cycle of high-intensity selection for high BW (Deeb and Cahaner, 2002), the higher GR of the selected group under normal AT resulted from similarly higher feed consumption, as compared with the non-selected group. These differences were halved under high AT from 17 to 28 days of age, and reversed from 28 to 42 days of age, where the selected broilers consumed significantly less feed and consequently gained less BW than the non-selected ones. Also the reduction in metabolic heat production under high AT was more pronounced in fast-growing broilers than in their slow-growing counterparts (Sandercock et al., 1995). The role of feed intake is important because the genetic potential for more rapid growth is based on enhanced tendency to consume feed. The higher rate of feed intake (per time unit) and metabolism in genetically high-GR broilers increase the amount of internally produced heat. Broilers must dissipate this excess of heat, but the efficiency of heat dissipation is significantly reduced by high AT. Therefore under hot conditions, today's fast-growing broilers cannot dissipate all the internal heat they generate; consequently many of them die due to lethal elevation in their body temperature (El-Gendy et al., 1992; Yaron et al., 2004), and the others avoid mortality by 'voluntarily' reducing feed intake and actual GR (Deeb and Cahaner, 2001b, 2002). However, these consequences increase rearing time to marketing and negatively affect feed conversion ratio, and thus significantly reduce the efficiency of broiler production in hot climates.

The Effects of Feather Coverage on Heat Tolerance in Broilers

Heat dissipation is hindered due to the insulation provided by the feather coverage. This insulation is advantageous in slow-growing chickens or when broilers are reared at low ambient temperatures. Under high AT, however, the feather coverage has a negative effect on thermoregulation (Deeb and Cahaner, 1999; Yunis and Cahaner, 1999). This negative effect can be countered by costly cooling and ventilation systems, which are not feasible or applicable in developing countries. The genetic alternative would be to introduce genes that reduce the feather coverage into the genetic makeup of fast-growing broiler stocks (Cahaner et al., 1994). Reduced feather mass may also contribute to increased meat yield, if the saved protein goes into building more muscles, as suggested by Cahaner et al. (1987).

The Effects of the Naked-neck Gene (*Na*) on Feather Coverage and Heat Tolerance in Broilers

Many studies have been conducted with the 'naked-neck' (*Na*) gene, which is quite common in rural chicken populations in hot regions (Merat, 1990). This gene reduces feather coverage by about 15–20% and 30–40% in heterozygous (*Na/na*) and homozygous (*Na/Na*) chickens, respectively (Cahaner et al., 1993; Yunis and Cahaner, 1999). It was suggested in the 1980s that heat tolerance of chickens can be improved by the *Na* gene (Merat, 1986, 1990). Compared to their normally feathered counterparts, naked-neck broilers were shown to have a higher rate of heat dissipation (Yahav et al., 1998) and better thermoregulation in hot conditions (Deeb and Cahaner, 1999), resulting in higher actual GR and meat production at controlled high AT (Cahaner et al., 1993; Deeb and Cahaner, 2001a) and natural hot conditions (Yalcin et al., 1997b). The effect of the *Na/na* genotype on performance at high AT differed in magnitude between commercial and non-commercial broiler lines, due to a large difference in potential GR (Yunis and Cahaner, 1999; Deeb and Cahaner, 2001a). In conclusion, many studies conducted under artificial high AT and in hot climates have demonstrated the advantage of naked-neck broilers over their normally feathered counterparts. However, although to a lesser extent than normally feathered broilers, their naked-neck counterparts also suffered substantially under heat stress. Enhancing the magnitude of reduction in feather coverage by combining the effects of the *Na* gene with those of the *F* gene for frizzled feathers was unsuccessful due to the latter's negative effects (Yunis and Cahaner, 1999). Therefore, it was hypothesized that a total elimination of feathers is required in order to maximize heat tolerance of fast-growing broilers under hot conditions (Cahaner et al., 2003).

Discovery of the Scaleless Gene (*Sc*), which Eliminates Feather Development

Featherless broilers were derived from the recessive spontaneous mutation called 'Scaleless', which was found in California in 1954 (Abbott and Asmundson, 1957). The mutated allele, designated *sc*, occurred in the New Hampshire egg-type breed, which is characterized by much lower GR and BW than contemporary commercial broilers. Due to their low GR, the original featherless mutants have not been considered for practical purposes (Somes, 1990). In the late 1970s, an attempt was made to derive featherless broilers from a cross between the scaleless mutant and a broiler stock of those days. Growth and carcass composition (Somes and Johnson, 1982) and cooking characteristics (Somes and Wiedenhefft, 1982) were tested. The results suggested an advantage of the featherless birds under hot conditions, but the effects were negligible because the GR of the birds used in these studies was very low, with maximum daily weight gain of 30 g (as compared with about 100 g/day in today's broilers). These birds reached an average body weight of about 1200 g at 8 weeks of age whereas today's broilers reach this body weight at about 4 weeks of age.

Therefore, for the research to be presented below, the original scaleless mutant line has been backcrossed repeatedly to contemporary fast-growing broilers. After several cycles of such backcrossing, accompanied with intensive selection on body weight, an experimental line was developed. The birds in this line are either normally feathered (+/*sc*, carriers of the *sc* allele) or featherless (*sc*/*sc*), all with a genetic potential GR similar to that of contemporary commercial broilers (Cahaner and Deeb, 2004; Cahaner and Druyan, 2007). At that stage, the featherless broilers and their feathered sibs (brothers and sisters, sharing the same genetic background) are capable of exhibiting economically important advantages when reared under warm or hot conditions.

Comparing Featherless and Naked-neck Broilers

A controlled trial was conducted with four experimental genetic groups (normally feathered, heterozygous naked-neck, homozygous naked-neck, and featherless), progeny of the same double-heterozygous parents (*Na/na* + /sc), and a commercial line (Ajuh et al., 2005). The birds from all five groups were brooded together. On day 21, each group was separated into two equal-size subgroups. Birds from one subgroup were pooled and reared together under normal AT (constant 25°C), and the other subgroups were pooled and reared together under high AT (constant 35°C). Live body weight and body temperature were recorded from hatch to slaughter – when carcass yields were recorded – at 46 and 52 days of age for birds at normal and hot conditions, respectively.

Only the featherless birds exhibited similar GR under the two temperatures, while the commercial broilers exhibited the highest and earliest depression in weight gains under hot versus AT. Mean breast yield of the naked-neck genotypes (*na/na*, *Na/na*, *Na/Na*) increased as their feather coverage decreased from 100% to 80% and 60%, respectively. This trend was lower under hot conditions. Mean breast meat yield of the featherless birds was much higher than their sibs, and also higher than the commercial broilers under hot conditions and similar to them under normal conditions. The elevation in body temperature due to rearing under hot versus normal conditions was highest (+2.0°C) in the commercial broilers. Significantly less elevation (1.4°C) was exhibited by the fully feathered (*na/na*) experimental birds (relatively slow-growing), and it was lower (1.3°C and 1.2°C) in the heterozygous and homozygous naked-neck birds. Only in the featherless broilers was body temperature the same under normal and hot conditions, as was their GR and breast meat yield. Thus, the effects of being featherless were in the same direction as those reported previously for naked-neck broilers, but much larger in magnitude. The results suggested that introduction of the *sc* gene into fast-growing commercial broiler lines is a promising approach to improve production of broiler meat in hot conditions. Accordingly, the consequent trials with featherless broilers, always compared with their feathered sibs and to commercial broilers as industry reference, covered all aspects of broiler production, from welfare and livability under heat, to GR, feed consumption and conversion, stocking density, meat yield and meat quality.

Heat Tolerance of Featherless Broilers versus their Feathered Counterparts

Featherless broilers and their feathered sibs were kept in an AT-controlled chamber (Yaron *et al.*, 2004). After brooding, AT of 28°C and 30°C was maintained up to 42 and 46 days of age, respectively. On day 47, when body weights averaged 1600 g, AT was elevated to 35°C, leading to an increase in body temperature (BT) of the feathered birds to 41.9°C, whereas the BT of the featherless birds remained unchanged (41.2°C). A second day (day 48) at an AT of 35°C further increased BT of the normally feathered birds to an average of 42.8°C, and only to 41.4°C in the featherless birds. On days 49 to 52, AT was reduced to 30°C and then increased to 36°C on day 53, when body weight averaged 1900 g. This 'heatwave' led to the death of 17 feathered birds (35%), with the BT of the surviving ones averaging 43.2°C. In contrast, mean BT of the featherless birds remained at 41.4°C and only one of them died. These results clearly indicate that being featherless assures the welfare and livability of broilers under acute hot conditions (Cahaner, 2007a; Cahaner and Druyan, 2007). Maintaining normal BT even under extreme AT is probably the key for the heat resistance of the featherless broilers. Elevated BT under heat stress was shown to negatively affect weight gain, feed consumption and feed utilization in standard broilers (Cooper and Washburn, 1998), suggesting that the prevention of reduced feed consumption and growth of featherless broilers under hot conditions is due to the capacity to dissipate the internally produced heat and maintain normal BT (Yadgari et al., 2006; Cahaner, 2007a).

In a trial in June 2007, featherless broilers and their feathered sibs were kept together on litter under constant hot conditions (32°C +/– 1°C between days and nights). When the birds were 41 days old, with mean body weight of about 1750 g, the AT in one of the trial's rooms was accidentally elevated to 38°C for about 5 h. Consequently, 50 out of 100 feathered birds in this room died, as compared with only 2 (also from 100) dead featherless birds. This event, although unfortunate for the birds that died and for the trial, proved that featherless birds have nearly complete resistance to heat.

Superior Performance under High AT of Featherless Broilers Compared with their Feathered Counterparts

When reared in a hot environment (constant 35°C), the featherless broilers and their feathered sibs exhibited similar GR up to 3 weeks of age, but at the 4th week the featherless broilers exhibited an advantage of 7.7% in GR over their feathered counterparts (Yaron-Azoulay, 2007). This advantage increased with age, and on day 53 (the trial's end) mean body weight of the featherless broilers was 46% higher (800 g) than that of their feathered counterparts. This difference reflected a substantial negative effect of the hot conditions on the growth of the feathered broilers compared with their counterparts under temperate environment (constant 25°C) in the same trial. The featherless broilers exhibited similar GR and mean body weight in both environments. Under hot conditions, the feed conversion ratio of the featherless broilers was also much better than that of their feathered counterparts. This is probably because the latter used more nutrients and dietary energy to develop feathers and to actively maintain normal body temperature, and thus hardly gained weight.

The featherless broilers also differed from their feathered counterparts in the size (relative to body weight) of the heart and liver, the breast meat, and the wings (Yaron-Azoulay, 2007). Heart and liver were larger in the featherless broilers, indicating a higher metabolism and activity of the blood system. The larger breast muscle in the featherless broilers, 20% of live body weight as compared with 16% in their feathered sibs, may also reflect higher metabolism and better supply of nutrients. The latter possibility is supported by the more reddish colour of the meat in featherless broilers (Cahaner,

2007a; Okere et al., 2007; Yadgari et al., 2007a). The wings of the featherless broilers were smaller, due to lower weight (per area) of their skin, which lacks the mass of feather follicles and the fat that adheres to them in the skin of feathered birds.

Featherless Broilers Produce More High-quality Meat in Hot Conditions and High Stocking Densities

A series of similar trials were conducted, each consisting of three groups: featherless broilers, their feathered sibs, and feathered commercial broilers as industry reference (Yadgari et al., 2006, 2007a; Cahaner, 2007 a,b). In each trial, all the chicks were hatched and brooded together in a room with air temperature declining from 36°C (day 1) to about 32°C on day 16, and maintained at that level thereafter, without forced ventilation. The rooms were divided by metal-screen fences into six or eight pens, with the area ranging between 1.75 and 2.7 m^2, depending on the trial. In the first trial, three stocking densities were applied, each in two pens: 12 birds/m^2 (medium), 17 birds/m^2 (high), and 22 birds/m^2 (very high). In each pen, one-third of the chicks were featherless, one-third feathered sibs, and one-third feathered commercial. In the second trial, featherless and feathered broilers were reared in separate pens. The feathered birds were housed at low, medium and high densities (7, 12 and 17 per m^2) whereas the featherless birds were housed at medium, high and very high densities (12, 17 and 22 per m^2). In a third trial, the air temperature was higher (around 34°C) and hence lower stocking densities were applied: eight pens with featherless birds (14/m^2) and eight pens with feathered birds (7/m^2). Four pens from each phenotype were fed with standard diets, and the other four pens received diets with 10% less protein and energy.

Body weight of each individual bird was measured twice a week, and feed intake per pen was measured daily (third trial). On day 44, samples of birds from all pens were taken to a commercial slaughterhouse, where their body weight was determined just before killing, after 10 h of feed withdrawal. After standard killing and evisceration, carcasses were air chilled for 24 h, then cut up and deboned by a professional. The weight of carcass, deboned breast and thigh meat, wings, and drumsticks were recorded. Three colour parameters (L^* = lightness, a^* = redness and b^* = yellowness) of the breast meat were determined by a Minolta colorimeter, and pH was measured. Breast meat was held at 5°C for 48 h, and drip loss was determined from the weight reduction. Weight was recorded for the wings' three sections (upper wing, lower wing, and wing tip), and for the skin of the lower wing (middle section).

As expected, GR of the feathered broilers was depressed by the trials' hot conditions, and the effect increased with stocking density and with age. Mean body weight at 44 days of age ranged from 2.4 kg (7 birds/m^2) to 1.8 kg (17 birds/m^2). In contrast, GR of the featherless broilers was hardly affected by density, with mean body weight of 2.4 kg (12 birds/m^2), 2.2 kg (17/m^2), and 2.1 kg (22/m^2), with the latter resulting in live weight production of 46 kg/m^2. The GR and body weight were not affected in featherless broilers that were fed the low-protein/low-energy diet, whereas a 7% reduction in performance was observed in their feathered counterparts. Apparently, the protein used to build the feathers and the energy used to maintain normal body temperature of the feathered broilers in hot conditions are saved in featherless broilers, thus allowing them to perform normally with low-density diets. About 15% of the feed nutrients, mainly the amino acids, are used to build the feathers, which are later discarded as waste at the slaughterhouse. These nutrients are saved in featherless broilers, thus contributing to their higher meat yield. Moreover, a 10% reduction in dietary nutrients in a recent nutritional trial led to a proportional reduction in the performance of feathered broilers but did not reduce growth rate, FCR and meat yield in featherless broilers (Cahaner, 2007a).

The stress of heat and density reduced the breast meat yield of the feathered broilers to 15%, with the meat being pale (L* = 50, a* = 4) and exudative (4% drip loss). In the featherless broilers, breast meat yield averaged 19% in all stocking densities, with the meat being darker (L* = 44 and a* = 5), and with less than 2% drip loss (Cahaner, 2007a,b; Okere *et al.*, 2007; Yadgari *et al.*, 2006, 2007a,b). Thus, in contrast to the reported negative association between muscle growth and meat quality (Dransfield and Sosnicki, 1999; Lu *et al.*, 2007), especially under heat stress (Mitchell and Sandercock, 1995b), which leads to the development of PSE syndrome (pale, soft, exudative) with all its negative effects on meat quality (Barbut, 1997; Woelfel *et al.*, 2002; Barbut *et al.*, 2005), the featherless broilers were also capable of producing high-quality, large breast muscles under hot conditions (Cahaner and Druyan, 2007).

Wings were smaller in the featherless broilers than in their feathered counterparts: 7.1 versus 8.1% of body weight and 9.3 versus 11.1% of carcass weight, due to lower fat content in the skin of featherless broilers. Wing tips (mainly bone and skin) were 12.2 versus 10.5% of wing weight in the feathered and featherless broilers, respectively. Similarly, the skin (plus its fat) was 31% of the weight of the wing's middle section in the feathered broilers, versus 23% only in the featherless broilers. Thus, the featherless broilers produced wings of higher nutritive quality, with less fat and more meat (Yadgari *et al.*, 2007a,b).

Summary

Broilers cannot be bred to exhibit high growth rate and large body weight at marketing under hot conditions without using costly housing and cooling systems. This, however, can possibly be achieved genetically by reduction or elimination of the feather coverage, to allow efficient dissipation of excessive internally produced heat without artificial ventilation.

The benefits of reduced feather coverage, mainly less heat-induced growth depression and higher breast meat yield, were demonstrated in many trials with naked-neck broilers during the last two decades. However, the magnitude of heat alleviation due to the naked neck was not large enough to justify commercialization. Thus, featherless broilers, homozygous *sc/sc* in the 'scaleless' gene, were employed in order to completely remove the feather coverage and thus increase the magnitude of heat alleviation.

Results from a series of trials over the last 4 years show that fast-growing featherless broilers substantially improve the cost-effectiveness of poultry meat production in hot and humid climates. Due to lower costs of feed and labour, the majority of broilers worldwide are reared under hot climates in tropical and subtropical regions of Asia, America and Africa, and the southern regions of the USA and China. Contemporary commercial broilers require low environmental temperatures in order to fully express their genetic potential of fast growth rate and high meat yield and quality. In hot conditions, the feathers of standard broilers prevent efficient dissipation of internal heat; consequently their growth rate and meat yield are dramatically reduced and many of them die before marketing. Currently, these negative consequences can be countered only by very costly, energy-dependent cooling systems that increase costs and reduce the competitiveness and sustainability of broiler production in hot climates.

A unique stock of fast-growing featherless broilers was developed by a complex breeding programme that used the *sc* gene, a well-documented natural mutation. Without feathers, these broilers easily dissipate excessive body heat, even under very hot conditions and at high stocking densities, and they do not need any artificial cooling or ventilation.

Featherless broilers in hot conditions

- Are resistant to heat stresses, which causes substantial growth depression and mortality among feathered broilers.
- Do not suffer from heat-related mortality or welfare problems.
- Fully express their genetic potential for rapid growth and heavy body, thus facilitating industrial slaughtering and marketing of processed or cooked products, which are safer than 'live marketing'.
- Require fewer days than their feathered counter parts to reach marketing body weight (about 2.25 kg at 6 weeks).
- Higher breast meat yield: about 20% of live weight (much higher than standard feathered broilers, which produce only about 14% breast meat in hot conditions, and about 18% in normal conditions).
- Higher quality of breast meat: not pale, not soft, and better water-holding capacity.
- Exhibit these advantages even under high stocking densities (over 40 kg/m^2), yet with high-quality broilers that enjoy good welfare.
- Exhibit these advantages on low-protein low-energy feed that costs 10% less than regular feed.

Breeding and production

Standard grandparent stock (GPS) of any broiler breeding company can be used, with featherless males in both paternal and maternal sides producing heterozygous males and females for the parent stock (PS) flocks. The *sc* allele is recessive and hence these males and females are normally feathered, and therefore the management and reproductive performances are the same as in standard PS flocks. Among the boiler progeny of these parents, 25% are featherless and the 75% remaining ones are standard (feathered) broilers. If featherless males are used in the PS flocks, 50% of the broilers produced are featherless.

Featherless broilers are especially suitable for companies that own all the components of a fully integrated broiler-meat production operation: breeding flocks (grandparent or parent stock), hatchery, broiler-rearing facilities, and slaughterhouse (preferably with deboning and further-processing facilities). In the suggested breeding scheme, 75% (or 50%) of the broilers are with feathers, which can be used to supply the live-bird market or to grow in well-ventilated houses. The featherless broilers can be reared in non-ventilated houses, and supply industrial slaughterhouses and processing plants.

References

Abbott, U.K. and Asmundson, V.S. (1957) Scaleless, an inherited ectodermal defect in domestic fowl. *Journal of Heredity* 48, 63–70.

Ain Baziz, H., Geraert, P.A., Padilha, J.C.F. and Guillaumin, S. (1996) Chronic heat exposure enhances fat deposition and modifies muscle and fat partition in broiler carcasses. *Poultry Science* 75, 505–513.

Ait Boulahsen, A., Garlich, J.D. and Edens, F.W. (1993) Calcium deficiency and food deprivation improve the response of chickens to acute heat stress. *Poultry Science* 123, 98–105.

Ait Boulahsen, A., Garlich, J.D. and Edens, F.W. (1995) Potassium chloride improves the thermotolerance of chickens exposed to acute heat stress. *Poultry Science* 74, 75–87.

Ajuh, J., Siegmund-Schultze, M., Valle Zárate, A., Azoulay, Y., Druyan, S. and Cahaner, A. (2005) The effect of naked neck (*Na*) and featherless (*sc*) genes on performance losses of broilers reared under hot temperature. In: *4th European Poultry Genetic Symposium*, Dubrovnik, Croatia (abstract).

Barbut, S. (1997) Problem of pale, soft, exudative meat in broiler chickens. *British Poultry Science* 38, 355–358.

Barbut, S., Zhang, L. and Marcone, M. (2005) Effects of pale, normal, and dark chicken breast meat on microstructure, extractable proteins, and cooking of marinated fillets. *Poultry Science* 84, 797–802.

Cahaner, A. (1990) Genotype by environment interactions in poultry. *Proceedings of the 4th World Congress on Genetics Applied to Livestock Production*, Edinburgh, pp. 13–20.

Cahaner, A. (1996) Improving poultry production under climatic stress through genetic manipulation. *Proceedings of the 20th World's Poultry Congress*, New Delhi, India, Vol. 1, pp. 127–139.

Cahaner, A. (2007a) Featherless broilers facilitate industrial production of quality meat under hot conditions. *Proceedings of FAO Conference on Poultry in the 21st Century*, Bangkok (abstract).

Cahaner, A. (2007b) Being featherless (homozygous sc/sc) provides fast-growing high-yield broilers with genetic adaptation to hot conditions. *Proceedings of the 1st International Conference on Food Safety of Animal Products*, Amman.

Cahaner, A. and Deeb, N. (2004) Breeding broilers for adaptability to hot conditions. *Proceedings of the 22nd World Poultry Congress*, Istanbul, Turkey.

Cahaner, A. and Druyan, S. (2007) The effect of increased growth rate of broilers on their susceptibility to a-biotic stresses. *Proceedings of the 5th European Poultry Genetic Symposium*, Braedstrup (Denmark), pp 33–52.

Cahaner, A. and Leenstra, F. (1992) Effects of high temperature on growth and efficiency of male and female broilers from lines selected for high weight gain, favorable feed conversion, and high or low fat content. *Poultry Science* 71, 1237–1250.

Cahaner, A., Dunnington, E.A., Jones, D.E., Cherry, J.A. and Siegel, P.B. (1987) Evaluation of two commercial broiler male lines differing in efficiency of feed utilization. *Poultry Science* 66, 1101–1110.

Cahaner, A., Deeb, N. and Gutman, M. (1993) Effects of the plumage-reducing naked-neck (*Na*) gene on the performance of fast-growing broilers at normal and high ambient temperatures. *Poultry Science* 72, 767–775.

Cahaner, A., Yunis, R. and Deeb, N. (1994) Genetics of feathering and heat tolerance in broilers. *Proceedings of the 9th European Poultry Conference*, Glasgow, UK, Vol. II, pp. 67–70.

Cahaner, A., Pinchasov, Y., Nir, I. and Nitsan, Z. (1995) Effects of dietary protein under high ambient temperature on body weight, breast meat yield and abdominal fat deposition of broiler stocks differing in growth rate and fatness. *Poultry Science* 74, 968–975.

Cahaner, A., Deeb, N., Yunis, R. and Lavi, Y. (1998) Reduced stress tolerance in fast growing broilers. *Proceedings of the 10th European Poultry Conference*, Jerusalem, Israel, Vol. I, pp. 113–117.

Cahaner, A., Yunis, R., Lavi, Y. and Deeb, N. (1999) Interactions between single-genes, polygenes, genetic backgrounds and environments, and their implication to broiler breeding and production. *Proceedings of the 1st European Poultry Genetics Symposium*, Mariensee, Germany, pp. 50–58.

Cahaner, A., Druyan, S. and Deeb, N. (2003) Improving broiler meat production, especially in hot climates, by genes that reduce or eliminate feather coverage. *British Poultry Science* 44 (Suppl.), 22–23.

Cooper, M.A. and Washburn, K.W. (1998) The relationships of body temperature to weight gain, feed consumption, and feed utilization in broilers under heat stress. *Poultry Science* 77, 237–242.

De Basilio, V., Requena, F., Leon, A., Vilarino, M. and Picard, M. (2003) Early age thermal conditioning immediately reduces body temperature of broiler chicks in a tropical environment. *Poultry Science* 82, 1235–1241.

Deeb, N. and Cahaner, A. (1999) The effect of naked-neck genotypes, ambient temperature, feeding status and their interactions on body temperature and performance of broilers. *Poultry Science* 78, 1341–1346.

Deeb N. and Cahaner, A. (2001a) Genotype-by-environment interaction with broiler genotypes differing in growth rate: 1. The effects of high ambient temperature and naked-neck genotype on stocks differing in genetic background. *Poultry Science* 80, 695–702.

Deeb, N. and Cahaner, A. (2001b) Genotype-by-environment interaction with broiler genotypes differing in growth rate: 2. The effects of high ambient temperature on dwarf versus normal broilers. *Poultry Science* 80, 541–548.

Deeb, N. and Cahaner, A. (2002) Genotype-by-environment interaction with broiler genotypes differing in growth rate: 3. Growth rate, water consumption and meat yield of broiler progeny from weight-selected vs. non-selected parents under normal and high ambient temperatures. *Poultry Science* 81, 293–301.

Dransfield, E. and Sosnicki, A.A. (1999) Relationship between muscle growth and poultry meat quality. *Poultry Science* 78, 743–746.

Eberhart, D.E. and Washburn, K.W. (1993) Assessing the effects of the naked neck gene on chronic heat stress resistance in two genetic populations. *Poultry Science* 72, 1391–1399.

El-Gendy, E., Washburn, K.W. and Eberhart, D.E. (1992) Selection for heat tolerance in young chicken. *Proceedings of the 19th World's Poultry Congress*, Amsterdam, the Netherlands, Vol. 2, p. 65.

Emmans, G.C. and Kyriazakis, I. (2000) Issues arising from genetic selection for growth and body composition characteristics in poultry and pigs. In: Hill, W.G., Bishop, S.C., McGuirk, B., McKay, J.C., Simm, G. and Webb. A.J. (eds) *The Challenge of Genetic Change in Animal Production*. Occasional Publication No. 27, British Society of Animal Science, Edinburgh, UK, pp. 39–53.

Havenstein, G.B., Ferket, P.R., Scheideler, S.E. and Larson, B.T. (1994a) Growth, livability, and feed conversion of 1957 vs 1991 broilers when fed "typical" 1957 and 1991 broiler diets. *Poultry Science* 73, 1785–1794.

Havenstein, G.B., Ferket, P.R., Scheideler, S.E. and Rives, D.E. (1994b) Carcass composition and yield of 1991 vs. 1957 broilers when fed "typical" 1957 and 1991 broiler diets. *Poultry Science* 73, 1795–1804.

Havenstein, G.B., Ferket, P.R. and Qureshi, M.A. (2003a) Carcass composition and yield of 1957 vs. 2001 broilers when fed representative 1957 and 2001 broiler diets. *Poultry Science* 82, 1509–1518.

Havenstein, G.B., Ferket, P.R. and Qureshi, M.A. (2003b) Growth, livability, and feed conversion of 1957 vs. 2001 broilers when fed representative 1957 and 2001 broiler diets. *Poultry Science* 82, 1500–1508.

Jain, G.L. (2000) Breeding a broiler for the Indian market. *Proceedings of the 21st World Poultry Congress*, Montreal, Canada.

Jain, G.L. (2004) Breeding for special needs of developing countries in collaboration with international companies. *Proceedings of the 22nd World Poultry Congress*, Istanbul, Turkey.

Leenstra, F. and Cahaner, A. (1991) Genotype by environment interactions using fast-growing, lean or fat broiler chickens, originating from The Netherlands and Israel, raised at normal or low temperature. *Poultry Science* 70, 2028–2039.

Leenstra, F. and Cahaner, A. (1992) Effects of low, normal, and high temperatures on slaughter yield of broilers from lines selected for high weight gain, favorable feed conversion, and high or low fat content. *Poultry Science* 71, 1994–2006.

Lu, Q., Wen, J. and Zhang, H. (2007) Effect of chronic heat exposure on fat deposition and meat quality in two genetic types of chicken. *Poultry Science* 86, 1059–1064.

Mathur, P.K. and Horst, P. (1994) Genotype by environment interactions in laying hens based on relationship between breeding values of sires in temperate and tropical environments. *Poultry Science* 73, 1777–1784.

May, J.D., Lott, B.D. and Simmons, J.D. (2000) The effect of air velocity on broiler performance and feed and water consumption. *Poultry Science* 79, 1396–1400.

Merat, P. (1986) Potential usefulness of the *Na* (naked neck) gene in poultry production. *World's Poultry Science Journal* 42, 124–142.

Merat, P. (1990) Pleiotropic and associated effect of major genes. In: Crawford, R.D. (ed.) *Poultry Breeding and Genetics*. Elsevier, Amsterdam, the Netherlands, pp. 429–467.

Mitchell, M.A. and Sandercock, D.A. (1995a) Creatine kinase isoenzyme profiles in the plasma of the domestic fowl (*Gallus domesticus*): effects of acute heat stress. *Research in Veterinary Science* 59, 30–34.

Mitchell, M.A. and Sandercock, D.A. (1995b) Increased hyperthermia induced skeletal muscle damage in fast growing broiler chickens? *Poultry Science* 74 (Suppl. 1), 74.

Mitchell, M.A. and Sandercock, D.A. (1997) Possible mechanisms of heat stress induced myopathy in the domestic fowl. *Journal of Physiology and Biochemistry* 53, 75.

Okere, I., Siegmund-Schultze, M., Cahaner, A. and Valle Zárate, A. (2007) Better breast meat quality in featherless broilers than in their feathered sibs under hot temperature conditions. *Proceedings of the 1st International Conference on Food Safety of Animal Products*, Amman.

Sandercock, D.A., Mitchell, M.A. and Macleod, M.G. (1995) Metabolic heat production in fast and slow growing broiler chickens during acute heat stress. *British Poultry Science* 36, 868.

Sandercock, D.A., Hunter, R.R., Nute, G.R. and Mitchel, M.A. (2001) Acute heat stress induced alterations in blood acid base status and skeletal muscle membrane in broiler chickens at two ages: implications for meat quality. *Poultry Science* 80, 418–425.

Settar, P., Yalcin, S., Turkmut, L., Ozkan, S. and Cahaner, A. (1999) Season by genotype interaction related to broiler growth rate and heat tolerance. *Poultry Science* 78, 1353–1358.

Somes, R.G. (1990) Mutations and major variants of plumage and skin in chickens. In: Crawford, R.D. (ed.) *Poultry Breeding and Genetics*. Elsevier, Amsterdam, the Netherlands, pp. 169–208.

Somes, R.G. and Johnson, S. (1982) The effect of the scaleless gene, *sc*, on growth performance and carcass composition of broilers. *Poultry Science* 61, 414–423.

Somes, R.G. and Wiedenhefft, M. (1982) Cooked and organoleptic characteristics of scaleless broiler chickens. *Poultry Science* 61, 221–225.

Washburn, K.W., El-Gendy, E. and Eberhart, D.E. (1992) Influence of body weight on response to heat stress environment. *Proceedings of the 19th World's Poultry Congress*, Amsterdam, the Netherlands, Vol. 2, pp. 53–56.

Woelfel, R.L., Owens, C.M., Hirschler, E.M., Martinez Dawson, R. and Sams, A.R. (2002) The characterization and incidence of pale, soft, and exudative broiler meat in commercial processing plant. *Poultry Science* 81, 579–584.

Yadgari, L., Kinereich, R., Druyan, S. and Cahaner, A. (2006) The effects of stocking density under hot conditions on growth, meat yield and meat quality of featherless and feathered broilers. *World's Poultry Science Journal* 62 (Suppl.), 603–604. (Abstract in *Proceedings of the XII European Poultry Conference*, Verona, Italy.)

Yadgari, L., Astrachan, N. and Cahaner, A. (2007a) Meat quality of featherless vs. feathered broilers under heat. *Proceedings of the XVIII European Symposium on Poultry Meat Quality*, Prague (abstract).

Yadgari, L., Kinereich, R., Astrachan, N., Hadad, Y., Druyan, S. and Cahaner, A. (2007b) Featherless broilers produce more high-quality meat in hot conditions and high stocking densities. *Proceedings of the VIII Asian-Pacific Poultry Conference*, Bangkok, Thailand.

Yahav, S., Luger, D., Cahaner, A., Dotan, M., Rusal, M. and Hurwitz, S. (1998) Thermoregulation in naked-neck chickens subjected to different ambient temperatures. *British Poultry Science* 39, 133–138.

Yalcin, S., Settar, P., Ozkan, S. and Cahaner, A. (1997a) Comparative evaluation of three commercial broiler stocks in hot versus temperate climates. *Poultry Science* 76, 921–929.

Yalcin, S., Testik, A., Ozkan, S., Settar, P., Celen, F. and Cahaner, A. (1997b) Performance of naked-neck and normal broilers in hot, warm, and temperate climates. *Poultry Science* 76, 930–937.

Yalcin, S., Özkan, S., Türkmut, L. and Siegel, P.B. (2001) Responses to heat stress in commercial and local broiler stocks. I. Performance traits. *British Poultry Science* 42, 149–152.

Yaron, Y., Hadad, Y., Druyan, S. and Cahaner, A. (2004) Heat tolerance of featherless broilers. *Proceedings of the 22nd World Poultry Congress*, Istanbul, Turkey.

Yaron-Azoulay, Y. (2007) The performance of featherless broilers with different genetic potential for rapid growth, reared under warm and moderate temperatures. MSc thesis submitted to the Hebrew University of Jerusalem (in Hebrew, with English abstract).

Yunis, R. and Cahaner, A. (1999) The effects of naked-neck (*Na*) and frizzle (*F*) genes on growth and meat yield of broilers, and their interactions with ambient temperatures and potential growth rate. *Poultry Science* 78, 1347–1352.

4 Behavioural, Physiological, Neuroendocrine and Molecular Responses to Heat Stress

R.J. Etches[1], T.M. John[2] and A.M. Verrinder Gibbins[3]

[1]Origen Therapeutics, 1450 Rollins Road, Burlingame, California, 94010, USA; Departments of [2]Zoology, and of [3]Animal and Poultry Science, University of Guelph, Guelph, Ontario, Canada N1G 2W1

Introduction	49
Heat stress and the maintenance of body temperature	49
Behavioural responses to heat stress	51
Physiological responses to heat stress	53
Acclimatization to high ambient temperature	53
Consumption of feed and water	53
Sensible heat loss through specialized heat exchange mechanisms	54
Sensible heat loss and feather cover in poultry	54
Changes in respiration rate and blood pH	55
Changes in plasma concentrations of ions	56
Heart rate, cardiac output, blood pressure and total peripheral resistance	57
Hormonal involvement in thermoregulation	58
Neurohypophyseal (posterior pituitary) hormones	58
Growth hormone (GH)	60
The hypothalamic–pituitary adrenal axis	60
Melatonin	62
Reproductive hormones	63
Thyroid hormones	63
Heat-shock proteins and heat stress	64
References	69

Introduction

The net energy stored in the tissues of a bird equals the difference between energy intake and energy loss. Metabolism of food and high environmental temperature are potential sources of energy while low environmental temperatures and the maintenance of normal body temperature are potential expenditures of energy. Excessive flow of energy into the body and excessive depletion of energy from the body both lead to death, although many birds can survive conditions in which the potential for energy flux is extreme by invoking various adaptive mechanisms that increase or decrease the flow of energy to or from the environment. The extreme examples of very cold environmental temperature and very hot environmental temperature both lead to death because the animal cannot cope with the excessive flow of energy out of or into, respectively, their body mass. In many parts of the world, particularly in warm tropical and subtropical regions, poultry are maintained in environmental temperatures which require the involvement of intricate molecular, physiological and behavioural changes that enable domestic birds to cope with the flux of energy into their tissues at high ambient temperatures. This chapter describes the range and complexity of molecular, physiological, neuroendocrine and behavioural responses that are invoked to maintain body temperature within the normal range at high ambient temperatures.

Heat Stress and the Maintenance of Body Temperature

Body temperature of domestic chickens is maintained within a relatively narrow range that is usually reflected by the upper and lower limits of a circadian rhythm in deep body temperature. In well-fed chickens that are neither dissipating heat to the environment nor gaining heat from the environment, the upper limit of the circadian rhythm is usually about 41.5°C and the lower limit is about 40.5°C. When exposed to a hot environment and/or performing vigorous physical activity, body temperature might rise by 1 or 2°C as heat is stored. Heat storage cannot continue for extended periods before body temperature increases past the limit that is compatible with life. Conversely, when birds are exposed to very cold environments, heat escapes from the bird and, unless it is replenished by energy from the metabolism of food, body temperature will decline until the bird is incapacitated and dies. These general considerations of the effect of environmental temperature have been synthesized into terminology that is commonly used to discuss the response of homeothermic animals to changes in environmental temperature and they are illustrated in Fig. 4.1. In this diagram, body temperature is approximated as a constant that is maintained over a wide range of environmental temperatures, indicated as the *zone of normothermia*. The lower critical temperature ([a] in Fig. 4.1) is the minimum environmental temperature which, even if maintained over a period of days, is compatible

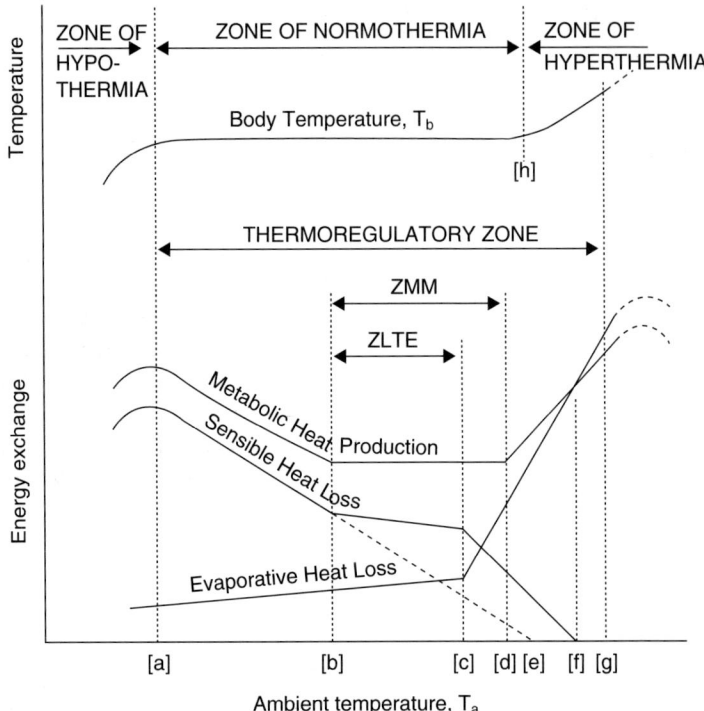

Fig. 4.1. Generalized schematic diagram illustrating T_b (body temperature) and partitioning of energy exchange through a wide range of T_a (ambient temperature). The zones and points are: ZMM, zone of minimum metabolism; ZLTE, zone of least thermoregulatory effort; point [a], lower critical temperature; point [b], critical temperature; point [c], temperature at which intense evaporative heat loss begins; point [d], upper critical temperature; point [e], T_a which equals normothermic T_b; point [f], where sensible heat loss is zero because metabolic heat production equals evaporative heat loss; point [g], critical thermal maximum; point [h], point of incipient hyperthermia. (From Hillman *et al.*, 1985.)

with life. When environmental temperature is less than the lower critical temperature, body temperature begins to decline and death occurs within the *zone of hypothermia*. At the upper end of the zone of normothermia, body temperature increases (indicated by [h] in Fig. 4.1) in the *zone of hyperthermia* until the critical thermal maximum ([g] in Fig. 4.1) is reached, above which the bird expires. Between the lower critical temperature and the critical thermal maximum, thermoregulatory processes are initiated to cope with the ambient temperature. Within the *zone of least thermoregulatory effort* (ZLTE in Fig. 4.1), metabolic heat production is at a minimum, sensible heat loss is relatively constant because physiological and behavioural responses limit the escape of heat throughout this range of environmental temperatures, and evaporative heat losses are limited to those occurring as a by-product of normal respiration and exposure of non-insulated areas of the body. The *zone of minimum metabolism* (ZMM in Fig. 4.1) extends throughout the range of the ZLTE and includes higher ambient temperatures (from [c] to [d] in Fig. 4.1) that can be accommodated by increasing both evaporative and sensible heat loss. Metabolic heat production increases as environmental temperature declines below the ZMM and ZLTE ([b] in Fig. 4.1) to provide energy to maintain body temperature, and increases above the ZMM ([d] in Fig. 4.1) to provide energy for panting. Evaporative heat loss is minimal at low ambient temperatures and increases rapidly as soon as thermoregulation is required to alleviate an increase in ambient temperature, i.e. at the transition from the ZLTE to the ZNM (indicated by [c] in Fig. 4.1). Sensible heat transfer (designated as sensible heat loss in Fig. 4.1), which is cumulative heat transfer from the bird by radiation, conduction and convection, is negative when the environment is colder than the bird and positive when the bird is colder than the environment (indicated by [e] in Fig. 4.1). As metabolic heat production increases in the zone of hyperthermia, body temperature rises and therefore sensible heat losses can increase if the ambient temperature is lower than body temperature. The point at which hyperthermic body temperature is equal to the ambient temperature, and above which sensible heat transfer increases body temperature further into the zone of hyperthermia, is indicated by [f] in Fig. 4.1.

As ambient temperature rises and falls, a wide variety of physiological, behavioural, neuroendocrine and molecular responses are initiated to maintain body temperature within the normal limits. In some instances, the responses are short-term measures that are invoked to withstand a brief period of extreme temperature. The responses can also be invoked to develop a longer-term response aimed at acclimatizing the bird to ambient temperatures that fall within the upper regions of the thermoregulatory zone. Finally, the responses can be terminal reactions that can be sustained for only brief periods; these responses are initiated to cope with extreme and life-threatening environmental conditions.

Behavioural Responses to Heat Stress

During thermal stress, birds alter their behaviour to help maintain body temperature within the normal limits. Behavioural adjustments can occur rapidly and at less cost to the bird than most physiological adjustments (Lustick, 1983), although they are preceded by the molecular response to heat stress that is mediated by the heat-shock proteins (see below).

As ambient temperature increases above the comfort zone, chickens devote less time to walking and standing (see Mench, 1985; McFarlane *et al.*, 1989). During exposure to high temperature, chickens consume less feed and more water (May and Lott, 1992) to compensate for water lost through evaporative cooling (see Mench, 1985), although a reduction in drinking time was observed when heat stress was applied concurrently with other stressors (McFarlane *et al.*, 1989). When exposed to high temperatures, domestic fowl may splash water on their combs and wattles in order to increase evaporative cooling from these surfaces (see Whittow, 1986). Heat-stressed birds also spend relatively less time engaging in social behaviour and in changing their posture. When maintained in cages, heat-stressed chickens tend to distance themselves from each other, pant, and often stand with their wings drooped and lifted slightly from the body to maximize sensible heat loss (see Mench, 1985).

In recent years, it has been shown that chickens will select their thermal environment to increase heat gain if the effective environmental temperature induces heat loss or to increase heat loss if the effective environmental temperature induces heat gain. In a natural environment, the hen would move to a shady area or seek a microenvironment that avoided the environmental extreme. In confinement, chickens will choose a preferred environment using operant control mechanisms (Richards, 1976; Morrison and Curtis, 1983; Morrison and McMillan, 1985, 1986; Morrison et al., 1987a,b; Laycock, 1989; Hooper and Richards, 1991). Operant control over the environment is achieved when the bird learns to perform a simple behaviour (usually pecking a switch) to choose a preferred environmental condition. Operant conditioning can be used to determine the optimum effective environmental temperature, which may be a combination of several factors such as temperature, air speed and humidity. In a study on young chicks, Morrison and McMillan (1986) observed that the birds responded rapidly to changing ambient temperatures when given the opportunity to press a microswitch to provide themselves with supplementary heat from a 250 W infrared bulb. As environmental temperature was reduced in increments of 1°C below an ambient temperature of 20°C, the chicks increased the request for supplementary heat by 1.6 min/h. Although it is often suggested that evaporative cooling is unimportant to birds because they have no sweat glands, when the ambient temperature is 40°C mature domestic hens will request a 30 s stream of air at a dry bulb temperature of 22°C using an operant control (Richards, 1976).

Operant control of the environment has been used to evaluate the thermal environment of mature laying hens (Laycock, 1989). Cooler cage temperatures are requested by hens during the night and during periods of low feeding activity, indicating that their thermal requirements are diminished during these periods. A similarity between the diurnal rhythm of operant-controlled supplemental heat and metabolic heat production was also observed in these birds, which might indicate that circadian rhythms of activity, feeding, basal metabolic rate and thermal requirements are interrelated.

Physical factors in the environment can also influence the effective environmental temperature, and operant control of supplemental heating sources can be used to evaluate these factors. For example, the influence of flooring was evaluated by Morrison and Curtis (1983) and Morrison and McMillan (1985), who observed that chicks raised on wire flooring requested 4.22 min/h more supplemental heat than those on floors with shavings. These data indicate that heat loss is greater on wire and consequently chicks experience cooler temperature on wire, although the ambient temperature was identical on both types of flooring. The influence of feather loss has also been investigated using operant control of environmental temperature, and, in the absence of their natural thermal insulation, domestic hens increase their request for supplemental heat (Horowitz et al., 1978). Studies with hens (Horowitz et al., 1978) as well as with pigeons (Necker, 1977) have demonstrated that the greatest demand for supplemental heat was produced when the bird's back was exposed, indicating that this region is the most thermally sensitive part of the bird.

From their studies on domestic fowl, Hooper and Richards (1991) concluded that the relative contribution of operant behaviour to overall temperature regulation was different both qualitatively and quantitatively under heat load and cold load temperatures. Under heat load, the preferential method of thermoregulation included an operant response, whereas under cold exposure the autonomic response (as indicated by increases in oxygen consumption and heart rate) appeared to be the major contributor to thermoregulation.

Physiological Responses to Heat Stress

The physiological responses to heat stress in birds involve the functional integration of several organs to meet the metabolic needs of birds that are trying to dissipate heat and maintain homeostasis. Responses that can be described in anatomical terms and involve the whole animal are discussed below, and integration of the physiological responses by endocrine systems are described in the following section.

Acclimatization to high ambient temperature

Exposure of chickens to high environmental temperatures produces an initial increase in the temperature of peripheral tissues and subsequently in core body temperature (Fronda, 1925; Heywang, 1938; Thornton, 1962; Kamar and Khalifa, 1964; Boone and Hughes, 1971a; Wang et al., 1989). Boone (1968) observed that the body temperature of chickens began increasing when the ambient temperature rose above 30°C if the rate of increase in ambient temperature was rapid. On the other hand, if the ambient temperature rose more slowly (Boone and Hughes, 1971b), the birds maintained their normal body temperature until the ambient temperature reached 33°C. Using the length of time required for body temperature to become constant as a measure of acclimatization, Hillerman and Wilson (1955) reported that adult chickens required 3 to 5 days to acclimatize to both hot and cold environments. Ambient temperatures above 32°C produce transient hyperthermia in turkeys (Wilson and Woodard, 1955), which can last for up to 21 days if the birds are exposed to 38°C (Parker et al., 1972).

Consumption of feed and water

As birds accumulate heat in their tissues, several responses to increase the dissipation of heat are invoked to reduce the heat load. Water consumption increases when chickens are exposed to high ambient temperature (North and Bell, 1990; Deyhim and Teeter, 1991; May and Lott, 1992), and survival in a hot environment is dependent upon the consumption of large volumes of water (Fox, 1951). Voluntary feed consumption is diminished in response to high environmental temperatures (Otten et al., 1989) and fasting for 1–3 days has been shown to progressively increase survival time of chicks exposed to heat stress (McCormick et al., 1979). However, since heavier broilers are more susceptible to heat stress (Reece et al., 1972), it is uncertain whether the effect of depressed feed consumption is due to the reduction in feed consumption or to a reduction in body weight resulting from low food intake. The increase in water consumption occurs immediately, whereas the reduction in food consumption is delayed until several hours after the birds have experienced high temperatures (May and Lott, 1992). The immediate increase in water consumption meets the immediate demands of evaporative cooling from respiratory surfaces, and the associated decline in food consumption reduces the contribution of metabolic heat to the total heat load that requires dispersion.

Sensible heat loss through specialized heat exchange mechanisms

Heat is dispersed through anatomical specializations in birds that provide increased blood flow to surfaces that can effectively transfer heat by radiation and conduction. The vascular system in the legs and feet of many birds, including domestic fowl, contains arteriovenous heat-exchange mechanisms that facilitate the dispersal of heat through these uninsulated surfaces. The volume of blood that flows through the arteriovenous network, which serves as a heat exchanger, is regulated by shunts in the vascular system (Midtgard, 1989). At high ambient temperatures, these shunts bring cool venous blood in close proximity to arterial blood to dissipate the maximum amount of heat to the environment. The loss of heat by sensible transfer through feet is adopted quickly by hens given the opportunity to roost on pipes in which cold water is circulating (Otten et al., 1989). While cooled roosts have not been extensively tested in poultry production systems, it would appear that they have considerable potential in the management of hens under heat stress.

In many birds, heat exchange can be increased through the reteopthalmicum, an arteriovenous heat exchanger situated between the optic cavity and the brain (Midtgard, 1989). This anatomical specialization can be used to dissipate heat through the cornea, the eye, the buccal cavity, the beak and the nasal passages. The extent to which domestic birds can disperse heat through this mechanism is not yet established.

Sensible heat loss and feather cover in poultry

Partial feather loss is not uncommon in hens, particularly in those housed in battery cages during the laying cycle. Feather loss occurs mainly from the neck, back and breast regions, and is believed to be related to cage shape, cage size, crowding of hens (Hill and Hunt, 1978) and feather pecking (Hughes and Wood-Gush, 1977; Hughes, 1978). In some breeds, the absence of feathers in the neck is controlled by a single autosomal dominant gene designated as naked neck (*Na*). In addition to eliminating plumage from the neck, the naked-neck gene suppresses 30–40% of the plumage in all of the other feather tracts. Regardless of the cause of poor feather covering, sensible heat loss is substantially increased (Richards, 1977). In normal environments, food consumption is increased to offset the increase in heat dissipation (Emmans and Charles, 1977; Gonyou and Morrison, 1983) and, consequently, the feed efficiency of poorly feathered birds is decreased (Leeson and Morrison, 1978). At ambient temperatures above 25°C, however, where the ability to dissipate heat is an asset, naked-neck (*Na Na*) chickens possessed superior growth rate, viability, egg weight and female reproductive performance (Merat, 1990).

Changes in respiration rate and blood pH

Panting is one of the visible responses of poultry during exposure to heat. This specialized form of respiration dissipates heat by evaporative cooling at the surfaces of the mouth and respiratory passageways. Hens may begin panting at an ambient temperature of 29°C (North, 1978) after 60 min of exposure to 37°C and 45% relative humidity (Wang *et al.*, 1989) or when their body temperature reaches 42°C (Hillman *et al.*, 1985). Panting enables hens to increase the rate of water evaporation from 5 to 18 g/h in response to a change in ambient temperature from 29 to 35°C with relative humidity of 50–60% (Lee *et al.*, 1945). However, it is believed that at an ambient temperature of 32°C and relative humidity of 50–60% hens reach the maximal ability to lose heat through evaporation (Barrot and Pringle, 1941; Wilson, 1948).

Panting increases the loss of carbon dioxide from the lungs, which leads to a reduction in the partial pressure of carbon dioxide (Wang *et al.*, 1989), and thus bicarbonate, in blood plasma. In turn, the lowered concentration of hydrogen ions causes a rise in plasma pH (Mongin, 1968; Richards, 1970), a condition generally referred to as alkalosis. In laying hens, the reduction in the plasma concentrations of bicarbonate compromises eggshell formation by limiting the availability of the anion required during formation of $CaCO_3$ crystals in the shell (Mongin, 1968).

The occurrence of respiratory alkalosis in response to thermal stress has not been consistently observed in all studies in poultry. For example, acute hyperthermia in female turkeys produced alkalosis (Kohne and Jones, 1975a), chronic hyperthermia in female turkeys had no effect on plasma pH (Kohne and Jones, 1975b) and thermal stress (37.8°C, 7 days) in male turkeys was associated with a reduction in blood pH (Parker and Boone, 1971). In Leghorn hens exposed to increasing heat, Darre *et al.* (1980) noted a curvilinear increase in blood pH, whereas in broilers reared under continuous heat (35°C) Siegel *et al.* (1974) and Vo and Boone (1975) were unable to detect any significant change. On the other hand, Bottje *et al.* (1983) and Raup and Bottje (1990) reported that blood pH in cockerels and broilers was increased at higher ambient temperature. Such discrepancies are presumed to be due to variations in the degree of thermal stress, the length of thermal stress period and the degree to which the birds had been acclimatized to the conditions (see Teeter et al., 1985). Variation in plasma pH noted during thermal stress may also have been a consequence of the time at which blood was sampled relative to bouts of panting and normal breathing that accompany exposure to chronically high ambient temperatures, since blood pH of panting chicks was elevated while that of non-panting birds was not significantly altered (Teeter et al., 1985).

The physiological mechanisms that are invoked by birds exposed to high temperatures must meet the opposing demands of thermoregulation and respiratory alkalosis. Dissipation of heat by evaporative cooling demands an increase in respiration, while respiratory alkalosis demands a decrease in respiration. Gular flutter and utilization during panting of respiratory passages that are not involved in gas exchange (e.g. nasal cavities, nasopharynx, larynx and trachea) are known to reduce respiratory alkalosis in heat-stressed birds (see Hillman et al., 1985). Gular flutter involves the rapid and sometimes resonant vibration of the upper throat passages, driven by the hyoid apparatus (Welty and Baptista, 1988). In pigeons, panting is superimposed on slower, deeper breathing, a system which may minimize respiratory alkalosis (Ramirez and Bernstein, 1976). The air sacs of birds are also utilized during panting to move air over surfaces that limit the exchange of gas between blood and air while facilitating evaporative heat loss (Whittow, 1986).

Respiratory alkalosis can also be combated nutritionally by providing a source of anion via feed or water. For example, Teeter and Smith (1986) have shown that supplemental ammonium chloride in drinking water of chronically heat-stressed birds can return blood pH to normal and enhance weight gain. During acute heat stress, the provision of ammonium chloride (Branton et al., 1986) or carbonated water (Bottje and Harrison, 1985) has been found to decrease blood pH.

Changes in plasma concentrations of ions

The normal functions of tissues are dependent upon the stability of the total osmolarity of intracellular and extracellular fluids. The major ions of the plasma are sodium, chloride, potassium, calcium, phosphate, sulfate and magnesium. The plasma concentration of each ion normally varies only within a remarkably small range; a substantial shift in their concentration can cause serious disturbance to cells since these ions and the plasma proteins play a major role in establishing the osmotic balance between plasma and fluids bathing the cells. The concentrations of the major ions are also important in determining the pH of the body fluids.

Elevation of body temperature to 44.5–45.0°C by exposing chickens to 41°C ambient temperature has been associated with increased plasma sodium and chloride and decreased plasma potassium and phosphate (Ait-Boulahsen et al., 1989). In normally hydrated fowls, however, heat stress (35–45°C for

10–12 h) produced no significant changes in the serum concentrations of sodium, potassium, chloride and calcium, or in serum osmolality, although serum phosphate declined (Arad *et al.*, 1983). Subjecting domestic fowl to 37°C and approximately 45% relative humidity for a period of 150 min did not significantly alter the plasma osmolality (Wang *et al.*, 1989). The different responses observed in these studies may be attributed to the differences in the extent and duration of heat exposure, and to the fact that the birds used by Arad *et al.* (1983) were acclimatized to high ambient temperature for a long period prior to the experiment.

The availability of water to heat-stressed hens is essential to support evaporative cooling from the respiratory surfaces. Whereas laying hens maintained within the thermoneutral zone will drink approximately 200 ml of water per day, hens at 40°C will consume approximately 500 ml per day (North and Bell, 1990). Arad *et al.* (1983) observed that water deprivation for 48 h, which included 24 h without food, of fowls maintained at 25°C was associated with an increase in serum concentrations of sodium and in serum osmolality. The increase in serum concentration of sodium and in serum osmolality were exacerbated and associated with a decline in serum concentrations of phosphate when dehydrated birds were subjected to 35–45°C for 10–12 h. In pigeons subjected to heat stress and water deprivation, serum concentrations of sodium increased while serum potassium levels declined (John and George, 1977). This response has been attributed to the release of neurohypophyseal hormones in heat-stressed pigeons (John and George, 1977; George, 1980), since it is known that injection of neurohypophyseal hormones into hens results in an elevation of blood sodium level and a decrease in blood potassium level (Rzasa and Neizgoda, 1969).

Heart rate, cardiac output, blood pressure and total peripheral resistance

Exposure to high ambient temperature is associated with a decline in blood pressure, an increase in cardiac output and a decrease in peripheral resistance (Weiss *et al.*, 1963; Whittow *et al.*, 1964; Sturkie, 1967; Darre and Harrison, 1987). As birds become acclimatized to elevated ambient temperature, however, cardiac output decreases, blood pressure increases and peripheral resistance returns to normal (Vogel and Sturkie, 1963; Sturkie, 1967).

Although in some experiments the heart rate of chickens exposed to acute heat stress declined (Darre and Harrison, 1981, 1987), an increase in heart rate was apparent in other experiments (Darre and Harrison, 1987; Wang *et al.*, 1989). In pigeons exposed to ambient temperatures between 6°C and 34°C, John and George (1992) observed that heart rate was significantly lower at ambient temperatures above 28°C when compared with temperatures below 28°C. In part, the variation in heart rate following exposure to heat stress may be the consequence of the trauma of repeated blood sampling, which would increase heart rate in some instances, overriding the inhibitory effect of thermal stress (Darre and Harrison, 1987). In both of the above experiments with chickens (Darre and Harrison, 1987; Wang et al., 1989), in which an increase in heart rate was observed following heat exposure, the birds had been subjected to repeated blood samplings.

The flow of blood from the body core to the periphery plays a significant role in the transfer of heat from deep body tissues to the peripheral tissues that are capable of dissipating heat to the environment (see Darre and Harrison, 1987). In chickens exposed to high ambient temperature, blood flow through the comb, wattles and shanks is increased due to peripheral vasodilation, and excess heat is dissipated to the surrounding air (Whittow et al., 1964; Darre and Harrison, 1981). During acute heat stress, the cardiovascular system distributes blood to functions related to thermoregulation, giving only secondary importance to other functions such as those related to the exchange of respiratory gases and digestion (see Darre and Harrison, 1987). In heat-exposed chickens, for example, Bottje and Harrison (1984) demonstrated that blood flow to the viscera was reduced by 44%.

Based upon their observation of cardiovascular response of chickens to acute mild hyperthermia, Darre and Harrison (1987) proposed that the thermoregulatory response starts with a decreased heart rate and peripheral vasodilation, which leads to decreased blood pressure, decreased peripheral resistance and compensatory increases in stroke volume and cardiac output. It is suggested that the large increase in cardiac output during heat stress demonstrates the intense demand placed upon the cardiovascular system to dissipate heat from the bird. Darre and Harrison (1987) concluded that the fine-tuning of the body temperature is accomplished primarily by cardiovascular adjustments to prevent major overshoots or undercontrol.

Hormonal Involvement in Thermoregulation

Hormones are produced by endocrine tissues and transported through the circulatory system to their target tissues. They provide an important link in the flow of information among cells and tissues in an animal to initiate and maintain the physiological and behavioural responses to heat stress. As a bird attempts to cope with heat stress, an intricate series of changes that is mediated by many, if not all, hormonal systems is initiated. The relative importance of each of these systems and the extent to which they are called upon depend on the severity of the heat stress. The following paragraphs describe the major physiological adaptations to heat stress that require the participation of endocrine systems.

Neurohypophyseal (posterior pituitary) hormones

Arginine vasotocin (AVT)

The principal neurohypophyseal hormone in birds, AVT, is an antidiuretic hormone in non-mammalian vertebrates (Munsick et al., 1960; Ames et al., 1971).

AVT is released in response to dehydration and stimulates the resorption of water by the kidney. However, AVT is believed to play a role in heat dissipation that is independent of its role in osmoregulation in chickens (Robinzon *et al.*, 1988; Wang *et al.*, 1989) and pigeons (John and George, 1992). In non-heat-acclimatized fowl, plasma concentrations of AVT increased after 90 min of exposure to 32°C (Wang, 1988) and after 60 min of exposure to 37°C (Wang *et al.*, 1989), without a significant change in plasma osmolality. Similar observations have been reported by Azahan and Sykes (1980), although in some cases plasma AVT levels increased only after 48 h of dehydration (Arad *et al.*, 1985). These observations have led to the suggestion that, in non-heat-acclimatized fowl, heat stress alone could increase AVT levels, while in heat-acclimatized birds, an increase in plasma osmolality could also be necessary (Wang *et al.*, 1989). The thermoregulatory role of AVT is shown further by the observation that injections of AVT decrease shank and comb temperatures in fowl (Robinzon *et al.*, 1988) and induce a drop in cloacal and foot temperature in heat-stressed pigeons (John and George, 1992). In pigeons, but not in chickens, AVT has also been implicated in the control of basal metabolic rate and respiratory rate (Robinzon *et al.*, 1988; John and George, 1992).

The release of AVT in response to heat stress has also been implicated in the mobilization of free fatty acids (FFA). In pigeons subjected to heat stress and dehydration, serum triglyceride levels dropped to less than half those in controls (John and George, 1977), whereas the plasma FFA levels showed a more than twofold increase (John *et al.*, 1975), suggesting that the increase in FFA is due to the breakdown (lipolysis) of blood triglycerides. In normothermic pigeons, intravenous injection of AVT (400 mU per pigeon) brought about a highly significant increase in FFA at 30 min post-injection (John and George, 1973), although injection of AVT into immature female chickens was without effect (Rzasa *et al.*, 1971). *In vitro* studies of pigeon adipose tissue have revealed that AVT induces FFA release (John and George, 1986). It is well established that fat is the major substrate for sustained muscular activity in birds (George and Berger, 1966) and a ready supply of FFA could meet the increased energy requirement of the respiratory muscles as panting is initiated in heat-stressed birds.

In a recent study with pigeons (John, T.M. and George, J.C., unpublished), increases in AVT were associated with increases in plasma thyroxine (T_4) level and concomitant decreases in triiodothyronine (T_3) level. Basal metabolic rate is determined by plasma T_3 in birds (see 'Thyroid hormones' below), and therefore a drop in T_3 level would reduce metabolic heat production to alleviate heat stress.

Mesotocin (MT)

Mesotocin, the avian analogue of oxytocin, has recently been implicated in thermoregulation in domestic fowl (Robinzon *et al.*, 1988; Wang *et al.*, 1989). Heat stress suppressed the circulating level of MT (Wang *et al.*, 1989), but it is not clear what role MT plays in thermoregulation. Since AVT could suppress MT release (Robinzon *et al.*, 1988), the decrease in MT may be the result of elevated concentrations of AVT. As in amphibians (Stiffler *et al.*, 1984), MT could be a diuretic hormone in birds (Wang et al., 1989). This suggestion is supported by the observation that MT release is stimulated by hypotonic saline infusion (Koike et al., 1986) and is positively correlated with avian renal blood flow (Bottje et al., 1989). It has been suggested that if MT has a diuretic function in fowl then the suppression of MT with a concomitant increase in AVT would be a useful mechanism to aid in the conservation of body fluids during heat stress (Wang et al., 1989). Infusion of MT has been shown to produce a dramatic increase in respiratory rate (Robinzon et al., 1988), which could enhance evaporative heat loss during heat stress. It is suggested that the thermoregulatory function of MT is carried out via either the central nervous system or peripheral mechanisms (Robinzon et al., 1988).

Growth hormone (GH)

In pigeons deprived of drinking water and subjected to high ambient temperature for 3 days (28°C, 31°C and 36.5°C, respectively), the plasma levels of growth hormone (GH) increased significantly (John et al., 1975). This increase in GH levels is believed to play a role in fatty acid mobilization since GH is a major lipolytic hormone in birds (John et al., 1973). Although the sequence of events that lead to the release of GH is not clearly understood, the importance of GH in diverting metabolism to provide a high-energy substrate for muscle metabolism is believed to contribute to the support of panting-related muscular activity during heat exposure.

The hypothalamic–pituitary adrenal axis

Corticosterone

Corticosterone is the principal steroid hormone of the avian adrenal cortex (Holmes and Phillips, 1976). Heat stress stimulates the release of corticosterone from the adrenal glands (Edens, 1978) and increases plasma concentrations of corticosterone in chickens (Edens and Siegel, 1975; Ben Nathan et al., 1976), turkeys (El Halawani et al., 1973) and pigeons (Pilo et al., 1985). Heat stress has also produced adrenal enlargement in ducks (Hester et al., 1981) and quail (Bhattacharyya and Ghosh, 1972). Since increased levels of circulating corticosterone have been observed under various stress situations including cold exposure (Etches, 1976; Pilo et al., 1985), the response to heat exposure is considered primarily as a reaction to stress.

The release of corticosterone from the adrenal cortex is mediated by the hypothalamus and the pituitary gland. Both neural and endocrine inputs to the central nervous system stimulate the production of corticotrophin-releasing factor (CRF) from neurons within the median eminence of the hypothalamus. CRF is released into the hypothalamic portal vascular system and transported to the pituitary gland, where it stimulates the production of adrenocorticotrophic hormone (ACTH). ACTH is released into the general circulatory system and is transported to its major target tissue, the adrenal cortex. Under ACTH stimulation, the adrenal cortex increases the production and release of all of the adrenocortical hormones, although the major hormones are corticosterone and aldosterone. As the primary mediators of stress responses, ACTH, aldosterone and corticosterone have widespread effects on many target tissues throughout the body. None of these effects are specific to heat stress, but all are initiated as a bird mounts a coordinated physiological response to cope with increasing environmental temperature.

Plasma concentrations of corticosterone increase in response to heat stress, but high levels of corticosterone can only be maintained for short periods to cope with acute exposure to the high temperatures. When birds are chronically exposed to high temperatures, plasma concentrations of corticosterone will decline after the initial surge and, unless other physiological and/or behavioural responses can be implemented to alleviate the heat stress, the bird will become hyperthermic and die. For example, in young chickens exposed to high ambient temperature (43°C), plasma corticosterone increased within 30 min, but dropped below pre-exposure levels within 120 min (Edens, 1978). This drop in circulating levels of corticosterone was accompanied by low plasma concentrations of glucose, phosphate and Na^+, and elevated plasma pH. In response to this severe acute thermal stress, therefore, the chickens exhibited acute adrenal cortical insufficiency within 120 min, which, in conjunction with a massive secretion of catecholamines, resulted in cardiovascular failure and death. Prevention of a decline in corticosterone by pretreatment of birds with reserpine, propanolol or dihydroergotamine sustained the adrenal response and reduced mortality during heat stress (Edens and Siegel, 1976). Hydrocortisone and cortisone therapy have also been found to reduce mortality in birds exposed to high temperatures (Burger and Lorenz, 1960; Sammelwitz, 1967), providing further evidence that corticosteroids can protect against the lethal effects of high ambient temperature.

Although the acute effects of aldosterone, corticosterone and ACTH are believed to initiate or support a number of physiological changes that delay incipient hyperthermia, the details of these effects are remarkably sparse. Increased plasma levels of aldosterone act in concert with AVT to promote renal absorption of water to prevent dehydration as evaporative cooling is utilized for thermoregulation.

Changes in the plasma concentrations of corticosteroids and ACTH affect the lymphoid tissues and consequently the ability of chickens to mount an immune response. For example, a diminution in the mass of the thymus, spleen and bursa of Fabricius (Garren and Shaffner, 1954, 1956; Glick, 1957, 1967; Siegel and Beane, 1961), and a decrease in the number of circulating lymphocytes and an increase in neutrophilic or heterophilic granulocytes (Dougherty and White, 1944; Gross and Siegel, 1983) have been reported following administration of corticosteroids. Corticosteroids bind to specific cytoplasmic receptors in lymphatic cells to redirect differentiation and metabolism of this cell lineage (Thompson and Lippman, 1974; Sullivan and Wira, 1979), and both ACTH administration and exposure to high temperature increase corticosteroid binding to cells in the lymphoid system

(Gould and Siegel, 1981). The precise effect of corticosteroids, however, on this diverse cell lineage depends on both the severity of persistence of the stress and the role of the affected cells in both humoral and cell-mediated immunity.

Catecholamines

The catecholamines, epinephrine (E) and norepinephrine (NE), are synthesized and released from adrenal chromaffin cells. Their secretion in response to stress is similar to the corticosterone response, since both adrenal cortical hormones and ACTH stimulate the release of both E and NE (see Harvey et al., 1986). For example, a substantial increase in the circulating levels of both E and NE were observed over a 140-min period in 8-week-old broilers exposed to 45°C (Edens and Siegel, 1975). Chronic exposure of adult roosters to an ambient temperature of 31°C produced little change in blood concentrations of E and NE (Lin and Sturkie, 1968). The increase in plasma concentrations of E and NE in response to acute heat stress are probably as transitory as that of corticosterone, since in 9-week-old male turkeys the adrenal concentrations of E and NE were not significantly altered when examined 6 h after exposure to 32°C (El Halawani et al., 1973). Catecholamines may also exert an influence on body temperature more directly than through their role as mediators of a general response to heat stress, since infusion of catecholamines into the hypothalamus of both young (Marley and Stephenson, 1970) and adult (Marley and Nistico, 1972) chickens lowers body temperature.

The turnover rate in brain tissue of NE increased and of E decreased upon acute exposure of turkeys to 32°C (El Halawani and Waibel, 1976). However, chronic exposure to the same temperature reduced the increase in NE turnover while not affecting the decrease in E turnover. Since similar observations have been reported for Japanese quail subjected to heat stress, Branganza and Wilson (1978a) suggested that acute heat stress increases central noradrenergic neuronal activity, which returns to normal following acclimatization. Branganza and Wilson (1978b) also reported that, while high ambient temperature (34°C) for 6 h increased brain NE level and NE turnover rate, it decreased heart NE levels and enhanced the turnover rate of NE in the heart. Although chronic exposure for 5 weeks to high temperatures did not increase NE turnover in the brain, NE turnover in the heart was increased. Thus, NE appears to play a role in heat stress in at least two organs and the effects in each of them appear to be dissimilar.

Melatonin

Melatonin (N-acetyl-5-methoxytryptamine) is an indoleamine that has been recognized as the major pineal hormone (Quay, 1974). In birds, as in many other vertebrates, melatonin is also produced in several extrapineal tissues (Pang et al., 1977; Cremer-Bartels et al., 1980; Ralph, 1981) such as the retina and the Harderian gland. Melatonin has been implicated in thermoregulation in birds (Binkley et al., 1971; John et al., 1978; John and George, 1984, 1991; George and John, 1986) and may regulate the circadian rhythm in body temperature since, in pigeons, body temperature is relatively low in the night, when both plasma and pineal levels of the melatonin are high (John et al., 1978). Conversely, body temperature is higher in the day, when the melatonin levels are low. In the presence of high concentrations of melatonin, heat dissipation by peripheral tissues is enhanced by vasodilation and blood flow, particularly to the foot, which is an important site for heat dissipation in birds (Jones and Johansen, 1972). Furthermore, melatonin may act centrally by lowering the set-point of the main 'thermostat', which is believed to be present in the hypothalamus (see John and George, 1991). A more general role for the pineal gland in the regulation of body temperature in chickens was suggested by Cogburn et al. (1976), who observed that the return to normal body temperature was delayed in pinealectomized birds.

Reproductive hormones

The effect of heat stress on reproductive performance in chickens has been indicated by reduced egg production. The diminished egg production is suspected to be at least partly influenced by the ovulatory hormones. In the hen, heat stress reduces serum luteinizing hormone (LH) levels, hypothalamic content of luteinizing hormone-releasing hormone (LHRH) (Donoghue et al., 1989) and the preovulatory surges of plasma LH and progesterone (Novero et al., 1991). Since the preovulatory surges of LH and progesterone are controlled in a positive feedback loop (Etches and Cunningham, 1976; Wilson and Sharp, 1976; Williams and Sharp, 1978; Johnson et al., 1985) and since both hormone levels are depressed concomitantly (Novero et al., 1991), it is difficult to identify the site or sites of action of heat stress. However, the hypothalamus could be a primary target for heat stress because it receives both neural and endocrine inputs that could be translated into general inhibition of the reproductive system. For example, corticosterone is known to have a suppressive effect on circulating LH levels in birds (Wilson and Follett, 1975; Deviche et al., 1979; Etches et al., 1984; Petitte and Etches, 1988) and the high concentration of circulating corticosterone during thermal stress (Donoghue et al., 1989) could promote the decline of LHRH. Brain monoamines have also been suspected of being involved in inhibiting hypothalamic function (Donoghue et al., 1989). Since heat stress has been reported to alter brain monoamine levels and turnover rate (El Halawani and Waibel, 1976; Branganza and Wilson, 1978a,b) and since monoamines are considered to be putative hypothalamic regulators of gonadotrophin secretion (El Halawani et al., 1982), monoamine turnover could be involved in bringing about the reduction in hypothalamic LHRH content that is associated with heat stress.

Thyroid hormones

The importance of the thyroid gland in adaptation to heat stress is related to the central role that thyroid hormones play in the regulation of metabolic rate of birds (Bellabarba and Lehoux, 1981, 1985; McNicholas and McNabb, 1987; McNabb, 1988). This effect has been demonstrated by surgical or chemical thyroidectomy of chickens, which produces a decrease in metabolic rate (Winchester, 1939; Mellen and Wentworth, 1962) and body temperature (Nobukumi and Nishiyama, 1975; Davison et al., 1980; Lam and Harvey, 1990), and by thyroid hormone administration, which stimulates heat production (Mellen and Wentworth, 1958; Singh et al., 1968; Arieli and Berman, 1979). In chickens, thyroid hormone secretion is depressed as ambient temperature increases (Reineke and Turner, 1945; Hahn et al., 1966) and heat tolerance improves as thyroid function is reduced. For example, radiothyroidectomy (Bowen et al., 1984) or chemical thyroidectomy by thiouracil (Fox, 1980; Bowen et al., 1984) increased survival time in heat-stressed chickens, while thyroid hormone administration decreased the survival time (Fox, 1980; May, 1982; Bowen et al., 1984).

The two active forms of thyroid hormones are T_4 and T_3, and the inactive form is reverse triiodothyronine (r-T_3). The selective peripheral conversion of T_4 to T_3 or r-T_3 is believed to play an important role in thermoregulation in domestic fowl (Kuhn and Nouwen, 1978; Decuypere et al., 1980; Rudas and Pethes, 1984); when chickens are exposed to warm temperatures, T_4 is inactivated by conversion into r-T_3, whereas during cold exposure T_4 is converted into T_3, which stimulates metabolic rate.

While it is generally accepted that T_3 stimulates metabolic rate and that both T_3 and T_4 are depressed following heat stress, this pattern is not universally observed. In Japanese quail and pigeons, for example, plasma T_3 and T_4 concentrations have been reported to increase, decrease or remain unchanged following heat stress (John and George, 1977; Bobek et al., 1980; Bowen and Washburn, 1985; Pilo et al., 1985). Heat stress experiments conducted on heat-acclimatized or non-acclimatized broilers of different lines also failed to elicit a consistent pattern in thyroid hormone response to heat stress (May et al., 1986), leading to the conclusion that the complex physiological response to heat stress does not consistently affect plasma concentrations of thyroid hormones.

Heat-shock Proteins and Heat Stress

A response of all organisms – animal, plant or microbe – to elevated temperature is the increased synthesis of a group of proteins known as the heat-shock proteins (reviewed by Lindquist and Craig, 1988; Pardue et al., 1989; Morimoto et al., 1990; Craig and Gross, 1991; Nover, 1991). The universal response to heat stress is the most highly conserved genetic system known, and some of the heat-shock proteins are not only the most abundant proteins found in nature but are the most highly conserved proteins that have been analysed. This degree of conservation indicates the fundamental role that these proteins must play in restoring normal function to cells or whole organisms that are exposed to potentially damaging stimuli. As discussed below, heat-shock proteins play an essential role by associating with a variety of proteins and affecting their conformation and location. In a heat-shocked cell, the heat-shock proteins may bind to heat-sensitive proteins and protect them from degradation, or may prevent damaged proteins from immediately precipitating and permanently affecting cell viability. Heat-shock proteins are members of a larger family of stress proteins, some of which may be synthesized because of nutrient deprivation, oxygen starvation or the presence of heavy metals, oxygen radicals or alcohol. These stimuli can also cause the synthesis of some of the heat-shock proteins. An important aspect of these stressful circumstances is that cells or organisms that have recovered from a mild stressful episode and are expressing elevated levels of stress proteins can exhibit tolerance to doses of the stress-causing agent, including heat, that would normally cause developmental abnormalities or death (e.g. see Mizzen and Welch, 1988; Welch and Mizzen, 1988).

The most commonly found forms of heat-shock proteins have relative molecular masses of approximately 10,000–30,000, 70,000, 90,000 and 100,000–110,000, and so are referred to as HSP70, HSP90 and so on. Some of these proteins, or their close relatives, apparently fulfil vital functions in normal cells, whereas some are required for growth at the upper end of the normal temperature range, and some are required to withstand the toxic effects of extreme temperatures. Strictly speaking, the term 'heat-shock' protein should apply only to those proteins that are synthesized in a cell in response to a heat shock, but currently it is unclear whether these proteins carry out any specialized function that cannot be performed by their constitutively expressed relatives present under normal conditions. These latter proteins are sometimes named 'heat-shock cognates' (HSCs). For the purposes of this chapter, and as is usually the case, the 'heat-shock' protein designations will be used broadly to refer to any members of these families, whether inducible or constitutive.

In general, heat-shock proteins are involved in the assembly or disassembly of proteins or protein-containing complexes during the life and death of a normal cell (the role of heat-shock proteins in protein folding is reviewed by Gething and Sambrook, 1992). Each type of heat-shock protein may interact with a specific group of molecules – for example, members of the HSP70 family bind to immunoglobulin heavy chains, clathrin baskets and deoxyribonucleic acid (DNA) replication complexes, and may be involved either in maintaining these structures in partially assembled form until they are required or in degrading them after use. All members of the HSP70 family bind adenosine triphosphate (ATP) with high affinity, have adenosine triphosphatase (ATPase) domains (Flaherty et al., 1990) and probably undergo conformational changes using energy released on ATP hydrolysis; the conformation of structures with which the heat-shock proteins are interacting may then also change (Pelham, 1986). HSP90, whose family members are abundant at normal temperatures (up to 1% of the total soluble protein in the cytoplasm), interacts with steroid hormone receptors and apparently masks the DNA-binding region of the receptor until the receptor has bound to the appropriate steroid hormone; the receptor can then bind to DNA and activate expression of specific genes. Another profound influence of HSP90 is its modulation of phosphorylation activity within the cell as a consequence of it binding to cellular kinases. These molecular chaperon characteristics of HSP90 apparently depend on the ability of HSP90 to recognize and bind certain subsets of non-native proteins and influence their folding to the native state, which they can do in the absence of nucleoside triphosphates (Wiech et al., 1992). Members of the HSP70 family also act as molecular chaperons and can conduct certain proteins through intracellular membranes (Chirico et al., 1988; Deshaies et al., 1988); perhaps HSP70 binds to newly synthesized proteins that are just being released from ribosomes and prevents the proteins from aggregating and precipitating before they are properly folded, transported and incorporated into complexes or organelles. Members of the HSP70 family can certainly distinguish folded from unfolded proteins (Flynn et al., 1991). This activity might provide a clue to the role of the large amount of HSP70 that is synthesized following heat shock – HSP70 may bind to cellular proteins that have been denatured by the heat and may prevent their catastrophic precipitation (Finley et al., 1984). On return to normal environmental conditions, sufficient cellular protein might be released in an undamaged state to allow the cell to resume its activity, and protein that has been irreparably damaged can be removed gradually by the ubiquitin salvage pathway. Ubiquitin expression is also induced by heat, and ubiquitin is conjugated through its terminal glycine residue to cellular proteins prior to their selective degradation. A further feature of heat-shock proteins that is yet to be fully explored is the role that they might play in immunity and immunopathology (Young, 1990). Although excessive heat can have many deleterious effects on the structure and physiology of a cell, including impairment of transcription, ribonucleic acid (RNA) processing, translation, post-translational processing, oxidative metabolism, membrane structure and function, cytoskeletal structure and function, etc., it is not clear which of these is the most harmful (reviewed by Rotii-Roti and Laszlo, 1987) or which is alleviated by any particular heat-shock protein.

The response of cells or a whole organism to heat shock is extremely rapid, but transient, and involves the redistribution of preformed heat-shock proteins within the cell, as well as immediate translation of preformed messenger RNA (mRNA) into heat-shock proteins, immediate transcription of genes encoding heat-shock proteins and cessation of transcription or translation of other genes or mRNA (Yost et al., 1990). An example of the redistribution of preformed proteins on heat shock is provided by some members of the HSP70 family that immediately migrate from their normal location in the cytoplasm to the nucleus, where they associate with preribosomes in the nucleoli (Welch and Suhan, 1985). Interestingly, the damage to nucleolar morphology that can be seen in cells that have undergone a heat shock can be repaired more rapidly if the cells are artificially induced to overexpress HSP70 by transformation with an exogenous HSP70-encoding gene (Pelham, 1984). During recovery from heat shock, HSP70 migrates back to the cytoplasm.

Transcription of heat-shock-protein-encoding genes is regulated by a heat-shock factor (HSF), which interacts with a conserved DNA sequence, the heat-shock element (HSE), located in the 5' flanking regions of the genes. In vertebrates, binding of significant amounts of HSF to HSE sequences only occurs after a heat shock, and may result from the conversion of small, inactive oligomeric forms of the protein to an active multimer (Westwood *et al.*, 1991). The HSE consists of three repeats of a five-base sequence, arranged in inverted orientation, and multiple copies of the HSE may result in a cooperative increase in levels of transcription (Tanguay, 1988); high-affinity binding of the HSF to this complex may only occur if the HSF is in the multimeric form. The HSF appears to be prelocalized in the nucleus, perhaps to minimize the time for response to a heat shock, and the chromatin surrounding some heat-shock-protein-encoding genes is in an 'open' conformation, allowing immediate association with transcription factors following a heat shock. Also, RNA polymerase may in some cases already be interacting with the gene but is blocked from transcribing until the heat shock occurs. In addition to associating with HSE in heat-shock-protein-encoding genes, there is evidence that the activated multimeric HSF may bind to other regions of the genome, and may play a role in suppressing transcription of other genes during the period of stress (Westwood *et al.*, 1991).

Although selective transcription of heat-shock-protein-encoding genes takes place at elevated temperatures, various aspects of RNA metabolism under these circumstances also result in the preferential synthesis of heat-shock proteins (Yost *et al.*, 1990). Most eukaryotic genes are transcribed to give complex RNA structures that require processing by removal of introns before functional mRNA results. This splicing reaction is inhibited under conditions of heat stress, perhaps to prevent the production of a range of incorrectly spliced, mutant mRNAs. Most heat-shock-protein genes do not contain introns, however, and so produce mRNAs that are functional under heat stress. Indeed, there is evidence that heat-shock protein may contribute to the thermotolerance of cells or organisms that have been previously heat-shocked by protecting the splicing machinery and thus enabling the production of functional mRNAs from a range of genes – a primed protection of normal function of the cell. A further aspect of heat-shock-protein mRNA metabolism that is of interest is that some classes of these molecules appear to be stable in heat-stressed cells but are highly unstable in unstressed cells, indicating that a significant increase in heat-shock-protein synthesis may result from stabilization of pre-existing or newly formed mRNA. Other evidence (Theodorakis *et al.*, 1988), however, indicates that the levels of some classes of heat-shock-protein mRNA may not be significantly different between stressed and unstressed cells, but that the elongation rate is much lower during protein synthesis from these templates in unstressed cells than in stressed cells.

Much of the work on the molecular biology of heat-shock-protein expression in vertebrates has been done using the chicken as a model system. Several of the chicken heat-shock-protein-encoding genes have been cloned and characterized, including *bsp70* (Morimoto *et al.*, 1986), *hsp90* (Catelli *et al.*, 1985), *hsp108* (Kulomaa *et al.*, 1986; Sargan *et al.*, 1986) and *UbI* and *UbII*, which are ubiquitin-encoding genes (Bond and Schlesinger, 1986). The *hsp108* gene was first cloned because of its expression in the chick oviduct and was subsequently shown to have sequence homology with heat-shock proteins. HSP108 is found constitutively in all chicken tissues examined, but its levels are enhanced following heat shock. Surprisingly, the *bspl08* gene is steroid-regulated in chick oviduct (Baez *et al.*, 1987), and HSP108, like HSP90, is found associated with all types of steroid hormone receptors. The induction of heat-shock-protein synthesis has been studied in a range of avian cell types, including reticulocytes (Atkinson *et al.*, 1986; Theodorakis *et al.*, 1988), lymphoid cells (Banerji *et al.*, 1986; Miller and Qureshi, 1992) and macrophages (Miller and Qureshi, 1992), generally by exposing the cells to 43–45°C for 30–60 min, and usually resulting in the synthesis of proteins of approximately 23,000, 70,000 and 90,000 Da. An additional protein of 47,000 Da is synthesized in chicken embryo fibroblasts that are exposed to heat shock, and appears to have collagen-binding activity (Nagata *et al.*, 1986).

Most protein synthesis in the chick embryo recovered from the unincubated egg is still dependent on recruitment of maternal mRNA rather than on embryonic gene expression, but expression of several heat-shock-protein-encoding genes can apparently be induced by heat shock at this stage (Zagris and Matthopoulos, 1988). By stage XIII of embryonic development (Eyal-Giladi and Kochav, 1976), heat-shock-protein expression is regionally specific within the blastoderm (Zagris and Matthopoulos, 1986) and heat shock (isolated blastoderms subjected to a temperature of 43–44°C for 2.5 h) disrupts normal embryonic development – the blastoderm becomes approximately twice as large and the primitive streak fails to form.

Certain breeds or strains of poultry appear to survive heat stress more successfully than others, and birds that are acclimatized gradually to elevated temperatures are more resilient than those experiencing a sudden heat shock. Jungle fowl survive heat stress more successfully than do normal commercial strains of chicken, as do Bedouin fowl of the Israeli desert (Marder, 1973). A cross of Leghorn and Bedouin fowl produced offspring with improved heat tolerance relative to the Leghorn parents (Arad *et al.*, 1975), indicating a genetic component in this tolerance. It is tempting to speculate that part of the difference in heat tolerance of various breeds could be attributable to different alleles of heat-shock-protein-encoding genes, resulting in differing ability to respond rapidly to a heat stress, differing final concentrations of the relevant heat-shock proteins in stressed birds or differing ability of various heat-shock-protein isoforms to interact with their normal ligands in the cell. A thorough molecular characterization of the heat-shock response in a range of poultry would provide the basis for future genetic manipulation of the heat-shock response in a way that has not been possible before. In the long term, perhaps, levels of certain heat-shock proteins could be raised, as indicated by Pelham (1984), bearing in mind that ancillary proteins such as the HSF might also have to be present in increased amounts (see Johnston and Kucey, 1988). These manipulations might be achieved either by introgression and marker-assisted selection or by more radical means through the development of transgenic birds. An intriguing hint of the genetic plasticity of the heat-shock response is provided by experiments that resulted in significant improvement of the temperature-specific fitness of lines of the bacterium, *Escherichia coli* (which has a heat-shock response that is very similar to that of birds), maintained at 42°C for 200 generations (Bennett *et al.*, 1990).

References

Ait-Boulahsen, A., Garlich, J.D. and Edens, F.W. (1989) Effect of fasting and acute heat-stress on body temperature, blood acid-base and electrolyte status in chickens. *Comparative Biochemistry and Physiology* 94A, 683–687.

Ames, E., Steven, K. and Skadhauge E. (1971) Effects of arginine vasotocin on renal excretion of Na+, K+, Cl⁻ and urea in the hydrated chicken. *American Journal of Physiology* 221, 1223–1228.

Arad, Z., Moskovits, E. and Marder, J. (1975) A preliminary study of egg production and heat tolerance in a new breed of fowl (Leghorn × Bedouin). *Poultry Science* 54, 780–783.

Arad, Z., Marder, J. and Eylath, U. (1983) Serum electrolyte and enzyme responses to heat stress and dehydration in the fowl (*Gallus domesticus*). *Comparative Biochemistry and Physiology* 74A, 449–453.

Arad, Z., Arnason, S.S., Chadwick, A. and Skadhauge, E. (1985) Osmotic and hormonal responses to heat and dehydration in the fowl. *Journal of Comparative Physiology* 155 B, 227–234.

Arieli, A. and Berman, A. (1979) The effect of thyroxine on thermoregulation in the mature fowl (*Gallus domesticus*). *Journal of Thermal Biology* 4, 247–249.

Atkinson, B.G., Dean, R.L. and Blaker, T.W. (1986) Heat shock induced changes in the gene expression of terminally differentiating avian red blood cells. *Canadian Journal of Genetics and Cytology* 28, 1053–1063.

Azahan, E. and Sykes, A.H. (1980) The effects of ambient temperature on urinary flow and composition in the fowl. *Journal of Physiology, London* 302, 389–396.

Baez, M., Sargan, D.R., Elbrecht, A., Kulomaa, M.S., Zarucki-Schulz, T., Tsai, M.-J. and O'Malley, B.W. (1987) Steroid hormone regulation of the gene encoding the chicken heat shock protein Hsp 108. *Journal of Biological Chemistry* 262, 6582–6588.

Banerji, S.S., Berg, L. and Morimoto, R.I. (1986) Transcription and post- transcriptional regulation of avian HSP70 gene expression. *Journal of Biological Chemistry* 261, 15740–15745.

Barrot, H.G. and Pringle, E.M. (1941) Energy and gaseous metabolism of the hen as affected by temperature. *Journal of Nutrition* 22, 273–286.

Bellabarba, D. and Lehoux, J.-G. (1981) Triiodothyronine nuclear receptor in chick embryo: nature and properties of hepatic receptor. *Endocrinology* 109, 1017–1025.

Bellabarba, D. and Lehoux, J.-G. (1985) Binding of thyroid hormones by nuclei of target tissues during the chick embryo development. *Mechanisms of Ageing and Development* 30, 325–331.

Ben Nathan, D., Heller, E.D. and Perek, M. (1976) The effect of short heat stress upon leucocyte count, plasma corticosterone level, plasma and leucocyte ascorbic acid content. *British Poultry Science* 17, 481–485.

Bennett, A.F., Dao, K.M. and Lenski, R.E. (1990) Rapid evolution in response to high-temperature selection. *Nature* 346, 79–81.

Bhattacharyya, T.K. and Ghosh, A. (1972) Cellular modification of interregnal tissue induced by corticoid therapy and stress in three avian species. *American Journal of Anatomy* 133, 483–494.

Binkley, S., Kluth, E. and Menaker, M. (1971) Pineal function in sparrows: circadian rhythms and body temperature. *Science* 174, 311–314.

Bobek, S., Niezgoda, J., Pietras, M., Kacinska, M. and Ewy, Z. (1980) The effect of acute cold and warm ambient temperatures on the thyroid hormone concentration in blood plasma, blood supply, and oxygen consumption in Japanese quail. *General Comparative Endocrinology* 40, 201–210.

Bond, U. and Schlesinger, M.J. (1986) The chicken ubiquitin gene contains a heat shock promoter and expresses an unstable mRNA in heat-shocked cells. *Molecular and Cellular Biology* 6, 4602–4610.

Boone, M.A. (1968) Temperature at six different locations in the fowl's body as affected by ambient temperatures. *Poultry Science* 47, 1961–1962.

Boone, M.A. and Hughes, B.L. (1971a) Effect of heat stress on laying and non-laying hens. *Poultry Science* 50, 473–477.

Boone, M.A. and Hughes, B.L. (1971b) Wind velocity as it affects body temperature, water consumption and feed consumption during heat stress of roosters. *Poultry Science* 50, 1535–1537.

Bottje, W.G. and Harrison, P.C. (1984) Mean celiac blood flow (MBF) and cardiovascular response to a-adrenergic blockade or elevated ambient CO_2 (%CO_2) during acute heat stress. *Poultry Science* 63 (Suppl. 1), 68–69 (Abstract).

Bottje, W.G. and Harrison, P.C. (1985) The effect of tap water, carbonated water, sodium bicarbonate, and calcium chloride on blood acid-base balance in cockerels subjected to heat stress. *Poultry Science* 64, 107–113.

Bottje, W.G., Harrison, P.C. and Grishaw, D. (1983) Effect of an acute heat stress on blood flow in the coeliac artery of Hubbard cockerels. *Poultry Science* 62, 1386–1387 (Abstract).

Bottje, W.G., Holmes, K.R., Neldon, H.L. and Koike, T.I. (1989) Relationships between renal hemodynamics and plasma levels of arginine vasotocin and mesotocin during hemorrhage in the domestic fowl (*Gallus domesticus*). *Comparative Biochemistry and Physiology* 92A, 423–427.

Bowen, S.J. and Washburn, K.W. (1985) Thyroid and adrenal response to heat stress in chickens and quail differing in heat tolerance. *Poultry Science* 64, 149–154.

Bowen, S.J., Washburn, K.W. and Huston, T.M. (1984) Involvement of the thyroid gland in the response of the young chicken to heat stress. *Poultry Science* 63, 66–69.

Branganza, A. and Wilson, W.O. (1978a) Elevated temperature effects on catecholamines and serotonin in brains of male Japanese quail. *Journal of Applied Physiology* 45, 705–708.

Branganza, A. and Wilson, W.O. (1978b) Effect of acute and chronic elevated air temperatures, constant (34°) and cyclic (10–34°), on brain and heart norepinephrine of male Japanese quail. *General and Comparative Endocrinology* 36, 233–237.

Branton, S.L., Reece, F.N. and Deaton, J.W. (1986) Use of ammonium chloride and sodium bicarbonate in acute heat exposure. *Poultry Science* 65, 1659–1663.

Burger, R.E. and Lorenz, F.W. (1960) Pharmacologically induced resistance to heat shock. II. Modifications of activity of the central-nervous and endocrine systems. *Poultry Science* 39, 477–482.

Catelli, M.G., Binart, N., Feramisco, J.R. and Helfman, D.M. (1985) Cloning of the chick *hsp90* cDNA in expression vector. *Nucleic Acids Research* 13, 6035–6047.

Chirico, W.J., Waters, M.G. and Blobel, G. (1988) 70 K heat shock related proteins stimulate protein translocation into microsomes. *Nature* 332, 805–810.

Cogburn, L.A., Harrison, P.C. and Brown, D.E. (1976) Scotophase-dependent thermoregulatory dysfunction in pinealectomized chickens. *Proceedings of the Society of Experimental Biology and Medicine* 153, 197–201.

Craig, E.A. and Gross, C.A. (1991) Is hsp70 the cellular thermometer? *Trends in Biochemical Science* 16, 135–140.

Cremer-Bartels, G., Kuchle, H.J., Ludtmann, W. and Malten, K. (1980) Melatonin biosynthesis in avian pineal gland and retina. *Advances in Bioscience* 29, 47–56.

Darre, M.J. and Harrison, P.C. (1981) The effects of heating and cooling localized areas of the spinal cord and brain stem on thermoregulatory responses of domestic fowl. *Poultry Science* 60, 1644. (Abstract).

Darre, M.J. and Harrison, P.C. (1987) Heart rate, blood pressure, cardiac output, and total peripheral resistance of single comb white leghorn hens during an acute exposure to 35°C ambient temperature. *Poultry Science* 66, 541–547.

Darre, M.J., Odom, T.W., Harrison, P.C. and Staten, F.E. (1980) Time course of change in respiratory rate, blood pH, and blood PCO_2 of SCWL hens during heat stress. *Poultry Science* 59, 1598 (Abstract).

Davison, T.F., Misson, B.H. and Freeman, B.M. (1980) Some effects of thyroidectomy on growth, heat production and the thermoregulatory ability of the immature fowl (*Gallus domesticus*). *Journal of Thermal Biology* 5, 197–202.

Decuypere, E., Hermans, C., Michels, H., Kuhn, E.R. and Verheyen, J. (1980) Thermoregulatory response and thyroid hormone concentration after cold exposure of young chicks treated with iopanoic acid or saline. In: Pethes, G., Peczely, P. and Rudas, P. (eds) *Recent Advances of Avian Endocrinology*. Pergamon Press, Oxford/Akademiai Kiado, Budapest, pp. 291–298.

Deshaies, R.J., Koch, B.D., Werner-Washburne, M., Craig, E.A. and Schekman, R. (1988) A subfamily of stress proteins facilitates translocation of secretory and mitochondrial precursor polypeptides. *Nature* 332, 800–805.

Deviche P., Heyns, W., Balthazart, J. and Hendick, J. (1979) Inhibition of LH plasma levels by corticosterone administration in the male duckling. *IRCS Medical Science* 7, 622–630.

Deyhim, F. and Teeter, R.G. (1991) Research note: sodium and potassium chloride drinking water supplementation effects on acid-base balance and plasma corticosterone in broilers reared in thermoneutral and heat-distressed environments. *Poultry Science* 70, 2551–2553.

Donoghue, D., Krueger, B.F., Hargis, B.M., Miller, A.M. and El Halawani, M.E. (1989) Thermal stress reduces serum luteinizing hormone and bioassayable hypothalamic content of luteinizing hormone releasing hormone in the hen. *Biology and Reproduction* 41, 419–424.

Dougherty, T.F. and White, A. (1944) Influence of hormones on lymphoid tissue structure and function: the role of pituitary adrenotrophic hormone in the regulation of lymphocytes and other cellular elements of blood. *Endocrinology* 35, 1–12.

Edens, F.W. (1978) Adrenal cortical insufficiency in young chickens exposed to a high ambient temperature. *Poultry Science* 57, 1746–1750.

Edens, F.W. and Siegel, H.S. (1975) Adrenal responses in high and low ACTH response lines of chicken during acute heat stress. *General and Comparative Endocrinology* 25, 64–73.

Edens, F.W. and Siegel, H.S. (1976) Modification of corticosterone and glucose responses by sympatholytic agent in young chickens during acute heat exposure. *Poultry Science* 55, 1704–1712.

El Halawani, M.E. and Waibel, P.E. (1976) Brain indole and catecholamines of turkeys during exposure to temperature stress. *American Journal of Physiology* 230, 110–114.

El Halawani, M.E., Waibel, P.E., Appel, J.R. and Good, A.L. (1973) Effects of temperature stress on catecholamines and corticosterone of male turkeys. *American Journal of Physiology* 224, 384–388.

El Halawani, M.E., Burke, W.H., Dennison, P.T. and Silsby, J.L. (1982) Neuropharmacological aspects of neural regulation of avian endocrine function. In: Scanes, C.G., Ottinger, M.A., Kenny, A.D., Balthazart, J., Cronshaw, J. and Chester Jones, I. (eds) *Aspects of Avian Endocrinology. Practical and Theoretical Implications.* Texas Tech Press, Lubbock, pp. 33–40.

Emmans, G.C. and Charles, D.R. (1977) Climatic environment and poultry feeding in practice. In: Haresign, W., Swan, H. and Lewis, D. (eds) *Nutrition and the Climatic Environment.* Butterworths, London, pp. 31–48.

Etches, R.J. (1976) A radioimmunoassay for corticosterone and its application to the measurement of stress in poultry. *Steroids* 28, 763–773.

Etches, R.J. and Cunningham, F.J. (1976) The interrelationship between progesterone and luteinizing hormone during the ovulation cycle of the hen (*Gallus domesticus*). *Journal of Endocrinology* 71, 51–58.

Etches, R.J., Williams, J.B. and Rzasa, J. (1984) Effects of corticosterone and dietary changes in the domestic hen on ovarian function, plasma LH and steroids and the response to exogenous LHRH. *Journal of Reproduction and Fertility* 70, 121–130.

Eyal-Giladi, H. and Kochav, S. (1976) From cleavage to primitive streak formation: a complementary normal table and a new look at the first stages of the development of the chick. I. General morphology. *Developmental Biology* 49, 321–337.

Finley, D., Ciechanover, A. and Varshavsky, A. (1984) Thermolability of ubiquitin-activating enzyme from the mammalian cell cycle mutant ts85. *Cell* 37, 43–55.

Flaherty, K.M., Deluca-Flaherty, C. and McKay, D.B. (1990) Three-dimensional structure of the ATPase fragment of a 70 K heat-shock cognate protein. *Nature* 346, 623–628.

Flynn, G.C., Pohl, J., Flocco, M.T. and Rothman, J.E. (1991) Peptide-binding specificity of the molecular chaperone BiP. *Nature* 353, 726–730.

Fox, T.W. (1951) Studies on heat tolerance in domestic fowl. *Poultry Science* 30, 477–483.

Fox, T.W. (1980) The effects of thiouracil and thyroxine on resistance to heat shock. *Poultry Science* 59, 2391–2396.

Fronda, F.M. (1925) Some observations on the body temperature of Poultry. *Cornell Vet* 15, 8–20.

Garren, H.W. and Shaffner, C.S. (1954) Factors concerned in the response of young New Hampshires to muscular fatigue. *Poultry Science* 33, 1095–1104.

Garren, H.W. and Shaffner, C.S. (1956) How the period of exposure to different stress stimuli affects the endocrine and lymphatic gland weights of young chicks. *Poultry Science* 35, 266–272.

George, J.C. (1980) Structure and physiology of posterior lobe hormones. In: Epple, A. and Stetson, M.H. (eds) *Avian Endocrinology.* Academic Press, New York, pp. 85–115.

George, J.C. and Berger, A J. (1966) *Avian Myology.* Academic Press, New York.

George, J.C. and John, T.M. (1986) Physiological responses to cold exposure in pigeons. In: Heller, H.C., Musacchia, X.J. and Wang, C.H. (eds) *Living in the Cold: Physiological and Biochemical Adaptations.* Elsevier Science, New York, pp. 435–443.

Gething, M.J. and Sambrook, J. (1992) Protein folding in the cell. *Nature* 355, 33–45.

Glick, B. (1957) Experimental modification of the growth of the bursa of Fabricius. *Poultry Science* 36, 18–24.

Glick, B. (1967) Antibody and gland studies in cortisone and ACTH-injected birds. *Journal of Immunology* 98, 1076–1084.

Gonyou, H.W. and Morrison, W.D. (1983) Effects of defeathering and insulative jackets on production by laying hens at low temperatures. *British Poultry Science* 24, 311–317.

Gould, N.R. and Siegel, H.S. (1981) Viability of and corticosteroid binding in lymphoid cells of various tissues after corticotropin injection. *Poultry Science* 60, 891–893.

Gross, W.B. and Siegel, H.S. (1983) Evaluation of heterophile:lymphocyte ratio as a measure of stress in chickens. *Avian Diseases* 27, 972–979.

Hahn, D.W., Ishibashi, T. and Turner, C.W. (1966) Alteration of thyroid secretion rate in fowls changed from a cold to a warm environment. *Poultry Science* 45, 31–33.

Harvey, S., Scanes, C.G. and Brown, K.I. (1986) Adrenals. In: Sturkie, P.D. (ed.) *Avian Physiology*. Springer, New York, pp. 479–493.

Hester, P.Y., Smith, S.G., Wilson, E.K. and Pierson, F.W. (1981) The effect of prolonged heat stress on adrenal weight, cholesterol and corticosterone in white pekin ducks. *Poultry Science* 60, 1583–1586.

Heywang, B.W. (1938) Effect of some factors on the body temperature of hens. *Poultry Science* 17, 317–323.

Hill, A.T. and Hunt, J.R. (1978) Layer cage depth effects on nervousness, feathering, shell breakage, performance and net egg returns. *Poultry Science* 57, 1204–1216.

Hillerman, J.P. and Wilson, W.O. (1955) Acclimation of adult chickens to environmental temperature changes. *American Journal of Physiology* 180, 591–595.

Hillman, P.E., Scott, N.R. and van Tienhoven, A. (1985) Physiological responses and adaptations to hot and cold environments. In: Yousef, M.K. (ed.). *Stress Physiology in Livestock*, Vol. 3, *Poultry*. CRC Press, Boca Raton, Florida, pp. 1–71.

Holmes, W.N. and Phillips, J.G. (1976) The adrenal cortex of birds. In: Chester Jones, I. and Henderson, I.W. (eds) *General, Comparative and Clinical Endocrinology of the Adrenal Cortex*. Academic Press, London, pp. 293–420.

Hooper, P. and Richards, S.A. (1991) Interaction of operant behaviour and autonomic thermoregulation in the domestic fowl. *British Poultry Science* 32, 929–938.

Horowitz, K.A., Scott, N.R., Hillman, P.E. and Van Tienhoven, A. (1978) Effects of feathers on instrumental thermoregulatory behaviour in chickens. *Physiology and Behaviour* 21, 233–238.

Hughes, B.O. (1978) The frequency of neck movements in laying hens and the improbability of cage abrasion causing feather wear. *British Poultry Science* 19, 389–393.

Hughes, B.O. and Wood-Gush, D.G.M. (1977) Agonistic behaviour in domestic hens: the influence of housing method and group size. *Animal Behaviour* 25, 1056–1062.

John, T.M. and George, J.C. (1973) Influence of glucagon and neurohypophysial hormones on plasma free fatty acid levels in the pigeon. *Comparative Biochemistry and Physiology* 45 A, 541–547.

John, T.M. and George, J.C. (1977) Blood levels of cyclic AMP, thyroxine, uric acid, certain metabolites and electrolytes under heat-stress and dehydration in the pigeon. *Archives Internationales de Physiologie et Biochimie* 85, 571–582.

John, T.M. and George, J.C. (1984) Diurnal thermal response to pinealectomy and photoperiod in the pigeon. *Journal of Interdisciplinary Cycle Research* 15, 57–67.

John, T.M. and George, J.C. (1986) Arginine vasotocin induces free fatty acid release from avian adipose tissue *in vitro*. *Archives Internationales de Physiologie et Biochimie* 94, 85–89.

John, T.M. and George, J.C. (1991) Physiological responses of melatonin-implanted pigeons to changes in ambient temperature. In: Riklis, E. (ed.) *Photobiology*. Plenum Press, New York, pp. 597–605.

John, T.M. and George, J.C. (1992) Effects of arginine vasotocin on cardiorespiratory and thermoregulatory responses in the pigeon. *Comparative Biochemistry and Physiology* 102 C, 353–359.

John, T.M., McKeown, B.A. and George, J.C. (1973) Influence of exogenous growth hormone and its antiserum on plasma free fatty acid level in the pigeon. *Comparative Biochemistry and Physiology* 46A, 497–504.

John, T.M., McKeown, B.A. and George, J.C. (1975) Effect of thermal stress and dehydration on plasma levels of glucose, free fatty acids and growth hormone in the pigeon. *Archives Internationales de Physiologie et Biochimie* 83, 303–308.

John, T.M., Itoh, S. and George, J.C. (1978) On the role of pineal hormones in the thermoregulation in the pigeon. *Hormone Research* 9, 41–56.

Johnson, P.L., Johnson, A.L. and van Tienhoven, A. (1985) Evidence for a positive feedback interaction between progesterone and luteinizing hormone in the induction of ovulation in the hen, *Gallus domesticus*. *General and Comparative Endocrinology* 58, 478–485.

Johnston, R.N. and Kucey, B.L. (1988) Competitive inhibition of *hsp70* gene expression causes thermosensitivity. *Science* 242, 1551–1554.

Jones, D.R. and Johansen, K. (1972) The blood vascular system of birds. In: Farner, D.S. and King, J.R. (eds) *Avian Biology*, Vol. 2. Academic Press, New York, pp. 257–285.

Kamar, G.A.R. and Khalifa, M.A.S. (1964) The effect of environmental conditions on body temperature of fowls. *British Poultry Science* 5, 235–244.

Kohne, H.J. and Jones, J.E. (1975a) Changes in plasma electrolytes, acid-base balance and other physiological parameters of adult female turkeys under conditions of acute hyperthermia. *Poultry Science* 54, 2034–2038.

Kohne, H.J. and Jones, J.E. (1975b) Acid-base balance, plasma electrolytes and production performance of adult turkey hens under conditions of increasing ambient temperature. *Poultry Science* 54, 2038–2045.

Koike T.I., Neldon, H.L., McKay, D.W. and Rayford, P.L. (1986) An antiserum that recognizes mesotocin and isotocin: development of a homologous radioimmunoassay for plasma mesotocin in chickens (*Gallus domesticus*). *General Comparative Endocrinology* 63, 93–103.

Kuhn, E.R. and Nouwen, E.J. (1978) Serum levels of triiodothyronine and thyroxine in the domestic fowl following mild cold exposure and injection of synthetic thyrotropin releasing hormone. *General and Comparative Endocrinology* 34, 336–342.

Kulomaa, M.S., Weigel, N.L., Kleinsek, D.A., Beattie, W.G., Conneely, O.M., March, C., Zarucki-Schulz, T., Schrader, W.T. and O'Malley, B.W. (1986) Amino acid sequence of a chicken heat shock protein derived from the complementary DNA nucleotide sequence. *Biochemistry* 25, 6244–6251.

Lam, S.K. and Harvey, S. (1990) Thyroid regulation of body temperature in anaesthetized chickens. *Comparative Biochemistry and Physiology* 95A, 435–439.

Laycock, S.R. (1989) Operant conditioning techniques and their use in evaluating the thermal environment of the mature laying hen. MSc Thesis, University of Guelph, Ontario, Canada.

Lee, D.H., Robinson, K.W., Yeates, N.T.M. and Scott, M.I.R. (1945) Poultry husbandry in hot climates – experimental enquiries. *Poultry Science* 24, 195–207.

Leeson, S.A. and Morrison, W.D. (1978) Effect of feather cover on feed efficiency in laying birds. *Poultry Science* 57, 1094–1096.

Lin, Y.C. and Sturkie, P.D. (1968) Effect of environmental temperatures on the catecholamines of chickens. *American Journal of Physiology* 214, 237–240.

Lindquist, S. and Craig, E.A. (1988) The heat-shock proteins. *Annual Review of Genetics* 22, 631–677.

Lustick, S.I. (1983) Cost-benefit of thermoregulation in birds: influences of posture, microhabitat selection, and color. In: Aspey, W. and Lustick, S.I. (eds) *Behavioral Energetics*. Ohio State University Press, Columbus, Ohio, pp. 265–294.

McCormick, C.C., Garlich, J.D. and Edens, F.W. (1979) Fasting and diet affect the tolerance of young chickens exposed to acute heat stress. *Journal of Nutrition* 109, 1797–1809.

McFarlane, J.M., Curtis, S.E., Shanks, R.D. and Carmer, S.G. (1989) Multiple concurrent stressors in chicks. 1. Effect on weight gain, feed intake, and behaviour. *Poultry Science* 68, 501–509.

McNabb, F.M.A. (1988) Peripheral thyroid hormone dynamics in precocial and altricial avian development. *American Zoologist* 28, 427–440.

McNicholas, M.J. and McNabb, F.M.A. (1987) Influence of dietary iodine availability. *Journal of Experimental Zoology* 244, 263–268.

Marder, J. (1973) Temperature regulation in the bedouin fowl (*Gallus domesticus*). *Physiology and Zoology* 46, 208–217.

Marley, E. and Nistico, G. (1972) Effects of catecholamines and adenosine derivative given into the brain of fowls. *British Journal of Pharmacology* 46, 629–636.

Marley, E. and Stephenson, J.D. (1970) Effects of catecholamines infused into the brain of young chickens. *British Journal of Pharmacology* 40, 639–658.

May, J.D. (1982) Effect of dietary thyroid hormone on survival time during heat stress. *Poultry Science* 61, 706–709.

May, J.D. and Lott, B.D. (1992) Feed consumption patterns of broilers at high environmental temperatures. *Poultry Science* 71, 331–336.

May, J.D., Deaton, J.W., Reece, F.N. and Branton, S.L. (1986) Effect of acclimation and heat stress on thyroid hormone concentration. *Poultry Science* 65, 1211–1213.

Mellen, W.J. and Wentworth, B.C. (1958) Studies with thyroxine and triiodothyronine in chickens. *Poultry Science* 37, 1226.

Mellen, W.J. and Wentworth, B.C. (1962) Observations on radiothyroidectomized chickens. *Poultry Science* 41, 134–141.

Mench, J. (1985) Behaviour and stress. *Maryland Poultryman*, July, 1–2.

Merat, P. (1990) Genes que majeurs chez la poule (*Gallus gallus*): autres genes que ceux affectant la taille. *INRA Production Animal* 3, 355–368.

Midtgard, U. (1989) Circulatory adaptations to cold in birds. In: Bech, C. and Reinertsen, R.E. (eds) *Physiology of Cold Adaptation in Birds*. Plenum Press, New York, pp. 211–222.

Miller, L. and Qureshi, M.A. (1992) Molecular changes associated with heat-shock treatment in avian mononuclear and lymphoid lineage cells. *Poultry Science* 71, 473–481.

Mizzen, L.A. and Welch, W.J. (1988) Characterization of the thermotolerant cell. I. Effects on protein synthesis activity and the regulation of heat-shock protein 70 expression. *Journal of Cell Biology* 106, 1105–1116.

Mongin, P.E. (1968) Role of acid-base balance in the physiology of egg-shell formation. *World's Poultry Science Journal* 24, 200–230.

Morimoto, R.I., Hunt, C., Huang, S.-Y., Berg, K.L. and Banerji, S.S. (1986) Organization, nucleotide sequence, and transcription of the chicken HSP70 gene. *Journal of Biological Chemistry* 261, 12692–12699.

Morimoto, R.I., Tissieres, A. and Georgopoulos, C. (eds) (1990) *Stress Proteins in Biology and Medicine*. Cold Spring Harbor Laboratory, Cold Spring Harbor, New York.

Morrison, W.D. and Curtis, S. (1983) Observations of environmental thermoregulation by chicks. *Poultry Science* 62, 1912–1914.

Morrison, W.D. and McMillan, I. (1985) Operant control of the thermal environment by chicks. *Poultry Science* 64, 1656–1660.

Morrison, W.D. and McMillan, I. (1986) Response of chicks to various environmental temperatures. *Poultry Science* 65, 881–883.

Morrison, W.D., McMillan, I. and Amyot, E. (1987a) Operant control of the thermal environment and learning time of young chicks and piglets. *Canadian Journal of Animal Science* 61, 343–347.

Morrison, W.D., McMillan, I. and Bate, L.A. (1987b) Effect of air movement on operant heat demand of chicks. *Poultry Science* 66, 854–857.

Munsick, R.S., Sawyer, W.H. and Van Dyke, H.B. (I960) Avian neurohypophysial hormones: pharmacological properties and tentative identification. *Endocrinology* 66, 860–871.

Nagata, K., Saga, S. and Yamada, K.M. (1986) A major collagen-binding protein of chick embryo fibroblasts is a novel heat shock protein. *Journal of Cell Biology* 103, 223–229.

Necker, R. (1977) Thermal sensitivity of different skin areas in pigeons. *Journal of Comparative Physiology* 116, 239–246.

Nobukumi, K. and Nishiyama, H. (1975) The influence of thyroid hormone on the maintenance of body temperature in male chicks exposed to low ambient temperature. *Japan Journal of Zootechnical Science* 46, 403–407.

North, M.O. (1978) *Commercial Chicken Production Manual*. AVI Publishing Company, Westport, Connecticut.

North, M.O. and Bell, D.D. (1990) *Commercial Chicken Production Manual*, 4th edn. Van Nostrand Reinhold, New York, 262 pp.

Nover, L. (1991) *Heat Shock Response*. CRC Press, Boca Raton, Florida.

Novero, R.P., Beck, M.M., Gleaves, E.W., Johnson, A.L. and Deshazer, J.A. (1991) Plasma progesterone, luteinizing hormone concentrations and granulosa cell responsiveness in heat-stressed hens. *Poultry Science* 70, 2335–2339.

Otten, L., Morrison, W.D., Braithwaite, L.A. and Smith, J.H. (1989) Development of cooled roosts for heat-stressed poultry. Presentation at the Meeting of the Canadian Society of Agricultural Engineering and American Society of Agricultural Engineers held in Quebec, PQ, Canada, 25–28 June 1989. Paper No. 89–4081, 13 pp.

Pang, S.F., Brown, G.M., Grota, L.J., Chambers, J.W. and Rodman, R.L. (1977) Determination of N-acetylserotonin and melatonin activities in the pineal gland, retina, Harderian gland, brain and serum of rats and chickens. *Neuroendocrinology* 23, 1–13.

Pardue, M.L., Feramisco, J.R. and Lindquist, S. (eds) (1989) *Stress-induced Proteins*. Alan R. Liss, New York.

Parker, J.T. and Boone, M.A. (1971) Thermal stress effects on certain blood characteristics of adult male turkeys. *Poultry Science* 50, 1287–1295.

Parker, J.T., Boone, M.A. and Knechtges, J.F. (1972) The effect of ambient temperature upon body temperature, feed consumption, and water consumption, using two varieties of turkeys. *Poultry Science* 51, 659–664.

Pelham, H.R.B. (1984) Hsp70 accelerates the recovery of nucleolar morphology after heat shock. *EMBO Journal* 3, 3095–3100.

Pelham, H.R.B. (1986) Speculations on the functions of the major heat shock and glucose-regulated proteins. *Cell* 46, 959–961.

Petitte, J.N. and Etches, R.J. (1988) The effect of corticosterone on the photo-periodic response of immature hens. *General and Comparative Endocrinology* 39, 424–430.

Pilo, B., .John, T.M., George, J.C. and Etches, R.J. (1985) Liver Na^+K^+-ATPase activity and circulating levels of corticosterone and thyroid hormones following cold and heat exposure in the pigeon. *Comparative Biochemistry and Physiology* 80A, 103–106.

Quay, W.B. (1974) *Pineal Chemistry*. Charles C. Thomas, Illinois.

Ralph, C.L. (1981) Melatonin production by extrapineal tissues. In: Birau, N. and Schloot, W. (eds) *Melatonin – Current Status and Perspectives*. Pergamon Press, New York, pp. 35–46.

Ramirez, J.M. and Bernstein, M.H. (1976) Compound ventilation during thermal panting in pigeons: a possible mechanism for minimizing hypocapnic alkalosis. *Federation Proceedings, Federation of the American Society of Experimental Biology* 35, 2562–2565.

Raup, T.J. and Bottje, W.G. (1990) Effect of carbonated water on arterial pH, $P\ CO_2$, and plasma lactate in heat-stressed broilers. *British Poultry Science* 31, 377–384.

Reece, F.N., Deaton, J.W. and Kubena, L.F. (1972) Effects of high temperature and humidity on heat prostration of broiler chickens. *Poultry Science* 51, 2021–2025.

Reineke, E.P. and Turner, C.W. (1945) Seasonal rhythm in thyroid hormone secretion of the chick. *Poultry Science* 24, 499–504.

Richards, S.A. (1970) Physiology of thermal panting in birds. *Annals of Biology, Animal Biophysics* 10, 151–168.

Richards, S.A. (1976) Behavioural temperature regulation in the fowl. *Journal of Physiology* 258, 122P–123P.

Richards, S.A. (1977) The influence of loss of plumage on temperature regulation in laying hens. *Journal of Agriculture Science, Cambridge* 89, 393–398.

Robinzon, B., Koike, T.I., Neldon, H.L., Kinzler, S.L., Hendry, I.R. and El Halawani, M.E. (1988) Physiological effects of arginine vasotocin and mesotocin in cockerels. *British Poultry Science* 29, 639–652.

Rotii-Roti, J.L. and Laszlo, A. (1987) The effects of hyperthermia on cellular macro molecules. In: Urano, M. and Douple, A. (eds) *Hyperthermia and Oncology*, Vol. 1. VNU Scientific Publishers, the Netherlands, pp. 13–56.

Rudas, P. and Pethes, G. (1984) The importance of peripheral thyroid hormone deiodination in adaptation to ambient temperature in the chicken (*Gallus domesticus*). *Comparative Biochemistry and Physiology* 77A, 567–571.

Rzasa, J. and Neizgoda, J. (1969) Effects of the neurohypophysial hormones on sodium and potassium level in the hen's blood. *Bulletin of the Academy of Polish Science CI. II, Ser. Science Biology* 17, 585–588.

Rzasa, J., Skotinicki, J. and Niezgoda, J. (1971) Metabolic effects of neuro-hypophysial hormones in the chicken. *Bulletin of the Academy of Polish Science CI. II, Ser. Science Biology* 19, 431–434.

Sammelwitz, P.H. (1967) Adrenocortical hormone therapy of induced heat stress mortality in broilers. *Poultry Science* 46, 1314.

Sargan, D.R., Tsai, M.J. and O'Malley, B.W. (1986) Hspl08, a novel heat shock inducible protein of chicken. *Biochemistry* 25, 6252–6258.

Siegel, H.S. and Beane, W.L. (1961) Time responses to single intramuscular doses of ACTH. *Poultry Science* 40, 216–219.

Siegel, H.S., Drury, L.N. and Patterson, W.C. (1974) Blood parameters of broilers grown in plastic coops and on litter at two temperatures. *Poultry Science* 53, 1016–1024.

Singh, A., Reineke, E.P. and Ringer, R.K. (1968) Influence of thyroid status of the chick on growth and metabolism, with observations on several parameters of thyroid function. *Poultry Science* 47, 212–219.

Stiffler, D.F., Roach, S.C. and Pruett, S.J. (1984) A comparison of the responses of the amphibian kidney to mesotocin, isotocin, and oxytocin. *Physiological Zoology* 57, 63–69.

Sturkie, P.D. (1967) Cardiovascular effects of acclimation to heat and cold in chickens. *Journal of Applied Physiology* 22, 13–15.

Sullivan, D.A. and Wira, C.R. (1979) Sex hormone and glucocorticoid receptor in the bursa of Fabricius of immature chickens. *Journal of Immunology* 122, 2617–2623.

Tanguay, R.M. (1988) Transcriptional activation of heat-shock genes in eukaryotes. *Biochemistry and Cell Biology* 66, 584–593.

Teeter, R.G. and Smith, M.O. (1986) High chronic ambient temperature stress effects on broiler acid-base balance and their response to supplemental ammonium chloride, potassium chloride, and potassium carbonate. *Poultry Science* 65, 1777–1781.

Teeter, R.G., Smith, M.O., Owens, F.N., Arp, S.C., Sangiah, S. and Breazile, J.E. (1985) Chronic heat stress and respiratory alkalosis: occurrence and treatment in broiler chicks. *Poultry Science* 64, 1060–1064.

Theodorakis, N.G., Banerji, S.S. and Morimoto, R.I. (1988) HSP70 mRNA translation in chicken reticulocytes is regulated at the level of elongation. *Journal of Biological Chemistry* 263, 14579–14585.

Thompson, E.B. and Lippman, M.E. (1974) Mechanism of action of glucocorticoids. *Metabolism* 23, 159–202.

Thornton, P.A. (1962) The effect of environmental temperature on body temperature and oxygen uptake by the chicken. *Poultry Science* 41, 1053–1062.

Vo, K.V. and Boone, M.A. (1975) The effect of high temperatures on broiler growth. *Poultry Science* 54, 1347–1348 (Abstract).

Vogel, J.A. and Sturkie, P.D. (1963) Cardiovascular responses of the chicken to seasonal and increased temperature changes. *Science* 140, 1404–1406.

Wang, S. (1988) Arginine vasotocin and mesotocin response to acute heat stress in domestic fowl. MS thesis, University of Arkansas.

Wang, S., Bottje, W.G., Kinzler, S., Neldon, H.L. and Koike, T.I. (1989) Effect of heat stress on plasma levels of arginine vasotocin and mesotocin in domestic fowl (*Gallus domesticus*). *Comparative Biochemistry and Physiology* 93A(4), 721–724.

Weiss, H.S., Frankel, H. and Hollands, K.G. (1963) The effect of extended exposure to a hot environment on the response of the chicken to hyperthermia. *Canadian Journal of Biochemistry* 41, 805–815.

Welch, W.J. and Mizzen, L.A. (1988) Characterization of the thermotolerant cell. II. Effects on the intracellular distribution of heat-shock protein 70, intermediate filaments, and small nuclear ribonucleoprotein complexes. *Journal of Cell Biology* 106, 1117–1130.

Welch, W.J. and Suhan, J.P. (1985) Morphological study of the mammalian stress response: characterization of changes in cytoplasmic organelles, cytoskeleton, and nucleoli, and appearance of intranuclear actin filaments in rat fibroblasts after heat-shock treatment. *Journal of Cell Biology* 101, 1198–1211.

Welty, J.C. and Baptista, L. (1988) *The Life of Birds*, 4th edn. Saunders College Publishing, Toronto, p. 138.

Westwood, J.T., Clos, J. and Wu, C. (1991) Stress-induced oligomerization and chromosomal relocalization of heat-shock factor. *Nature* 353, 822–827.

Whittow, G.C. (1986) Regulation of body temperature. In: Sturkie, P.D. (ed.) *Avian Physiology*. Springer, New York, pp. 221–252.

Whittow, G.C., Sturkie, P.D. and Stein, G., Jr (1964) Cardiovascular changes associated with thermal polypnea in the chicken. *American Journal of Physiology* 207, 1349–1353.

Wiech, H., Buchner, J., Zimmerman, R. and Jakob, U. (1992) Hsp90 chaperones protein folding *in vitro*. *Nature* 358, 169–170.

Williams, J.B. and Sharp, P.J. (1978) Control of the preovulatory surge of luteinizing hormone in the hen (*Gallus domesticus*): the role of progesterone and androgens. *Journal of Endocrinology* 11, 57–65.

Wilson, F.E. and Follett, B.K. (1975) Corticosterone-induced gonadosuppression in photostimulated tree sparrows. *Life Science* 17, 1451–1456.

Wilson, S.C. and Sharp, P.J. (1976) Induction of luteinizing hormone release by gonadal steroids in the ovariectomized domestic hen. *Journal of Endocrinology* 71, 87–98.

Wilson, W.O. (1948) Some effects of increasing environmental temperatures on pullets. *Poultry Science* 27, 813–817.

Wilson, W.O. and Woodard, A. (1955) Some factors affecting body temperature of turkeys. *Poultry Science* 34, 369–371.

Winchester, C.F. (1939) Influence of thyroid on egg production in poultry. *Endocrinology* 27, 697–703.

Yost, H.J., Petersen, R.B. and Lindquist, S. (1990) RNA metabolism: strategies for regulation in the heat shock response. *Trends in Genetics,* 6, 223–227.

Young, R.A. (1990) Stress proteins and immunology. *Annual Review of Immunology* 8, 401–420.

Zagris, N. and Matthopoulos, D. (1986) Differential heat-shock gene expression in chick blastula. *Roux's Archives of Developmental Biology* 195, 403–407.

Zagris, N. and Matthopoulos, D. (1988) Gene expression in chick morula. *Roux's Archives of Developmental Biology* 197, 298–301.

5 Poultry Housing for Hot Climates

M. CZARICK III AND B.D. FAIRCHILD

The University of Georgia, Athens, Georgia 30602, USA

Introduction	81
Definitions	81
Bird heat loss	83
Naturally ventilated houses	86
House width	86
House length	87
Side-wall openings	87
Local obstructions	88
Ridge openings	89
House orientation	91
Roof overhang	92
Roof slope	93
Roof insulation	94
Side-wall and end-wall insulation	97
Roof coatings	97
Roof sprinkling	98
Circulation fans	98
Layer houses	99
Power-ventilated houses	100
House construction	101
Air exchange	101
Air inlet systems	104
Types of inlet ventilation systems	106
Air movement in inlet-ventilated houses	108
Tunnel ventilation systems	109
Poultry-house exhaust fans	117

© CAB International 2008. *Poultry Production in Hot Climates* 2nd edition (ed. N.J. Daghir)

Evaporative cooling	121
Fogging systems	122
Sprinkling systems	124
Pad systems	124
Summary	128
References	131

Introduction

In hot weather conditions it is critical that birds are kept cool. Mammals and avian species have the ability to maintain body temperature by losing any extra heat that is produced or by generating heat if the body temperature is too low. If the body temperature of a bird, which normally runs between 39.4°C and 40°C, is allowed to increase the bird will not perform well. The thermoneutral zone, which varies with age, is the temperature where the bird is not using any energy to lose or gain heat. For adult chickens the thermoneutral zone is between 18°C and 23.9°C (Sturkie, 1965), while for adult broilers it is between 26°C and 27°C (Van Der Hel et al., 1991). When birds' body temperatures are in their thermoneutral zone, the energy from the feed can be directed to growth, immune system development and reproduction. When temperature goes below the thermoneutral temperature, energy from the feed is used to generate heat rather than for growth and development. If the temperature increases above thermoneutral temperature, body temperature will begin to rise and the bird will use a portion of the energy to lose heat, which tends to result in reduced feed conversion efficiency. Another disadvantage to increasing body temperature is that the birds' feed consumption will decrease, resulting in reduced weight gains and overall performance.

The results of heat stress on performance are dependent on the degree and the length of the time that the temperature is above the thermoneutral zone. Birds that are heat stressed experience reduced feed consumption, reduced weight gain, and poor feed conversion (Van Kampen, 1981). Heat stress results in increased respiratory rate (panting) and degradation of shell quality and egg production and increased mortality and morbidity (Reece et al., 1972; Scott and Balnave, 1988). The bird's body makes several physiological changes to maintain body temperature, including electrolyte depletion and reduced immune response. Any and all of these corrective measures results in the shunting of energy away from growth and egg production, resulting in poor performance.

Definitions

Sensible Heat:
Sensible heat is heat that when absorbed by a substance causes a rise in temperature of the object or when lost from a substance causes a decrease in the

temperature of the object. In poultry houses it is the sensible heat produced by the birds that causes a rise in house air temperature.

Latent Heat:
Latent heat is the amount of heat that a substance may absorb with a change in temperature which is used in the conversion of a solid to liquid state or from a liquid to gaseous state. Bird latent heat is the heat lost from a bird due to the evaporation of water from its body that does not increase house temperature but does increase house humidity.

Evaporative Cooling:
A cooling process by which the air is cooled through the evaporation of water into the air. The drier the air the greater the amount of water that can be evaporated into the air and the greater the amount of cooling produced. Though evaporative cooling is very effective in decreasing house temperatures it also leads to an increase in relative humidity, which can be problematic, especially in humid climates.

Static Pressure:
Static pressure typically refers to the pressure differential created by exhaust fans in a power-ventilated poultry house. Static pressure is typically measured in pascals or inches of water column. The lower amount of opening available to exhaust fans the higher the static pressure and the faster the air will try to enter a house through available openings. High entrance velocities at air inlets during cool weather can help to direct the incoming air towards the ceiling and away from the birds. Excessive static pressures will dramatically reduce the air-moving capacity of exhaust fans as well as increase electricity usage.

Radiation:
The transfer of heat from a warm object to a cold object through electromagnetic waves. In most poultry houses birds lose very little heat through radiation owing to the fact that very few surfaces within the house have a lower temperature than the birds' surface temperature. Poorly insulated ceilings and side-wall curtains typical radiate heat to the birds during hot weather.

Conduction:
The transfer of heat through a solid medium. This method of heat loss is fairly minimal in most houses during hot weather. Most bedding materials are good insulators, which reduce conductive heat loss from the bird to the floor. An exception would be birds in houses with concrete floors with no or minimal bedding material, or birds in cages.

Convection:
The transfer of heat through a moving fluid. Air movement over birds during hot weather transfers heat from the birds to air.

Thermoneutral Temperature:
A range in environmental temperature where birds can maintain their body temperature without utilizing excess energy. If temperature drops below or rises above the thermoneutral zone, energy will be shunted away from

important economic functions, such as growth, development, and egg and semen production, to increase or decrease their body temperature. The range of temperatures will vary with bird breed and age.

Upper and Lower Lethal Temperature:
This is the deep body temperatures at which chickens begin to die from excessive hot or cold environmental temperatures.

Bird Heat Loss

The primary methods of heat loss during hot weather are through evaporation and convection. Evaporation (or latent heat loss) is a very important method of heat dissipation at all ages and temperatures. Even at thermoneutral temperatures evaporative heat loss typically accounts for approximately 60% of a bird's total heat loss. As house temperatures rise a bird will begin to pant in an effort to increase evaporative heat loss. Bird panting is undesirable in poultry production as it results in dehydration if adequate water supplies are not provided. Bird energy is directed away from growth and development as the bird pants and it can result in respiratory alkalosis, which can result in the drop in blood pH as carbon dioxide is given off. Evaporative/latent heat loss depends on the evaporation of water off a bird's respiratory system to produce cooling and is therefore adversely affected by high humidity. This is problematic for two reasons. First, most poultry-producing areas of the world would be considered humid climates. Second, evaporative cooling is the primary method of air-temperature reduction used by poultry producers during hot times of the year, which reduces house temperature but increases humidity. So though the temperature reduction produced through the evaporation of water into the air in a poultry house increases convective cooling it can reduce latent heat loss from the bird.

Increasing the amount of air movement over a bird is one of the most effective methods that producers can use to increase heat removal from birds. Air movement provides a number of benefits that help to cool the birds during hot weather. First, it increases convective heat loss from the birds in a house, thereby lowering the effective air temperature. Second, it helps to remove trapped heat from between the birds as well as from cages. Last, but not least, it lessens the adverse effects of high-humidity environments.

Movement of air over the birds removes heat (from the bird), which helps keep the birds' body temperature at or near thermoneutral. As air speed increases, the amount of heat removal increases and the bird cooling is enhanced. The effect of this is to transfer the bird heat loss from latent heat loss (heat loss through panting) to sensible heat loss (convective heat loss), so as wind speed increases more sensible heat loss occurs than latent heat loss (Mitchell, 1985; Simmons et al., 1997) (Fig. 5.1). This will result in less panting allowing the nutrient energy to be used for growth and development.

Tunnel ventilation on commercial broiler farms results in better bird performance, including increased weight gain and improved feed conversion.

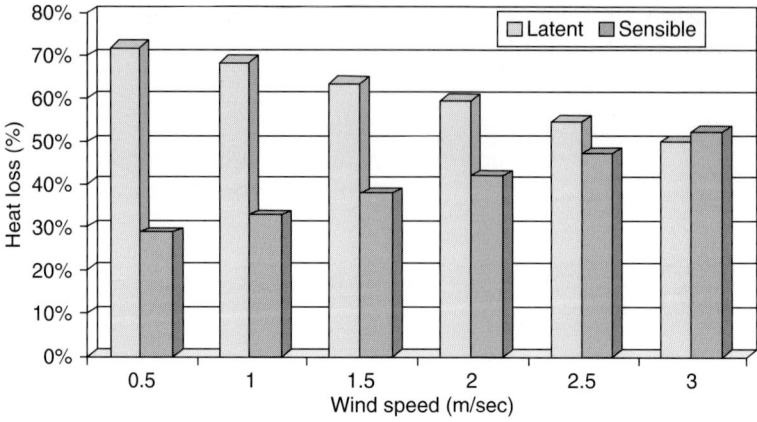

Fig. 5.1. Effect of wind speed on sensible heat loss.

Table 5.1. Air-velocity effects on male broiler body-weight gain (g) in cyclic temperatures of 25–30–25°C.

	<0.25 m/s (still air)	2 m/s	3 m/s
4 wk	526[a]	545[a]	552[a]
5 wk	579[a]	653[b]	666[b]
6 wk	489[a]	620[b]	650[b]
7 wk	366[a]	504[b]	592[c]

[a–c] Values within a row without common letter are significantly different ($P < 0.05$)
* Table adapted from Simmons *et al.*, 2003.

Even with the greater dependence on electricity for cooling, income is generally greater on tunnel-ventilated farms than naturally ventilated farms (Lacy and Czarick, 1992). A study conducted with rearing male broilers in a moderate cyclic temperature environment of 25–30–25°C compared wind speeds of still air (< 0.25 m/s), 2 and 3 m/s. The results of this study demonstrated that the benefit of air speed increased as the birds got older and bigger. Therefore, 6-week-old birds benefited from higher wind speeds more than the 4-week-old birds. Wind speeds of 2 and 3 m/s had no effect on body weight gain or feed conversion of 4-week-old chickens when compared to still air. However, 5- and 6-week-old birds had significantly better weight gain in the 2 and 3 m/s treatments than in the still air treatment, while 7-week-old birds in the 3 m/s treatment were significantly heavier than either in the 2 m/s and still air treatments (Table 5.1).

As temperatures increase, the impact of wind speed can be observed earlier. In a study conducted with male broilers reared from 4 to 7 weeks of age in a cyclic temperature of 25–35–25°C over a 24-h period, difference in body weight gain, feed consumption and feed conversion (Table 5.2) between the treatments were noted as early as 4 weeks of age (Dozier *et al.*, 2005a).

Table 5.2. Air-velocity effects on male broiler body-weight gain (g) in cyclic temperatures of 25–35–25°C.

	Still air	2 m/s	3 m/s
4 wk	500[c]	546[b]	588[a]
5 wk	476[c]	572[b]	630[a]
6 wk	340[c]	468[b]	574[a]
7 wk	187[c]	300[b]	381[b]

[a–c] Values within a row without common letter are significantly different ($P < 0.05$)
* Table adapted from Dozir et al., 2005[a].

This indicates that rearing the current high meat yielding strains in higher environmental temperatures is more detrimental to bird performance compared with similar studies conducted in 1997.

Water consumption is important to bird performance, not only as an essential nutrient but also when water consumption is reduced feed consumption is reduced (Pesti et al., 1985). Nipple drinkers are the industry standard for poultry-house drinking systems, resulting in cleaner drinking water, reduced water spillage and reduced labour for drinker cleaning. However, May et al. (1997) demonstrated a reduction in water consumption by chickens drinking from nipple drinkers compared with those drinking from open water systems during an elevated temperature of 21–35–21°C diurnal cycle. When the two drinker systems were compared under air velocity treatments of 0.25 m/s and 2.1 m/s, birds in the higher air velocity with nipple drinkers did not differ from those on open water drinkers, but experienced increased weight gain and better feed conversion than birds at the low air velocity (Lott et al., 1998).

Bird age is a factor in determining the air velocity needed to cool the birds. In general, older birds require more air velocity than younger birds. A recent study (Dozier et al. 2005b) examined the differences between still, constant (2 m/s) and variable (1.5 m/s from 28 to 35 days of age; 2 m/s from 36 to 42 days of age, 3 m/s from 43 to 49 days of age) air speeds when birds were exposed to a cyclic diurnal temperature of 25–30–25°C. Body weights were similar among the air velocity treatments through 42 days of age, but increasing the air velocity to 3 m/s in the last week improved broiler growth rate (Table 5.3).

Thus, removing the heat from the birds and the house is crucial to obtaining optimum performance. The amount of air velocity required to keep bird body temperature at normal levels is dependent on several factors, which include bird density, bird age, and the local climate. As bird density increases, higher air velocity is required to cool the birds. As birds get older, their body weight increases and feathers provide a layer of insulation that will trap the heat. High air velocity is more crucial for older birds than younger birds. In areas with low humidity, lower air velocity can be effective. However, in areas with high humidity, low air speed can be deadly to birds. Even with high air speeds, high humidity is a problem, but air speed, owing to the cooling effect it provides, can be managed to where birds' performance

Table 5.3. Still, constant and variable air-velocity effects on male broiler body-weight gain (g) in cyclic temperatures of 25–30–25°C.

	Still air	Constant**	Variable***
5 wk	549[b]	644[a]	628[a]
6 wk	511[b]	637[a]	643[a]
7 wk	378[c]	493[b]	599[c]

[a–c] Values within a row without common letter are significantly different ($P < 0.05$)
* Table adapted from Dozir et al., 2005b.
** Constant = 2 m/s for entire study.
*** Variable = 1.5 m/s from 28 to 35; 2 m/s from 36 to 42, 3 m/s from 43 to 49 days of age.

can be optimized. In hot weather, cooling birds to maintain body temperatures or to return body temperature to normal after a heat stress period is crucial in optimizing bird performance.

Naturally Ventilated Houses

One of the keys to minimizing heat stress in naturally ventilated poultry houses during hot weather is making sure that outside air can easily flow into and out of the house. The easier it is for outside air to flow through a house the less likely there will be a detrimental build-up of heat within the house, minimizing inside to outside temperature differentials. Furthermore, increased air-exchange rates tend to result in increased air movement over the birds within the house, thus maximizing heat loss due to convection. Factors which affect heat build-up in a naturally ventilated house include house width, house length, side-wall openings, local obstructions, ridge openings, house orientation, roof overhang, roof slope, roof insulation, side-wall and end-wall insulations, roof coatings, roof sprinkling and circulation fans. The first four factors are the primary factors that determine the ease with which outside air moves through a house.

House width

Traditionally poultry producers have found that natural ventilation tends to be most effective in houses which are 12 m in width or less. Wider houses tend to have lower air-exchange rates and significantly less natural air movement towards the centre of the house. Since air-exchange rates tend to be reduced, significant temperature differences can occur between the upwind and downwind sides of the house, especially during cooler times of the year. Though the use of interior circulation fans in a house has been found to mitigate the magnitude of the problem, it is still advisable that house width should be kept to 12 m or less in most hot climates.

House length

Houses can be of any convenient length. If mechanical feeding, egg collection belts or manure belts are used, there is a practical limit on length for most systems (check with manufacturer for recommendations). In long buildings, doors are often placed in the side walls at intervals of 15–30 m (50–100 ft), to provide access for service and bird removal.

Side-wall openings

All naturally ventilated houses must be equipped with some type of adjustable side-wall curtains to control the flow of air into the house during cooler times of the year or when small birds are present. Some curtain materials allow light to enter the building. These are widely used in applications where the manager wants natural light to penetrate the building so that artificial light is not needed during daylight hours.

Black (opaque) curtains are used in applications where it is desirable to exclude all external light (e.g. to provide blackout housing for broiler breeder pullets). Insulated curtains are also available for use in areas where it is desirable to reduce heat loss from buildings during the cool season (Timmons et al., 1981). In order to facilitate the rapid exchange of air during hot weather, curtain openings should generally account for between 50 and 80% of the side-wall height. The hotter the climate, the wider the house, and the greater the percentage the curtain opening should account for in the side wall. It is very important that there are solid portions of wall above and below the curtain opening. The lower solid-wall portion of the wall serves a number of important functions. First, it helps to reduce the amount of rain that can enter the house during storms as well as prevent outside water from running into the house. It also reduces the likelihood of direct sunlight entering the house when the curtains are fully opened, and provides a surface for the bottom of the curtain to overlap and reduce drafts during cooler times of the year. Last, but not least, the solid lower wall helps to exclude rodents from the house and minimize contact between the birds in the house and birds outside the house. Though the optimum height of the lower solid wall can in general vary, it is best if it is a minimum of 40 cm (Fig. 5.2).

A solid wall above the curtain opening allows an adequate surface for the side-wall curtain to seal against during cold weather. Even in hot climates it is important to be able to seal a house relatively tightly during brooding periods to minimize heat loss as well as drafts. Furthermore, it is important to have a solid wall above the curtain opening to facilitate the installation of side-wall air inlets for use during cooler times of the year or during brooding. In general, it is best that the wall above the curtain opening is such that it will allow the side-wall curtains to overlap the wall by a minimum of 30 cm.

The side-wall curtain should be constructed of clear, non-breathable material to limit air exchange when heat conservation is desirable. A rigid curtain rod at the top of the curtain opening should be used to more precisely control

Fig. 5.2. 40-cm lower solid side wall.

the amount of curtain opening during winter, when heat conservation is required. It is best if the bottom of the curtain opening is also equipped with a curtain rod to minimize drafts when fully closed. Like the top of the curtain, the bottom of the curtain should overlap the lower solid wall by 30 cm or more.

Typically, translucent curtains are advisable to allow the maximum amount of sunlight to enter the house when the curtains are closed or partially closed. When black curtains are used to control light it is best that the outside surface is silver or white in colour to reflect solar radiation, thus minimizing solar heat gain. Black curtains are usually used during rearing periods to reduce high light intensity and consequently cannibalism of layer and breeder flocks.

Local obstructions

House spacing

House spacing can significantly affect the environment in poultry houses during hot weather (Fig. 5.3). A structure creates a zone on the downwind side where wind velocities are reduced. If an adjacent house is placed in this zone it can be subjected to decreased air-exchange rates, as well as heat, moisture, dust and microorganisms emanating from the upwind house. Though there are a number of factors that determine optimal house spacing (prevailing wind speed, direction, topography, etc.), a minimum recommended spacing can be calculated from the following formula (Timmons, 1989):

Fig. 5.3. Wide spacing between poultry houses.

$$D = 0.4 \times H \times L^{0.5}$$

Where:
D = Separation distance (ridge to the closest wall of the next house)
H = Height of obstructing building
L = Length of the obstructing building.

Vegetation

Vegetation height should be kept to a minimum around poultry houses, not only to discourage rodents but also to maximize air flow into the house and discourage wild bird nesting (Fig. 5.4). Close-cut green vegetation can be very beneficial compared with bare ground, owing to the fact that vegetation temperatures can be as much as 30°C cooler than bare ground. Furthermore, green vegetation reflects less solar heat into the house than does bare earth. Tall trees planted next to a poultry house can prove very beneficial during hot weather. If the tree's canopy is above a house's side wall, it shades a house's roof from direct sunlight, reduces heat gain from the ceiling and the ground surrounding the house, but it can also help to direct air into a house. In order not to impede the flow of air into and out of the house, the trees should have no branches below the eaves of the house (Fig. 5.5).

Ridge openings

Ridge openings for naturally ventilated houses are only effective in poultry houses with uninsulated roofs (Fig. 5.6). Air next to an uninsulated roof can

Fig. 5.4. Low-cut green grass between houses.

Fig. 5.5. Trees shading poultry house and ground near house.

Fig. 5.6. Ridge vent in uninsulated poultry house.

easily exceed 55°C. Since this hot air is much warmer than the rest of the air in the house, it will tend to rise to the peak of a sloped roof. If there is an opening at the ridge this superheated air will leave the house. Without the ridge opening this hot air would tend to accumulate, leading to increased house temperatures. In a poultry house with an insulated roof, air temperatures next to the ceiling are typically not much warmer than those next to the floor, so there is no great need to make a special effort to rid the house of the air next to the ceiling. Research has shown that ridge ventilation has very little effect on the overall air-exchange rates in most naturally ventilated houses that are insulated (Timmons *et al.*, 1986).

House orientation

Naturally ventilated houses should always be orientated in an east–west direction. The reason for this is to minimize the possibility of direct sunlight entering the house (Fig. 5.7). Direct sunlight striking upon a bird can dramatically increase the effective temperature a bird is experiencing. Direct sunlight can increase the surface temperature of a bird to well above 38°C, creating a heat stress situation at air temperatures that would not normally be thought of as problematic. What is more often the case when sunlight enters a house is that the birds will move away from the side wall where the sun is entering the house, thereby dramatically increasing the effective density of the birds. The higher density significantly decreases the amount of air movement between and over the bird's body as well as putting a bird in

Fig. 5.7. Sunlight entering broiler house.

direct contact with other hot birds. Though it is true that orientating a house in an east–west direction may not take full advantage of winds blowing from east or west, this is typically not a problem for narrow houses (12 m or less) with proper house spacing, curtain openings, and low vegetation.

Roof overhang

A properly designed roof overhang helps to reduce the possibility of both direct and indirect sunlight entering a house during hot weather. For most locations in the world the sun travels slightly to the north or south of a house orientated in the east–west direction. Without a proper roof overhang the sun would be able to shine directly on to one of the house's side walls. In a naturally ventilated house, this means that the sun will shine into the house, leading to an increase in heat-stress-related problems. The length of the overhang is a function of side-wall height and proximity of the side-wall opening to the ground. The taller a house's side wall, the longer a house's roof overhang should be to prevent sunlight from entering the house. The closer the side-wall opening is to the ground, the longer the roof overhang should be. Roof overhangs should typically be a minimum of 0.6 m in most instances, but some houses with taller side walls and large curtain openings could benefit from roof overhangs of 1.25 m or more (Fig. 5.8).

Roof overhangs can also help direct rain coming off the roof of a house away from the house, as well as keep rain from directly entering a house.

Fig. 5.8. 1.2-m roof overhang.

Roof slope

Though there are structural considerations related to the most desirable roof slope for a poultry house, the optimal roof slope is more often determined by the level of roof insulation. In houses with uninsulated roofs a steep roof slope (45°) is highly desirable for a number of reasons. First, a steep roof slope tends to collect less radiant heat from the sun than does a flat roof. Second, a steep roof maximizes the distance between the birds and the hot ceiling, which reduces the amount of radiant heat the birds receive from the hot, uninsulated roof. A steep roof also encourages the superheated air immediately next to the ceiling to quickly rise towards the peak of the ceiling far from the birds. If the house has some type of an open ridge, the heated air will quickly leave the house. Last but not least, a steep roof tends to create a more 'open' environment, making it easier for air to flow into and out of the house. If a ceiling is properly insulated it produces essentially no radiant heat, so the need for steeply sloped roofs is significantly reduced. It is still advisable in a naturally ventilated house to have a sloped roof to ease the flow of air into and out of the house and to allow the equipment to be raised out of the way (in between the flocks) to facilitate house clean-out.

Roof insulation

To minimize heat-stress-related problems during hot weather it is always beneficial to insulate poultry-house roofs/ceilings. Most poultry-house roofs are fabricated from galvanized steel, which will commonly reach temperatures of 50–70°C on a sunny day. The hot roof not only leads to increased house air temperatures but can also dramatically increase the amount of thermal radiation the birds are exposed to. Much like direct thermal radiation from the sun, which can increase the surface temperature of objects to 25°C or more above ambient air temperature, so can, to a lesser extent, the radiant heat produced by a hot roof. It is not uncommon to find nest systems, cages and other objects in a poultry house with an uninsulated ceiling to be 1–5°C above ambient air temperature.

The best way to eliminate heat from a ceiling is through insulation. Insulation acts as a thermal barrier, keeping heat from the hot roof from entering the house. A typical minimum level of ceiling insulation for a naturally ventilated house is R-value 1.25 m^2 C/W. Whereas houses that have high temperatures above 40°C or temperatures below 0°C typically require ceiling R-values of 2.25 m^2 C/W or more.

There are a variety of methods of insulating a poultry-house ceiling (Fig. 5.9).

Dropped ceiling

A dropped ceiling is most commonly installed in houses with a scissor truss. Metal is installed along the top cord of the truss and a strong vapour barrier supported by strap/strings is installed on the underside of the bottom cord of the truss. Fibreglass batt or blown cellulose insulation is placed on top of the vapour barrier.

Rigid board insulation

Rigid board insulation is typically constructed of polystyrene or polyurethane insulation. Rigid board insulation is available in thickness of 1.25 to 4 cm and comes in sheets 1.2 to 1.8 m in width in a variety of lengths. The board insulation is placed on the topside of the truss and the metal is placed on top of the board insulation. Thicker boards are available with 'tongue and grove' construction, which facilitates a tighter seal between the metal roof and the building space.

Sprayed polyurethane

Polystyrene board insulation

Fig. 5.9. Different types of roof insulation.

Fig. 5.9. Continued.

Reflective 'bubble' insulation

'Dropped ceiling' breeder house

Spray polyurethane insulation

Spray polyurethane insulation is applied on to the underside of a poultry house roof to form an airtight insulation barrier. Thickness typically ranges from 2 to 4 cm. Spray polyurethane is a popular method of insulating existing houses owing to the relatively easy application.

Reflective insulation

Reflective insulation consists of a single reflective film or two pieces of reflective film with some type of 'bubble' insulation between the two pieces of film. Reflective insulations are installed either on top of the truss, just beneath metal roofing, or on the underside of the bottom cord of a truss. Reflective insulation works by reflecting radiant energy from the roofing material back towards the roof and away from the birds. In order to be effective an air space between the reflective insulation and the hot roof is necessary. Furthermore, the insulation must stay clean. Dust accumulation on the surface of the reflective insulation can dramatically reduce its heat-reflecting ability. Reflective insulations have very limited ability to limit heat flow through conduction and therefore are generally not as effective at limiting the long-term heat flow from the ceiling as the more traditional forms of insulation but are superior to having no insulation. Reflective insulations have a very limited ability to reduce conductive heat loss during cold weather and are generally not recommended for installation in houses where low outside temperatures (< 30°C) are commonly experienced.

Side-wall and end-wall insulation

For naturally ventilated houses in hot climates insulating the side walls above and below the side-wall curtain opening, although advisable, is typically not a high priority. End walls should always be insulated because direct sunlight striking the eastern or western end walls in the morning or evening can lead to significant heat stress to those birds in the vicinity of the end walls. The level of insulation required depends on the severity of the climate. The hotter or colder the climate is, the greater is the level of insulation required.

Roof coatings

Though not a replacement for ceiling insulation, reflective roof coatings can prove effective in reducing roof temperatures during hot weather (Fig. 5.10). Reflective roof coatings reduce roof temperatures primarily by reflecting solar radiation away from the roof. Reflective roof paints have been shown to reduce roof temperature from 5°C to 32°C, thus dramatically reducing heat gain through roof surfaces (Bucklin *et al.*, 1993). The effectiveness of roof coatings is heavily dependent upon reflectivity. Application of paintings/coatings may sometimes be required on a yearly basis, since their effectiveness over time is reduced as they become dull.

Fig. 5.10. Reflective roof coating.

Owing to the fact that a properly insulated ceiling reduces heat flow through the ceiling by 90% or more, reflective coatings do not offer any significant benefits for houses with insulated ceilings. Roof coatings do little to stem heat loss from a poultry-house ceiling during cool weather.

Roof sprinkling

Roof sprinkling is another method of reducing the heat gain from an uninsulated roof. If sufficient water is distributed uniformly on the roof surface, the roof temperature can actually be decreased below outside air temperature owing to the evaporation of the water off the surface of the roof. Water needs to be sprinkled on the roof on a near constant basis during the hottest portion of the day, and as a result large volumes of water are used, which by far exceed bird water usage.

Roof staining or rusting can be a potential problem when using roof sprinklers. The staining can lead to reduced roof life and can lead to increased house temperatures due to heat from the sun being absorbed by the darkened roof and not reflected away.

Circulation fans

The primary purpose of circulation fans in a naturally ventilated house is not to bring air into the house but rather to produce air movement over the birds to increase convective cooling. When examining the use of circulation fans, it is important to realize that a circulation fan will only produce a significant amount of air movement over a limited area of specific dimensions. In general, most circulation fans will produce some air movement (0.5 m/s or greater) over an area 15 times the diameter of the fan in length by five times the diameter of the fan in width. The exact shape and size of the coverage area is affected by the type of fan. Small-diameter, high-speed fans produce longer but narrower coverage areas while large-diameter, low-speed fans tend to produce wider yet shorter coverage areas.

It is generally advisable to have a minimum of 50% of the floor air covered with air movement produced by circulation fans. When less than 50% of floor area is covered, not only is the number of birds receiving convective cooling limited but birds often crowd into the few areas covered by the circulation fans, thus reducing their effectiveness.

Generally, it is best if circulation fans are orientated to blow with the long axis of a house and positioned towards the centre of the house, where air movement tends to be most needed. Circulation fans should generally be installed in rows (Fig. 5.11). The use of a single row or multiple rows of circulation fans depends on the width of the house, type and size of fan, air movement, and the desired floor area to be covered. By installing circulation fans in rows such that they tend to act together as a system, air blown by one fan is picked up by the next fan and so on down the line. By acting together to move air down the house the total coverage area of the circulation fans can be increased far beyond what they do as individual units. Furthermore, with sufficient circulation and fan capacity, fans working as a system can actually help to draw air into the house on days with limited breeze.

To maximize air movement over the birds, circulation fans should generally be installed 1–1.5 m above the floor and tilted downward at approximately a 5° angle.

Layer houses

In the case of layer houses, the type of cage system used has a significant effect on how well the house can be naturally ventilated during hot weather. A traditional two-deck system lends itself well to natural ventilation owing to the ease with which outside winds can pass above and below the cage system (Fig. 5.12).

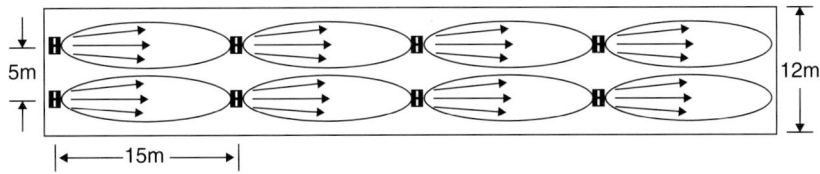

Fig. 5.11. 0.9-m circulation fan layout for a 12-m-wide house.

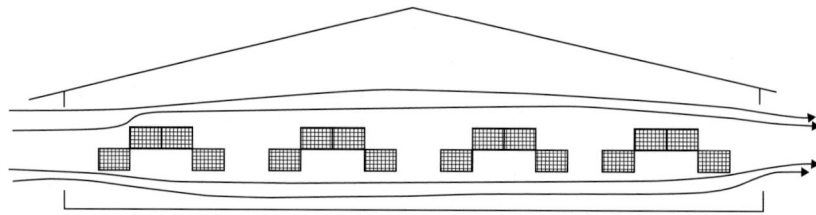

Fig. 5.12. Air flow through a 12-m-wide house with a two-deck cage system.

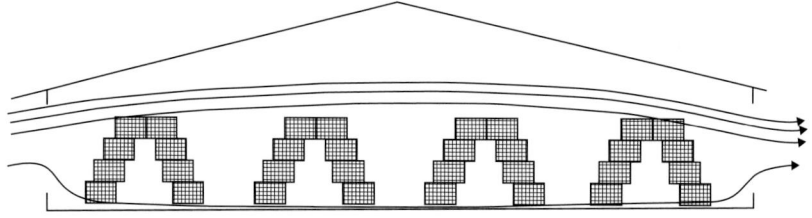

Fig. 5.13. Air flow through a 12-m-wide house with a four-deck, A-frame cage system.

Multiple rows of three- and four-deck systems make it increasingly difficult to naturally ventilate a house effectively. The three- and four-deck systems tend to act like walls, impeding natural air flow through a house, resulting in reduced air-exchange rates and low air velocities within the cage rows. This is especially true where the distance between the top cages and the ceiling is less than 1 m and/or there are more than three cage rows. Natural ventilation tends to be most effective in three- and four-deck systems where the number of cage rows is limited to three, the centre aisles are a minimum of 1.2 m in width, and circulation fans are placed down the two centre aisles, similar to that depicted in Fig. 5.13.

Power-ventilated Houses

Two types of power-ventilation system have been used to ventilate poultry houses, positive- or negative-pressure systems. With a negative-pressure ventilation system the air is exhausted from the building by fans and enters through inlet openings. Though natural ventilation can be used to control the environment within a poultry house during hot weather it has limitations to its effectiveness, especially when it comes to keeping older birds productive. The primary driving force for air exchange and air movement in a naturally ventilated house is the wind. Wind speeds and direction by their very nature are uncontrollable; therefore it is difficult to control air exchange and air movement within a poultry house. To gain more control over the environment during hot weather it is necessary to use a power-ventilation system, where exhaust fans and air inlets provide a very high level of control over both air exchange and air movement within a poultry house during hot weather. The vast majority of hot-weather power-ventilation systems are negative-pressure systems. In a negative-pressure system, exhaust fans are used to draw air out of the house, creating a low-pressure zone within the house. Since air pressure is greater outside the house than inside the house, air from outside rushes into the house through some type of air inlet system.

There are basically two types of negative-pressure systems used to ventilate poultry houses: inlet ventilation and tunnel ventilation. In an inlet ventilation system exhaust fans and air inlets tend to be uniformly distributed down the length of a house, while in a tunnel ventilation system air inlets are located at one end of the house and exhaust fans are located on the opposite end of the house. The tunnel ventilation system tends to create a more positive air exchange and greater air movement than does a traditional inlet ventilation system and as a result has proven superior in cooling birds during hot weather.

House construction

In order to be effectively power-ventilated it is crucial that a poultry house is well insulated and tightly constructed. The same methods of ceiling insulation discussed previously can be used, though higher insulation levels are typically suggested for power-ventilated houses. Side walls can be either of solid construction or equipped with adjustable side-wall curtains for use during mild times of the year or when there are young birds in the house or in emergency situations where there is a power failure. A common difference in power-ventilated houses with side-wall curtains is that the curtains are smaller since they are not typically used during hot weather. Ideally, power-ventilated houses are not equipped with side-wall curtains. Side-wall curtains have a very low insulation value, which can lead to greater heat gain. When solid walls are used it is important that they are properly insulated to minimize heat gain through the side walls during hot weather.

Air Exchange

During hot weather the air in a poultry house is heated by the heat flowing in through building surfaces as well as the heat produced internally by the lights, motors and the birds. A basic design goal for a power-ventilated house is that heat is exchanged quickly enough that the temperature of the house inside (without evaporative cooling) is no more than 2.8°C hotter than it is outside. In other words the air leaving the house is no more than 2.8°C hotter than the air entering the house. In order to determine the appropriate exhaust fan capacity, a heat balance of a poultry house must be conducted. In a heat balance, all the heat gains to the house are calculated and their sum is then balanced with the air-exchange rate required to maintain the desired 2.8°C differential between intake and exhaust air. Heat gain through a building surface is calculated from the following equation:

Building surface heat = $(A/R) \times (T_o - T_i)$

Where:
Heat = watts
A = area of building surface in m²
R = insulation value of wall (m² • °C/W)
T_o = temperature outside (°C)
T_i = temperature inside (°C)

For most building surfaces T_o is the hottest outside temperature expected for the house location. The exception is when calculating heat gain through the house's roof. If the house has an attic space it is generally recommended that an outside temperature of 55°C is used. This is because the 'outside surface' of the ceiling insulation is the attic space. For ceilings where the insulation is directly beneath the roofing material an outside temperature of 65°C is recommended, because the 'outside surface' of the ceiling insulation is in close proximity to the roofing material, which is often as hot as 65°C.

For inside temperature, it is best to assume a temperature of about 27°C. The house needs to be designed assuming that it is equipped with an evaporative cooling system, which when designed properly will typically lower house temperature to around 27°C. In some very dry climates it is possible to decrease house temperatures to below 27°C with a properly designed evaporative cooling system and therefore a design inside temperature of 24°C may be required.

The R-value of a building surface is the sum of the insulation values of the wall section. Therefore, if a wall is constructed of multiple components the total R-value of the wall is the sum of the individual R-values of the wall components.

Example 1:

12 m × 150 m totally enclosed broiler house with flat, dropped ceiling.
Side-wall height = 2.7 m
Ceiling insulation value = 2.1 m² • °C/W
Side-wall insulation value = 1.6 m² • °C/W
End-wall insulation value = 1.6 m² • °C/W
Maximum outside temperature = 38°C
Maximum attic space temperature = 55°C
Inside temperature = 27°C
Birds = 25,000 birds @ 2.3 kg/bird

The heat production of the birds in a house can vary significantly depending on size of bird, age, feed energy, house temperature, egg production, air movement, etc. As a result, coming up with a precise value of heat production of the birds in a house during hot weather can be difficult. From a design standpoint, we are most interested in the heat production of 'older' birds when heat production is at its greatest and heat stress is most likely to be a problem. Based on the latest research, total heat production (sensible + latent) for a market-age broiler is approximately 7.9 W/kg. For commercial layers, broiler breeders, pullets and turkeys it is approximately 5.1 W/kg. If it is assumed that 50% of the total heat production of a bird is in the form of sensible heat, the sensible heat production for broilers can be estimated at 4 W/kg, while for commercial layers, broiler-breeders, pullets, and turkeys, 2.6 W/kg is more appropriate (Chepete and Xin, 2001).

In the above example the total heat production of the birds can be calculated by the following equation:

Bird heat = total kg × 4 W/kg
= 25,000 birds × 2.3 kg/bird × 4 W/kg
= 230,000 W

The total heat added to the air in the house is the sum of the heat entering the house through building surfaces and the heat produced by the birds. Feed motors and lights are usually not considered, owing to the fact that their heat production is very small in comparison with the heat gain through building surfaces and heat production of the birds.

Total heat = building surface heat + bird heat
= 30,016 W + 230,000 W
= 260,016 W

The air-moving capacity required to maintain a 2.8°C temperature difference between intake and exhaust air can be calculated using the following equation:

Air-moving capacity = total heat load × 3.4/2.8°C
= 260,016 watts × 3.4/2.8°C
= 315,733 m^3/h

315,733 m^3/h is roughly an air exchange a minute, a commonly used rule of thumb to determine maximum air-exchange rate for broiler houses (Table 6.2).

Example 2:

12 m × 150 m curtain-sided broiler-breeder house with open ceiling.
Total side wall height = 2.7 m
Ceiling width = 13.4 m (pitched ceiling)
Curtain height = 1.5 m
Side wall without curtain = 1.2 m
Ceiling insulation value = 1.6 m^2 °C/W
Curtain insulation value = 0.3 m^2 °C/W

Table 5.4. Total building surface heat gain

Surface	Area	R-Value	Heat Gain
Ceiling	1,800 m^2 (12 m × 150 m)	2.1	24,000 watts (1,800/2.1) × (55C – 27C)
Side walls	810 m^2 (2.7 m × 150 m × 2)	1.6	5,569 watts (810/1.6) × (38C – 27C)
End wall	65 m^2 (12 m × 2.7 m × 2)	1.6	447 watts (65/1.6) × (38C – 27C)
Total			30,016 watts

Side-wall insulation value = 1.6 m² • °C /W
End-wall insulation value = 1.6 m² • °C/W
Maximum outside temperature = 38°C
Maximum roof temperature = 65°C
Inside temperature = 27°C
Birds = 10,000 birds @ 3.2 kg/bird

Total sensible heat production of the birds
= total kg × 2.6 W/kg
= 10,000 birds × 3.2 kg/bird × 2.6 W/kg
= 83,200 W

Total house sensible heat production
= building surfaces' heat production + bird heat production
= 67,160 W + 83,200 W
= 150,360 W

The air-moving capacity required to maintain a 2.8°C temperature difference between intake and exhaust air can be calculated using the following equation:

Required air-moving capacity = total heat load × 3.4/2.8°C
= 150,260 W × 3.4/2.8°C
= 182,580 m³ /h

(see Table 5.5)

For ease of calculating the required air-flow rate of a force-ventilated house, the following formula is adapted:

Air-flow rate
= cross-sectional area of the house × required or maximum speed desired.

Example:

A house whose dimensions are
= 15 m wide and 3 m high (at ceiling or deflectors level) and air speed of 2 m/s
= (15 × 3) m² × 2 m/s × 60 s/m × 60 m/h = 324,000 m³/h

Table 5.5. Total building surface heat gain

Surface	Area	R-Value	Heat Gain
Ceiling	2,010 m² (13.4 × 150)	1.6	47,738 watts (2,010/1.6) × (65C − 27C)
Side walls	360 m² (1.2 × 150 × 2)	1.6	2,475 watts (360/1.6) × (38C − 27C)
Curtains	450 m² (1.5 × 150 × 2)	0.3	16,500 watts (450/0.3) × (38C − 27C)
End wall	65 m² (12 × 2.7 × 2)	1.6	447 watts (65/1.6) × (38C − 27C)
Total			67,160 watts

Air inlet systems

There are a number of different types of negative-pressure air inlet systems that are used to control poultry-house environments during hot weather. Most of these inlet systems are the same ones used to maintain the proper house temperature and air quality during cold weather. Whereas a house's exhaust fans determine how much air enters a house, an inlet system allows the producer to control where the fresh air enters the house, how it moves once it enters the house, and the speed at which it enters the house.

Inlet air speed

With a negative-pressure system, the speed at which the air enters the house is determined by the pressure differential between inside and outside the house (Fig. 5.14) (Wilson, 1983). The higher the pressure differential, the faster the air tries to enter the house, and the lower the pressure difference, the slower the air enters the house. Pressure differential is determined by the relative relationship between the amount of air moved by a house's exhaust fans and the opening available to the fans. The greater the opening available to the fans, the lower the differential pressure, and the slower the air enters the house. The lower the amount of opening, the greater the pressure differential, and the faster the air enters the house. By manipulating the pressure differential it is possible to control air-flow patterns within the house.

Inlet area

In order to gain control over fresh air distribution and air movement within the poultry house the inlet area has to be matched to a house's exhaust fan capacity. Typically, the required inlet area is determined so that when all the exhaust fans are operating the static pressure is in the range of 12 to 25 Pa. Inlet manufacturers can generally provide the air-moving capacity of their inlets at various static

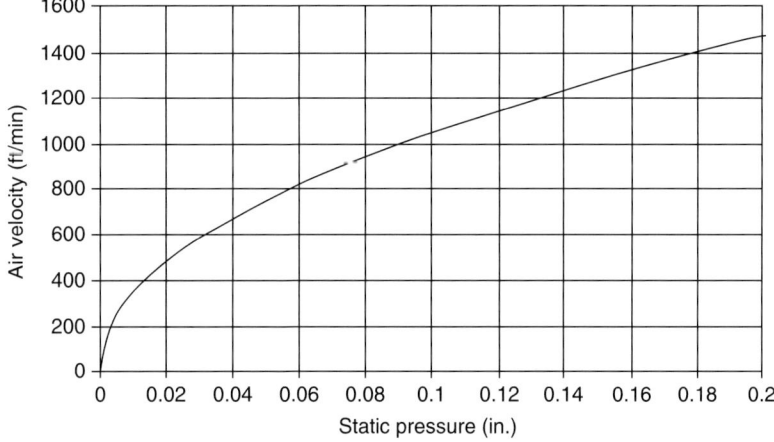

Fig. 5.14. Inlet air velocity versus static pressure.

pressures, so the number of inlets is easily determined by dividing the total air-moving capacity of the fans by the air-moving capacity of the inlets at the desired maximum static pressure. If there is no data on the air-moving capacity of the inlet then it is generally recommended that there should be a minimum of 1 m² of inlet area for every 14,000 m³/h of exhaust fan capacity.

Air inlet design

Ideally air inlets should be of a design that allows air to be directed away from the birds and towards the ceiling during cooler times of year or when the birds are young (less feather) during hot weather, and towards the floor during hot weather when the birds are old (more feather) to increase convective cooling. Owing to the fact that the objectives of an inlet system for use during cold weather are very different from what is desired during hot weather, some inlet ventilation systems actually consist of two separate sets of inlets. The first tends to be located near the ceiling and is designed to direct air along/toward the ceiling. The second inlet system is typically located lower in the side wall and is specifically designed to direct the air along the floor during hot weather.

Air inlet control

Whereas exhaust fans are controlled based on air temperature and air quality, control of inlet opening is based on the difference in static pressure between inside and outside the house. If the static pressure difference is too low, the air will not enter the house with sufficient velocity to produce the required air movement within the house. If the static pressure is too high, fan capacity will be reduced and air-exchange rates will be decreased. In general, the desired static pressure is between 12 and 25 Pa. Inlets are connected through cables or rods to an inlet machine that will open and close the inlets in the house to maintain the desired static pressure.

Types of inlet ventilation systems

Cross-ventilation

In the typical 'cross-ventilated' poultry house the exhaust fans are evenly spaced down one side of the house and inlets are installed along the other side of the house (Fig. 5.15). This form of inlet ventilation when used in hot climates is best suited for relatively narrow houses (i.e. less than 10 m). When used in wider houses air movement on the fan side of the house tends to be very low, leading to significant differences in environmental conditions across the width of the house.

Side-wall inlet ventilation

In the typical side-wall inlet-ventilated house, exhaust fans are spaced evenly down one or both side walls and inlets are installed along both side walls (Fig. 5.16). Care must be taken to make sure that there is at least a two-fan-diameter spacing between the exhaust fans and the air inlets, to ensure that there is no 'short circuiting' of inlet air. If the inlets direct air along the ceiling, the air jets will tend to collide in the centre of the house and move down to and then across the floor (Fig. 5.16). In order for this air pattern to occur the house needs to be fairly narrow (12 m or less). In wider houses the air jets may not make it to the centre of the house and the proper circulation pattern will not occur. In houses with air inlets that direct air along the floor the air will tend to move across the floor towards the centre of the house, collide, and then move upward towards the ceiling (Fig. 5.17). This flow pattern will tend to produce superior air movement along the floor than when the inlets are directed upward. Where inlets direct air along the

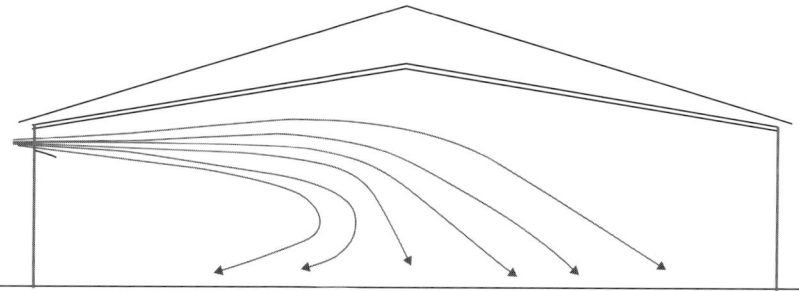

Fig. 5.15. Air-flow pattern in a cross-ventilated house.

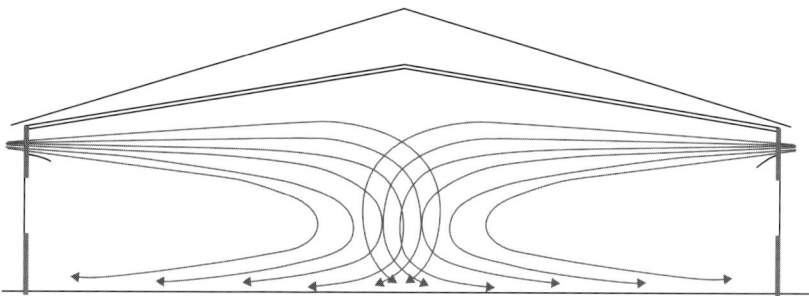

Fig. 5.16. Air-flow pattern in house with air inlets on both sides of the house.

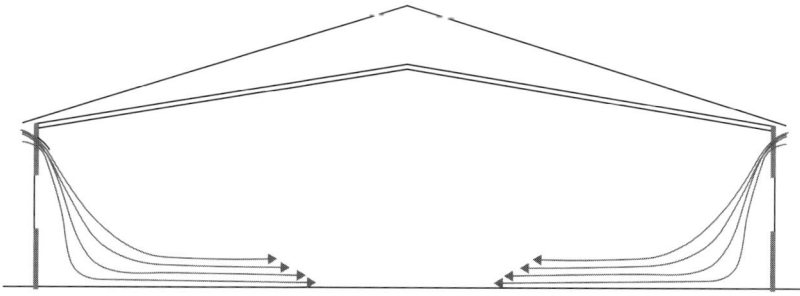

Fig. 5.17. Air-flow pattern in a house with air inlets directed downward.

ceiling, house width is very important. Wider houses typically require circulation fans to be installed towards the centre of the house.

Attic inlet ventilation

In an attic inlet ventilated house exhaust fans are spaced evenly down both side walls, and air inlets are installed in the ceiling to pull air from the attic (Fig. 5.18). For hot climates it is important that the ceiling between the birds is insulated but the attic space should be insulated from the roofing material. These systems are primarily used in wide commercial layer houses, where side-wall inlets tend to be less effective in distributing fresh air across the width of the house (Fig. 5.19).

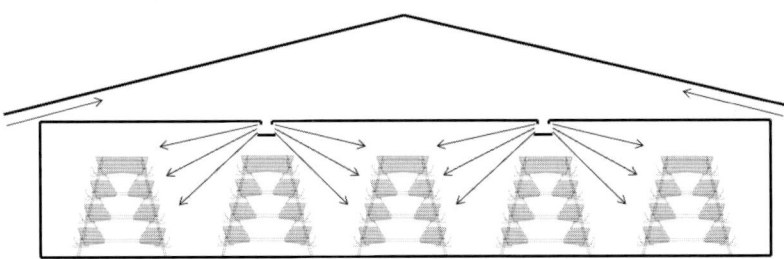

Fig. 5.18. Air-flow pattern in layer house with slot inlets.

Fig. 5.19. Slot attic inlet in commercial layer house.

Air movement in inlet-ventilated houses

The greatest challenge with any traditional inlet ventilation system is producing sufficient air movement to produce sufficient bird cooling during hot weather. Though the air may enter through an inlet at velocities between 3.5 and 6 m/s it quickly drops as the air jet expands as it moves away from the side wall. Depending on the type and size, the inlet air movement at bird level is typically well below 1 m/s in the majority of the house. For this reason circulation fans (in addition to exhaust fans) are often required in inlet-ventilated houses to produce the required air movement over the birds to minimize heat-stress-related problems during hot weather.

Inlet-ventilated houses are generally best suited to control the environment during cool to moderate weather conditions. They tend to produce minimum air movement at floor level and do not flush the air out of the house as well as a tunnel-ventilated house does.

Tunnel ventilation systems

When designing a tunnel ventilation system not only must it be designed to produce the proper air exchange but it must also be able to produce the desired air velocity. The desired design air velocity depends on the type of bird being raised, density, and climate. Table 5.6 below provides some general recommendations for the desired air speed for different types of housing.

In humid climates or where birds are placed at high densities, higher velocities tend to be used. In dry climates where evaporative cooling systems are more effective in decreasing house temperature or in a house where birds are placed at lower densities, lower air velocities can often be used.

Tunnel fan capacity and air velocity

A house's minimum fan capacity for a tunnel-ventilated house is determined by the same method as for inlet-ventilated houses, namely by conducting a heat balance. Using this method assures that there should be no more than a 2.8°C differential between the tunnel inlet and tunnel fans at the end of the house on the hottest day with mature birds. This is not necessarily the proper amount of air-moving capacity, because although a house may have the

Table 5.6. Recommended air velocities in tunnel-ventilated houses.

House Type	Air Speed
Broilers	2.5 – 3 m/s
Pullets	1.75 – 2.25 m/s
Broiler Breeders	2.25 – 3 m/s
Commercial Layers	2.5 – 3 m/s
Turkeys	2.5 – 3 m/s

proper air-exchange rate in a tunnel house, it may not be adequate to produce sufficient tunnel air velocity to achieve the desired level of convective cooling. The average air velocity in a tunnel house can then be determined from the following equation:

Air velocity
= Tunnel fan capacity/(cross-sectional area of the house × 3600)

Where:
Air velocity is in m/s
Tunnel fan capacity is in m^3/h
Cross-sectional area is in m^2

From the above equation, it can be seen that a major factor that determines air velocity in a tunnel house is its cross-sectional area (a plane containing width and height). Large cross-sectional areas tend to lead to lower air speeds and lower levels of convective cooling. A house can have enough fan capacity to obtain the proper air-exchange rate but due to its large cross-sectional area may not have sufficient air velocity for cooling. For this reason, it is very desirable for tunnel houses to have lower ceilings rather than higher ceilings. Another important characteristic is that for a given floor area it is better for a house to be long and narrow rather than short and wide. This goes back to cross-sectional area. Two houses with the same floor space (one long and narrow and the other shorter and wider) will require the same fan capacity for air exchange. If the calculated minimum tunnel fan capacity is insufficient to produce the desired air velocity, fan capacity will need to be increased. The following equation can be used to determine the tunnel fan capacity required to produce the desired tunnel air velocity.

Tunnel fan capacity
= desired air velocity × cross-sectional area × 3600

Where:
Desired air velocity is in m/s
Tunnel fan capacity is in m^3/h
Cross-sectional area is in m^2

Another method of increasing air speed is through the installation of air deflectors. Air deflectors temporarily reduce the cross-sectional area of a poultry house, thus increasing the speed of the air in the vicinity of the deflector. Air deflectors are made of curtain material and extend from the ceiling to within 2.5–3 m of the floor. The reduction in cross-sectional area increases the velocity of the air for a distance of approximately 1.2 m upwind of the deflector and between 6 and 9 m downwind of the deflector. If deflectors are placed at 8 m on centre the increase in air velocity will be fairly consistent throughout the house (Fig. 5.20).

Air deflectors are primarily used in houses with relatively large cross-sectional areas. Many of these houses often require nearly the same air-exchange rate as does a house with a lower ceiling, but since the cross-sectional area can be 15% greater, the resultant air velocity is reduced by 15%. This leaves two

Fig. 5.20. Air deflector in house with a large cross-sectional area.

options: increase fan capacity or install air deflectors. Caution should be exercised with installing air deflectors. If the cross-sectional area beneath a deflector is too small, resulting in an increase in static pressure, the fan performance and air-exchange rates will be reduced. To avoid these problems it is recommended that air deflectors should extend no closer than 2.5 m to the floor, or deflectors should not decrease the cross-sectional area of a house by more than 15%.

Air velocity distribution

Although air velocity within a tunnel house tends to be much more uniform than in traditional naturally ventilated houses, there are still some variations in air velocity within the house. For instance, air velocity will vary between the centre of the house and near the side walls from around 15 to 40%, depending on a variety of factors. The most important factor is side-wall roughness or side-wall obstructions. The 'rougher' the side wall the more likely air will be deflected off the side wall towards the centre of the house (Fig. 5.21). For this reason it is very desirable to have a smooth side wall (Fig. 5.22). The presence of nests or other objects on the side wall can also deflect air from the side walls to the centre of the house, creating significant difference between the air speed in the centre of the house and the side walls.

Bi-directional tunnel houses

In most cases it is best to install the tunnel fans on one end of the house and the tunnel inlet opening on the opposite end (Fig. 5.23). This arrangement produces the highest air speeds and fewer dead spots than bi-directional

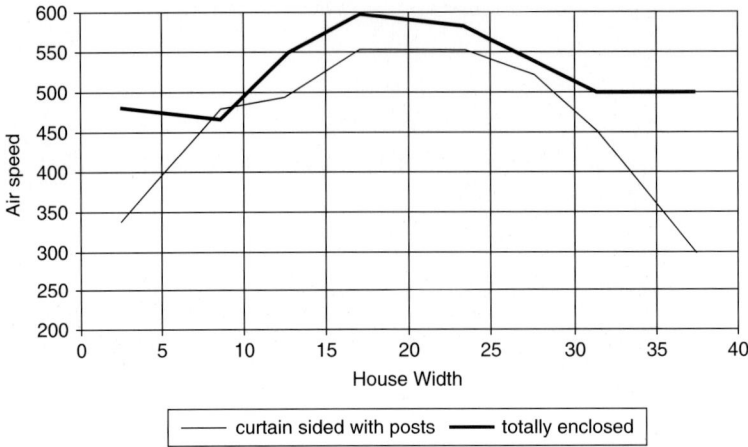

Fig. 5.21. Air velocity profile in house with rough and smooth side walls.

tunnel houses, where either the fans are placed in the centre of the house and the tunnel inlet opening near the end walls or the tunnel inlet opening is placed in the centre of the house and fans in the end walls (Fig. 5.24).

In a bi-directional tunnel house air velocity is cut in half because the air flows in two directions instead of one (Fig 5.24). For instance, when half the tunnel fans are placed in each end wall, the tunnel inlet opening is placed in the centre of the house and the amount of air flowing through the cross-sectional area of each end of the house is cut in half, which reduces the air speed to one half. The temperature difference between the inlet and the fan ends of the house will remain the same because the air-exchange rate has not changed, while the cooling effect produced by the air velocity will be dramatically reduced.

Bi-directional tunnel houses are only advisable when the house is very long (over 180 m) or when the velocity generated by tunnel ventilation in one direction would be excessive (over 3.5 m/s). This is often the case in high-density commercial layer houses. The tunnel fan capacity required to provide the proper air-exchange rate to maintain temperature uniformity can be excessive due to the relatively high bird density in many caged layer houses. In these cases it is best to place the tunnel inlet opening in the centre of the house and fans in the end walls (Fig. 5.25).

Tunnel fan placement

Tunnel fans can be installed in either the end wall of a house or the side walls near the end wall. There is essentially no difference in fan performance with tunnel fans installed in either location. When fans are installed in the side walls near the end wall there tends to be a small dead spot near the centre of the end wall. This spot tends to increase in size as house width increases, and for this reason it is generally recommended that a portion of the tunnel fans are installed in the end walls (Fig. 5.26).

Fig. 5.22. Examples of 'rough' and 'smooth' side walls.

Tunnel inlet opening

For tunnel houses without evaporative cooling pads, it is generally recommended that the tunnel inlet area should be approximately 10% greater than

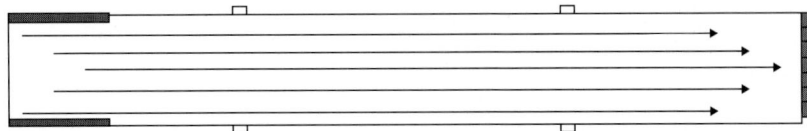

Fig. 5.23. Single-direction tunnel-ventilated house.

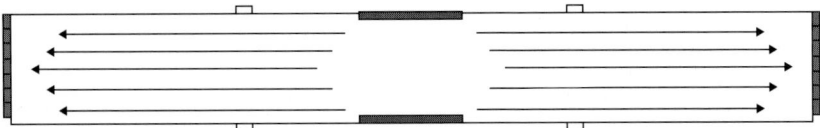

Fig. 5.24. Bi-directional tunnel-ventilated house.

the cross-sectional area of the house. For houses with evaporative cooling pads, the tunnel inlet area is determined by the appropriate amount of pads for the tunnel fan capacity installed, which depends on the type of pad installed. Tunnel inlet opening should be started as close to the end wall as possible. In houses over 15 m in width it is advisable to place some of the tunnel inlet openings in the end wall of the house to reduce the size of the dead spot that occurs near the end wall in wider houses. Tunnel inlet openings need to be built so as to be opened and closed through the use of either a tunnel curtain or tunnel door system (Fig. 5.27).

Cool weather inlet systems for tunnel-ventilated houses

It is important to realize that tunnel ventilation should only be used during hot weather, when the objective is to cool the birds. During cooler weather, if houses are tunnel ventilated, large difference in air temperature and air quality can occur between the tunnel inlet end and tunnel fan end of the house, due to the relatively low air-exchange rates typical during cold weather. To maintain uniform house conditions, houses should also be equipped with a conventional inlet system for use when bird cooling is not required. Traditional side-wall fans can be installed but it is possible to use a portion of the tunnel fans for ventilating a house during cooler weather. As a general rule, it is best if a house is designed so that a minimum of 60% of the tunnel fan capacity can be operated through a traditional inlet system before transitioning to tunnel ventilation (Fig. 5.28).

Poultry-house exhaust fans

Types of exhaust fans

EXTERIOR SHUTTER FANS. This is the most basic form of exhaust fan. The shutter is important so that when the fan is not operating outside air does not enter the house through the fan opening. Shutters typically consist of a series

Bi-directional tunnel-ventilated house with fans.

Bi-directional tunnel-ventilated house with evaporative cooling pads in the centre.

Fig. 5.25. Bi-directional tunnel houses.

Fig. 5.26. Tunnel fans in the side walls near the end wall.

of horizontal, hinged louvres, which gravity acts on to hold them closed. An exterior shutter tends to restrict air flow due to the fact that it impedes the spinning motion of the air as it comes off the fan blades (Fig. 5.29).

INTERIOR SHUTTER FANS. An interior shutter fan is on when the shutter is on the intake side of the fan. The interior shutter is an advantage over the exterior shutter for a number of reasons. First, it is less restrictive to air flow due to the fact that air flowing into the fans is not travelling in a circular motion as it moves through the horizontal louvre openings. Second, interior shutters tend to be larger than exterior shutters, which makes it easier for the fan to move air through the shutter. The fan shutters are easier to keep clean when they are on the intake side of the fan. The combination of these factors tends to increase air-moving capacity of a fan by 5–10% (Fig. 5.30).

DISCHARGE CONE FANS. Discharge cones on fans tend to increase fan performance 5–10% by easing the transition of the spinning air coming off the fan blades (Fig. 5.31).

BELT-DRIVE FANS. In a belt-drive fan the pulley on the fan motor drives a pulley on the fan-blade shaft through the use of a fan belt. To minimize problems associated with belt slippage it is generally recommended that belt-driven fans are equipped with an automatic belt tensioner.

DIRECT-DRIVE FANS. In a direct-drive fan the blades are attached directly to the fan motor shaft, thus eliminating the need for drive belts. Although they do have the advantage of potentially being able to be controlled through a variable-speed controller, they tend to be less energy efficient than belt-driven fans. This tends to be especially true of larger diameter fans (1.2 m or larger), which tend to be associated with hot-weather ventilation systems.

Open tunnel curtain.

Open tunnel door.

Fig. 5.27. Tunnel inlet openings.

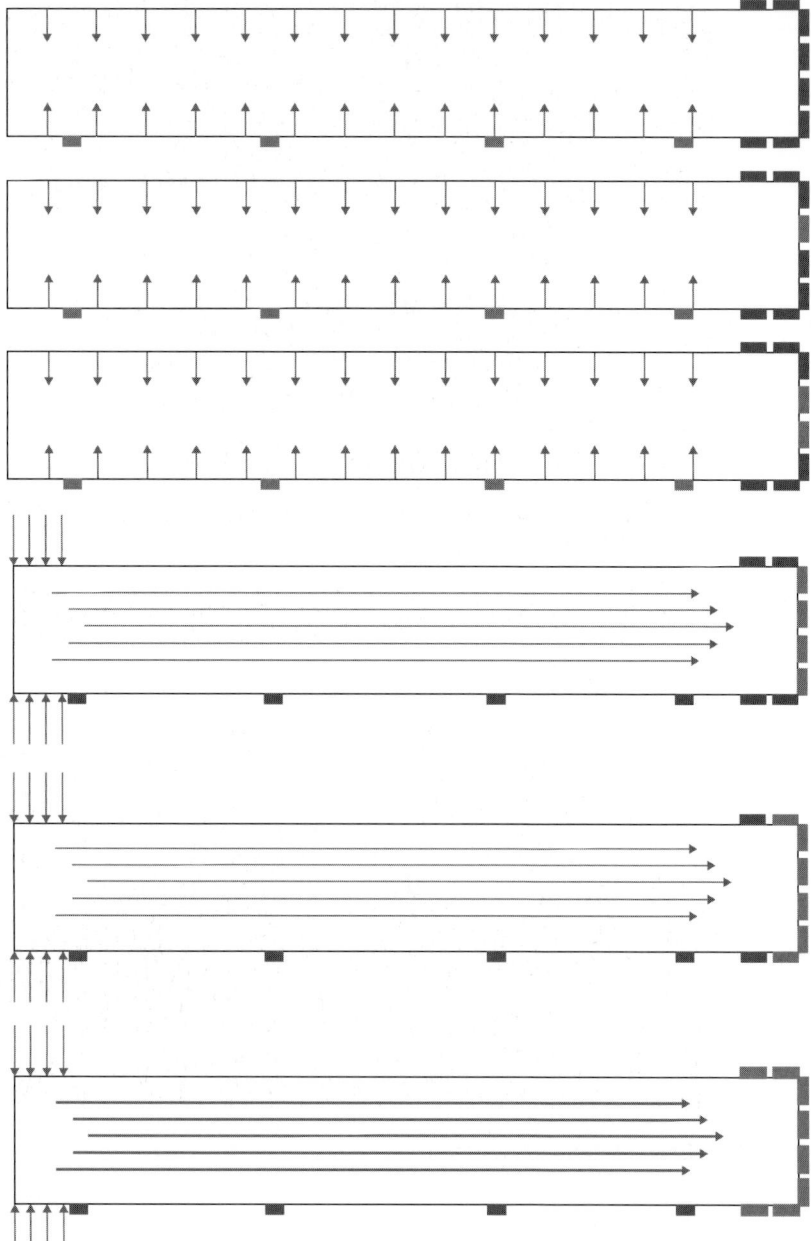

Fig. 5.28. Fan staging in a modern tunnel house.

Exhaust fan performance factors

To provide optimal conditions during hot weather, as well as to keep operating costs to a minimum, it is crucial that the right fan is installed. There are basically

Fig. 5.29. Exterior shutter fan.

three performance factors that should be used to evaluate poultry-house exhaust fans: the amount of air delivered at a static pressure of 25 Pa, energy-efficiency rating, and air-flow ratio. Fan numbers should be determined based on a fan's air-moving capacity at 25 Pa static pressure, because that is the typical static pressure the fans will be operating at or near when in operation. Determining the number of fans a house should have by a fan's air-moving capacity using a lower static pressure can lead to a house being significantly under-ventilated during hot weather.

Fig. 5.30. Interior shutter fan.

Fig. 5.31. Discharge cone fans.

A fan's energy-efficiency rating is similar to a car's mileage rating. Instead of speaking in terms of miles per gallon when comparing fans we look at how many cubic metres per minute the fan can move with a single watt of power ($m^3/h/W$). As with a car's mileage rating, the higher the m^3 per h per watt, the more energy efficient is the fan. A 3.5 $m^3/h/W$ difference between two fans typically results in an approximately 10% difference in electricity usage. In general, energy-efficiency ratings vary from around 26 $m^3/h/W$ to nearly 50 $m^3/h/W$. As with the case of a fan's air-moving capacity, comparing a fan's energy-efficiency rating should be done at a static pressure of 25 Pa. In general, it is best to select fans with an energy-efficiency rating of 35 $m^3/h/W$ or better at 25 Pa static pressure.

A fan's air-flow ratio is another important factor to consider when purchasing a fan. A fan's air-flow ratio is an indicator of how well the fan will hold up as static pressure increases due to factors such as dirty shutters, pads and baffle curtains, or tunnel curtains not fully opening. A fan's air-flow ratio is determined by dividing the amount of air it moves at 50 Pa static pressure by the amount of air it moves at a static pressure of 12 Pa. Air-flow ratios typically vary from 0.50 to 0.85. The higher the ratio, the less the fan is affected by high static pressures, and the more desirable is the fan. To put this into perspective, the air-moving capacity of a fan with an air-flow ratio of 0.50 will decrease by as much as 50% in a worse-case scenario in which the static pressure was very high (50 Pa) due to dirty pads and fan shutters; for a fan with an air-flow ratio of 0.85, air-flow capacity would only decrease 15%. For best results exhaust fans should have an air-flow ratio of 0.70 or better. Small fans, 0.9-m diameter or less, are generally best used for minimum ventilation during cold weather. During hot weather, when high volumes of air are required, it is best to use large-diameter fans (1.2 m or larger). Larger-diameter fans are not only less expensive to purchase per cubic metre of air moved but they tend to be more energy efficient than smaller fans, resulting in lower operating costs. Furthermore, using larger fans reduces the number of fans installed, thereby reducing installation and operating costs.

Fan safety guards

Fans that are within reach of personnel should always have safety guards attached. When the guard is installed within 10 cm of the moving parts, a woven wire mesh of at least 16 gauge with 1.3-cm openings should be used. If the guard can be installed further than 10 cm from the moving parts, the use of a 12 gauge with 5-cm mesh screen would give less air resistance.

Evaporative Cooling

An evaporative cooling system is an essential component of any hot-weather ventilation system. As house temperatures rise, the temperature difference between a bird's body and the ambient air decreases, which results in a decrease in heat loss from a bird. Although high air movement over the birds can help increase bird heat loss at high ambient air temperatures, its effectiveness decreases as air temperatures increase.

Evaporative cooling systems have proven to be very effective in reducing house air temperatures in virtually all hot climates. Though evaporative cooling systems can reduce house temperatures by 15°C or more in dry climates, even in the most humid climates evaporative cooling systems have proven to reduce house temperature by 3–6°C, which is often sufficient to limit heat-stress-related problems. The best type of evaporative cooling system depends on a variety of factors, including climate, housing type and ventilation system design. There are basically two types of evaporative cooling systems used in poultry houses: fogging systems and pad systems. While pad systems can only be installed in a power-ventilated house, fogging systems can be used in either power-ventilated or naturally ventilated housing to lower house temperatures during hot weather. A third system, sprinkling, has limited usage except in very dry climates.

Fogging systems

Fogging systems are generally divided into three groups, based on their operating pressure: low (< 700 kPa), medium (700–1400 kPa) and high pressure (3500–7000 kPa). Medium- and high-pressure 'fogging' systems generally produce cooling through evaporating water into the air, thereby reducing air temperature. Low-pressure 'misting' systems produce cooling primarily through bird wetting and the subsequent evaporation of water off the surface of the birds.

In general, the greater the water pressure produced by a fogging system the finer the mist produced. A smaller droplet has a higher surface area to volume ratio than a larger droplet, which increases the rate at which the droplet evaporates, which in turn results in more effective cooling of the air in a poultry house. Furthermore, smaller droplets tend to stay aloft longer than larger droplets. The longer a droplet stays aloft, the more likely the droplet will evaporate before coming in contact with the house surface, equipment or the birds, leading to wetting. High-pressure systems produce the smallest droplets (10 µm) and are therefore the most effective type of fogging system in reducing house temperature without causing house and equipment wetting. In comparison, medium-pressure nozzles typically produce droplets approximately 50 µm in diameter.

Another factor that affects droplet evaporation is air movement. Without air movement a droplet emitted from a nozzle may only stay suspended a few seconds before hitting the ground. Suspension time, and therefore cooling, can be increased dramatically by introducing fog into a moving airstream. It can be an airstream created by circulation fans or, in the case of a power-ventilated house, by the air inlets. Furthermore, by introducing fog into an airstream, droplets, and therefore cooling, can be better distributed throughout the house, further increasing the effectiveness of the fogging system. It is important to note that the larger the water droplet the greater the need to have good air movement to keep the droplets suspended.

In order to maximize cooling that a fogging system can produce while at the same time minimizing house wetting, the amount of moisture added to the air should be adjusted based on the relative humidity of the air in the house. The lower the relative humidity, the greater the amount of water which can be added and the more cooling that can be produced. On the other hand, only a limited amount of water can be added on very humid days. Ideally, a system would monitor house temperature as well as relative humidity. In general, a house relative humidity should not exceed 85% when using evaporative cooling. The output of a fogging system should be controlled either through the use of an interval timer or by adjusting the number of nozzles operating so as not to exceed 80–85% relative humidity.

Naturally ventilated houses

It is difficult to precisely determine the number of nozzles that should be installed in a naturally ventilated house due in large part to the unpredictability of air-exchange rates in naturally ventilated houses. As a general rule of thumb the typical poultry house should have a minimum of 0.35 l/h of fogging nozzle capacity for every square metre of floor space. This may prove insufficient in dry climates and excessive in humid climates during cooler weather. To minimize the possibility of house wetting problems the nozzles should be spread out evenly throughout the house and either installed in stages or controlled by an interval timer. Side-wall curtain openings may need to be adjusted to reduce air-exchange rates in dry climates or on windy days.

Inlet-ventilated houses

Fogging systems tend to be better suited to inlet-ventilated houses than pad systems owing to the simplicity of installation. Nozzles should be installed so that they cool the incoming air before it moves down to floor level. Typically, nozzles are placed near the side walls above the air inlets so that the fog that is produced drops into the airstream. By dropping the fog into a high-velocity airstream the droplets evaporate at a faster rate and the likelihood of floor wetting is minimized. One of the best methods of ensuring the incoming air is thoroughly cooled before moving down to floor level is by placing fogging nozzles outside the house inside an inlet hood. Typically the inlet hood extends 1 m or more below the inlet opening in the side wall. The fogging nozzles are placed near the intake to the hood so that the fog has the greatest likelihood of evaporating before entering the house.

The number of nozzles required depends on the capacity of the nozzles, climate, exhaust fan capacity and the amount of cooling desired. As a general rule of thumb, in order to produce 1°C cooling it requires the evaporation of 0.5 l/h of water for every 1,000 m^3/h of exhaust fan capacity. It is typically recommended that a fogging system be designed to produce a minimum of 5°C cooling. For dry climates a minimum of 10°C is typically recommended. Owing to the fact that the level of evaporation of the fog added to a house is dependent upon outside temperature and relative humidity, it is best that the fogging nozzles are split into stages or, at a minimum, controlled through an interval timer so that house wetting and humidity do not become excessive.

Tunnel-ventilated houses

Although pad systems are the most popular method of evaporative cooling, fogging systems have proven to be very effective in reducing house temperatures during hot weather in tunnel-ventilated houses. Fogging nozzles tend to be more effective in reducing house temperature in a tunnel-ventilated house than in naturally ventilated houses or inlet-ventilated houses, owing to the uniform air movement, which maximizes the suspension time of the water droplets. Whereas in naturally or inlet-ventilated houses the fogging nozzles tend to be installed lengthwise in a house, in tunnel-ventilated houses nozzles should be installed in lines running from side wall to side wall. The typical house might have between 10 and 20 cross lines of fogging nozzles installed so they can be turned on in a minimum of two stages. To maximize cooling of the incoming air, generally 50% of the nozzles are installed in the first 30–40 m of the house. No nozzles should be installed within 30 m of the exhaust fans, to minimize the possibility of fan wetting.

Sprinkling systems

Although bird-sprinkling systems do produce some cooling of the air in a poultry house they primarily act by bird surface wetting. Air flowing over the birds' body surfaces evaporates water off their bodies, producing cooling. Since the primary method of cooling is through bird wetting it is important that nozzles are spread out evenly throughout a house, so that all birds within the house are wetted when the nozzles are used. Sprinkler systems should operate off a 5- or 10-min interval timer to keep moisture levels within a house from becoming excessive. Sprinkling systems are generally best suited for drier climates, where excess moisture added to the house quickly evaporates.

Pad systems

Though it is possible to install pad systems in an inlet-ventilated house, generally it is not a common option owing to cost and maintenance issues. Unlike tunnel-ventilated houses, where pads are located in a limited area, in inlet-ventilated houses the pad system must be installed down the entire length of the building, so that air for all the house's inlets enters through the pads and is cooled before entering the house. Typically some type of plenum needs to be installed owing to the fact that the pad area typically needs to be two to three times the inlet area, adding additional cost.

In order to ensure maximum bird cooling during hot weather, it is crucial that evaporative cooling pads are properly sized. Insufficient pad area can lead to reduced exhaust fan performance and reduced cooling. Pad systems are generally designed so that the static pressure drop across the pads is no greater than 15 Pa when all fans are operating and the pad system has a cooling efficiency of between 70 and 80%.

Excessive static pressure drop across a pad system leads to reduced air-moving capacity of the fans, which in turn leads to reduced air-exchange rates, greater temperature differentials, and reduced air speeds. The static pressure drop across a pad is a function of the pad design and the velocity of the air through the pad. With most 15-cm corrugated-paper pad systems, a pad static pressure drop of 15 Pa is achieved at an air velocity of approximately 1.8 m/s. For 10-cm corrugated paper and some aspen pad systems, air velocity is approximately 1.25 m/s. For determining the static pressure drop across a pad system at different velocities, it is best to contact the manufacturer of the pad in question.

It is important to realize that the 15 Pa static pressure drop across the pad system is not the total static pressure that the exhaust fans are working against. Since the pad area tends to be greater than the cross-sectional area of the house, additional pressure is generated as the air velocity increases as it is pulled into the smaller cross-section area of the house. Furthermore, additional pressure can be generated as the air moves down the length of the house. The amount of pressure increase is a function of velocity and restrictions within the house (i.e. cages, nest boxes, side-wall and ceiling roughness). In most instances, although the static pressure drop across the pad system may be only 15 Pa, the total static pressure drop the fans are working against could be approximately 22–30 Pa. A pad system's cooling efficiency is basically a measure of the system's effectiveness at decreasing the dry bulb temperature to the wet bulb temperature. Most pads are designed for a cooling efficiency of approximately 75% at the recommended area. A cooling efficiency of 75% basically means that the pad system is 75% effective at decreasing the dry bulb temperature to the wet bulb temperature. Fig. 5.32 illustrates the incoming air temperature through a properly designed and maintained pad system that is 75% efficient with different combinations of temperature and relative humidity. For example, at 31°C and 60% RH, the incoming coming air temperature is 24.6°C.

Pad system installation

WATER-DISTRIBUTION SYSTEMS. To ensure that the entire pad remains wet and clean, it is vital that there is adequate water flowing over the pad. In general, it is recommended that the water-distribution system be capable of circulating a minimum of 10 l/min for every linear metre of pad length. It is important to realize that, at most, only about 10% of the water flowing over the pad is evaporated.

PAD ROOMS. For best results, evaporative cooling pads should be installed at least 0.5 m out from the side of a house (Fig. 5.33). Pad rooms allow for access to both sides of the pads to make it easier to clean the pad by eliminating the need to remove the pads from the system for cleaning. Furthermore, pad

%RH		23	25	27	29	31	33	35	37	39	41
	100	23.0	25.0	27.0	29.0	31.0	33.0	35.0	37.0	39.0	41.0
	95	22.4	24.4	26.3	28.3	30.3	32.3	34.2	36.2	38.2	40.2
	90	21.8	23.7	25.7	27.6	29.5	31.5	33.4	35.4	37.4	39.3
	85	21.1	23.0	25.0	26.9	28.8	30.7	32.6	34.6	36.5	38.4
	80	20.5	22.4	24.2	26.1	28.0	29.9	31.8	33.7	35.6	37.5
	75	19.8	21.6	23.5	25.4	27.2	29.1	30.9	32.8	34.6	36.5
	70	19.1	20.9	22.7	24.6	26.4	28.2	30.0	31.8	33.7	35.5
	65	18.4	20.2	22.0	23.7	25.5	27.3	29.1	30.9	32.7	34.5
	60	17.7	19.4	21.1	22.9	24.6	26.4	28.1	29.9	31.6	33.4
	55	16.9	18.6	20.3	22.0	23.7	25.4	27.1	28.8	30.5	32.3
	50	16.2	17.8	19.4	21.1	22.7	24.4	26.1	27.7	29.4	31.1
	45	15.4	17.0	18.6	20.2	21.8	23.4	25.0	26.6	28.2	29.8
	40	14.6	16.1	17.6	19.2	20.7	22.3	23.8	25.4	26.9	28.5
	35	13.7	15.2	16.7	18.2	19.7	21.1	22.6	24.1	25.6	27.1
	30	12.8	14.3	15.7	17.1	18.5	20.0	21.4	22.8	24.2	25.7
	25	12.0	13.3	14.7	16.0	17.4	18.7	20.1	21.4	22.8	24.1
	20	11.0	12.3	13.6	14.9	16.1	17.4	18.7	19.9	21.2	22.5
	15	10.1	11.3	12.5	13.7	14.9	16.1	17.2	18.4	19.6	20.7
	10	9.1	10.2	11.4	12.4	13.5	14.6	15.7	16.7	17.8	18.8

Temperature °C

Fig. 5.32. Incoming air temperature with a 15-cm pad under different weather conditions.

Fig. 5.33. Pad room.

Fig. 5.34. Restrictive pad cover.

rooms keep the pads away from the house and eliminate the possibility of water leaking from the pad system into the house. Pad rooms most importantly allow the tunnel curtain to be installed on the outside surface of the side wall. If the curtain is installed on the inside surface of the side wall it is very difficult to get it to seal tightly during cooler weather. When installed on the outside surface of the pad, over time raising and lowering the tunnel curtain will cause damage to the pad and prevent water (from the distribution system) from entering the house.

CIRCULATION PUMP LOCATION. To maximize water distribution uniformity, the circulation pump should be located in the centre for pad systems longer than 15 m. With longer pad systems, more water is often circulated on the pads nearest the pump than is circulated at the end of the system. By placing the circulation pump in the centre of the system, water distribution becomes much more uniform and the flow of water back to the circulation pump is improved. Furthermore, with the installation of shut-off valves, operators can conveniently turn off half the pad on each side of the house. This will allow the flexibility of providing limited cooling for younger birds, and for older birds during cool or humid weather.

PAD SYSTEM COVERS. In some instances it is desirable to protect the pads from direct sunlight, thus making the house darker in the vicinity of the pads and/or helping to minimize algae growth on the pads. In other cases farm managers want to build a structure to protect the pads from blowing sand, which can dramatically reduce the life of pads. In either case care must be taken not to restrict the flow of air into the evaporative cooling pads. To accomplish this, the opening created by the structure should be at least the same area of the pads they are protecting (Fig. 5.34).

CONSTRUCTION OF POWER-VENTILATED HOUSES. In order for any power-ventilated house to be properly ventilated, a house should be constructed such that it does not allow air seepage when closed from other than the desired inlets especially the pads. Air seepage from walls, ceiling or curtains will disrupt proper ventilation and cooling of the house, resulting in poor performance of birds housed.

Summary

Although heat loss from birds can occur in four ways (radiation, conduction, convection, and evaporation), the primary methods of heat loss during hot weather are through evaporation and convection. Evaporative heat loss, which produces cooling, is adversely affected by high humidity and therefore not very effective in very humid climates. Air movement provides a number of benefits that help to cool birds during hot weather. Tunnel ventilation on commercial broiler farms results in better bird performance than in naturally ventilated farms, and income from those farms is generally greater. Bird age is an important factor in determining air velocity needed to cool birds, older birds requiring more air velocity than younger ones. Removing heat from the poultry house is crucial for optimum bird performance. The air velocity required to remove heat from the house and keep bird body temperature normal depends on bird density, bird age and the environmental temperature. The higher the humidity in the house, the higher air velocity is needed.

Poultry houses are either naturally ventilated or power-ventilated. Naturally ventilated houses are very common in developing regions of the world and in small to medium-size poultry operations. In such houses it is important during hot weather to facilitate the flow of air into and out of the poultry house. Such houses should not be more than 12 m in width and even less in hot climates. All naturally ventilated houses should be equipped with some type of adjustable side-wall curtains. During hot weather, curtain openings should account for about 50–80% of the side-wall height. There should be solid portions of wall above and below the curtain opening. The side-wall curtain should be made of clear, non-breathable material to limit air exchange when heat conservation is needed. House spacing significantly affects the environment in poultry houses during hot weather and factors like prevailing wind speed, direction, topography, etc. determine the optimal house spacing. Vegetation height should be kept to a minimum around poultry houses to discourage rodents and maximize air flow into the house. Tall trees planted next to a poultry house can be beneficial during hot weather. These trees, however, should have no branches below the eaves of the house, which may impede the flow of air in and out of the house. Ridge openings are necessary for naturally ventilated houses and can be very effective in those houses with uninsulated roofs. Naturally ventilated houses should always be oriented in an east–west direction. This is to minimize direct sunlight from entering the house. Properly designed roof overhangs reduce both direct and indirect sunlight entering a house during hot weather. A roof overhang should typically be a minimum of 0.6 m, but may need to be longer with taller side walls and large curtain openings. A roof slope is often determined by the level of roof insulation. With uninsulated roofs, a steep roof slope (45°) is highly desirable. To reduce heat-stress-related problems, it is beneficial to insulate roofs in naturally ventilated houses. A variety of methods are used to insulate a poultry-house ceiling (dropped ceiling, rigid board insulation, spray polyurethane insulation and reflective insulation). Houses in high-temperature areas (40°C) require ceiling R-values of 2.25 $m^2 \bullet$ °C/W. Roof coatings can also help in reducing roof temperature during hot weather. They are heavily dependent on reflectivity and over time can dull, thus requiring reapplication on a yearly basis. They are not needed for insulated ceilings. Roof sprinkling is another method of reducing heat gain from an uninsulated roof. Circulation fans in a naturally ventilated house are mainly used to produce air movement over the birds, to increase convective cooling. The type of cage system used in a layer house can significantly affect natural ventilation in that house. The traditional two-deck system lends itself well to natural ventilation because outside winds can pass easily above and below the cage system. Multiple rows of three- and four-deck systems are difficult to naturally ventilate.

In areas where climates are harsh and temperature extremes exist, it is necessary to use power-ventilated systems, in which exhaust fans and air inlets provide a very high level of control over both air exchange and air movement within a poultry house. The majority of hot-weather power-ventilated systems are negative-pressure systems, and these are of two types; inlet ventilation and tunnel ventilation. Power-ventilated poultry houses should be well insulated and tightly constructed. The air in a poultry house is heated by the heat flowing in through building surfaces as well as heat produced internally by lights, motors and the birds. The air leaving the house is no more than 2.8°C hotter than that entering the house. In houses with evaporative cooling systems, when properly designed, house temperature should be lowered to 27°C. In such houses, in very dry climates temperature can be lowered to 24°C. R-values are calculated on the basis of total heat entering the house through building surfaces and heat produced by the birds. A number of different types of negative-pressure air inlet systems are used to control poultry-house environments in hot climates. With a negative-pressure system, the speed at which air enters the house is determined by the pressure differential between inside and outside the house. In order to gain control over fresh-air distribution and air movement within the poultry house, the inlet area has to match a house's exhaust fan capacity. Three types of inlet ventilation systems are in use. In a cross-ventilation poultry house, the exhaust fans are evenly spaced down one side of the house and inlets are installed along the other side. In a side-wall inlet-ventilated house, exhaust fans are spaced evenly down one or both side walls and inlets are installed along both side walls. In an attic inlet-ventilated house, exhaust fans are spaced evenly down both side walls and air inlets are installed in the ceiling to pull air in from the attic. The greatest challenge with any traditional inlet ventilation system is producing sufficient air movement to provide sufficient bird cooling during hot weather. In designing a tunnel ventilation system, not only must it be designed to produce the proper air exchange but also to produce the desired air velocity. Minimum fan capacity for a tunnel-ventilated house is determined by the same method as for inlet-ventilated houses, namely by conducting a heat balance. Although air velocity within a tunnel house tends to be much more uniform than within a traditional naturally ventilated house, there are still some variations in air velocity within the house. In a bi-directional tunnel, it is best to install the tunnel fans on one end of the house and the tunnel inlet opening on the opposite end. Tunnel fans can be installed in either the end wall of a house or the side walls near the end wall. For tunnel houses without evaporative cooling pads, it is recommended that the tunnel inlet area be approximately 10% greater than the cross-sectional area of the house.

There are several types of exhaust fans in use. The most basic form is the exterior shutter fan. An interior shutter fan has the shutter on the intake side of the fan. A discharge cone fan is, as the name indicates, equipped with a discharge cone, which tends to increase fan performance by 5–10% by easing the transition of the spinning air coming off the fan blades. In a belt-drive fan, the pulley on the fan motor drives a pulley or the fan blade shaft through the use of a fan belt. In a direct-drive fan, the blades are attached directly to the fan motor shaft, thus eliminating the need for drive belts. To provide optimal conditions during hot weather and to keep operating costs to a minimum, it is important that the right fan is installed. Three performance factors are used to evaluate poultry-house exhaust fans: the amount of air delivered at a static pressure of 25 Pa, energy-efficiency rating and air-flow ratio.

Evaporative cooling systems have proven to be very effective in reducing house temperatures in virtually all hot climates. In dry climates, such systems can reduce house temperature by 15°C, while in humid climates temperature can be reduced only by 3–6°C. The best type of evaporative cooling depends on a variety of factors such as climate, housing type and ventilation system design. Basically two types are used: the fogging system and the pad system. Pad systems are used only in power-ventilated houses, while fogging systems can be used in both power-ventilated and naturally ventilated houses. Evaporative cooling pads should be properly sized because insufficient pad area can lead to reduced exhaust fan performance and reduced cooling. Most pads are designed for a cooling efficiency of about 75%. This means that the pad system is 75% effective at decreasing the dry bulb temperature to the wet bulb temperature. It is recommended that the water-distribution system be capable of circulating a minimum of 10 l/min for every linear metre of pad length. Cooling pads should be installed at least 0.5 m out from the side of a house for maximum water-distribution uniformity; the circulation pump should be located in the centre of pad systems longer than 15 m.

References

Bucklin, R.A., Bottcher, R.W., VanWicklen, G.L. and Czarick, M. (1993) Reflective roof coatings for heat stress relief in livestock and poultry housing. *Applied Engineering in Agriculture* 9(1), 123–129.

Chepete, H.J. and Xin, H. (2001) Heat and moisture production of poultry and their housing systems – a literature review. *Proceedings of the 6th International Symposium,* Louisville, Kentucky. ASAE Publication Number 701P0201, 319–335.

Dozier, W.A., Lott, B.D. and Branton, S.L. (2005a) Growth responses of male broilers subjected to increasing air velocities at high ambient temperatures and a high dew point. *Poultry Science* 84, 962–966.

Dozier, W.A., Lott, B.D. and Branton, S.L. (2005b) Live performance of male broilers subjected to constant or increasing air velocities at moderate temperatures with a high dew point. *Poultry Science* 84, 1328–1331.

Lacy, M.P. and Czarick, M. (1992) Tunnel-ventilated broiler houses: broiler performance and operating costs. *Journal of Applied Poultry Research* 1, 104–109.

Lott, B.D., Simmons, J.D. and May, J.D. (1998) Air velocity and high temperature effects on broiler performance. *Poultry Science* 77, 391–393.

May, J.D., Lott, B.D. and Simmons, J.D. (1997) Water consumption by broilers in high cyclic temperatures: bell versus nipple waterers. *Poultry Science* 76, 944–947.

Mitchell, M.A. (1985) Effects of air velocity and convective and radiant heat transfer from domestic fowls at environmental temperatures of 20 and 30°C. *British Poultry Science* 26, 413–423.

Pesti, G.M., Armato, S.V. and Minear, L.R. (1985) Water consumption of broiler chickens under commercial conditions. *Poultry Science* 64, 803–808.

Reece, F.N., Deaton, J.W. and Kubena, L.F. (1972) Effects of high temperature and humidity on heat prostration of broiler chickens. *Poultry Science* 51, 2021–2025.

Scott, T.A. and Balnave, D. (1988) Comparison between concentrated complete diets and self-selection for feeding sexually maturing pullets at hot and cold temperatures. *British Poultry Science* 29, 613–625.

Simmons, J.D., Lott, B.D. and May, J.D. (1997) Heat loss from broiler chickens subjected to various wind speeds and ambient temperatures. *Applied Engineering in Agriculture* 13, 665–669.

Simmons, J.D., Lott, B.D. and Miles, D.M. (2003) The effects of high air velocity on broiler performance. *Poultry Science* 82, 232–234.

Sturkie, P.D. (1965) *Avian Physiology,* 2nd edn. Cornstock Press, Ithaca, New York.

Timmons, M.B. (1989) Improving ventilation in open-type poultry housing. *Proceedings of the 1989 Poultry Symposium,* University of California, pp. 1–8.

Timmons, M.B., Boughman, G.R. and Parkhurst, C.R. (1981) Development and evaluation of an insulated curtain for poultry houses. *Poultry Science* 60: 2585–2592.

Timmons, M.B., Baughman, G.R. and Parkhurst, C.R. (1986) Effects of supplemental ridge ventilation on curtain-ventilated broiler housing. *Poultry Science* 65, 258–261.

Van Der Hel, W., Verstegen, M.W.A., Henken, A.M. and Brandsma, H.A. (1991) The upper critical ambient temperature in neonatal chicks. *Poultry Science* 70, 1882–1887.

Van Kampen, M. (1981) Water balance of colostomized and non-colostomized hens at different ambient temperature. *British Poultry Science* 22, 17–23.

Wilson, J.D. (1983) Ventilation air distribution. In: *Ventilation of Agricultural Structures.* ASAE Monograph. American Society of Agricultural Engineers, St Joseph, Michigan.

6 Nutrient Requirements of Poultry at High Temperatures

N.J. DAGHIR

Faculty of Agricultural and Food Sciences, American University of Beirut, Lebanon

Introduction	133
General temperature effects	133
Energy requirements	135
Protein and amino acid requirements	137
Vitamins	142
Vitamin C	142
Vitamin A	143
Vitamins E and D_3	144
Thiamine	144
Vitamin B_6	145
Mineral requirements	145
Calcium	145
Phosphorus	147
Potassium	148
Zinc	149
Dietary electrolyte balance (DEB)	149
Non-nutrient feed additives	150
Antibiotics	150
Aspirin	150
Coccidiostats	151
Reserpine	151
Flunixin	152
Genestein	152
Conclusions and recommendations	152
References	153

Introduction

Nutrition under temperature stress is one area of research that has been receiving a great deal of attention for many years. There have been at least five reviews on this subject in the 1980s. The first was by Moreng (1980), in which he reviewed the effects of temperature on vitamin requirements of poultry. The second review was by the National Research Council (NRC, 1981), which appeared in the publication *Effect of Environment on Nutrient Requirements of Domestic Animals* and dealt mainly with feed intake as affected by temperature changes, efficiency of production, metabolizable energy (ME) requirements and water intake. The third review, by Austic (1985), was published as one of the chapters in *Stress Physiology in Livestock*, Vol. 3, *Poultry* and covered, in a concise and comprehensive way, energy, protein, amino acids, vitamins, minerals, essential fatty acids and water. The fourth review was by Leeson (1986), who dealt with various nutritional considerations during heat stress and gave some specific dietary manipulations that could lead to improvement in performance of broilers and layers under heat-stress conditions. The fifth review was by Shane (1988), who discussed the interaction of heat and nutrition, with special emphasis on the immune system. Several studies have been conducted in recent years on nutritional intervention in alleviating the effects of high temperature. Gous and Morris (2005) reviewed the work for interventions in broilers, while Balnave and Brake (2005) covered work on pullets and laying hens. Lin *et al.* (2006) discussed nutritional and feeding strategies to prevent heat stress in birds, as well as genetic and environmental strategies. Balnave (2004) discussed the challenges for accurately defining the nutrient requirements of heat-stressed poultry and concluded that the number of factors that influence nutrient requirements of heat-stressed poultry are much greater than those that influence the nutrient requirements of poultry at the thermoneutral temperature.

General Temperature Effects

Before getting into specific nutrient requirements, it should be stated that there is a great deal of disagreement as to what is the ideal temperature range for the different classes and age-groups of poultry. This is probably due to the fact that many factors influence the reaction of poultry to temperature changes. Humidity of the atmosphere, wind velocity and previous acclimatization of the bird are among the most important. Birds, in general, perform well within a relatively wide temperature range. This range, which extends between 10 and 27°C, is not too different for broilers, layers or turkeys (Milligan and Winn, 1964; de Albuquerque *et al.*, 1978; Mardsen and Morris, 1987).

Kampen (1984) found that the highest growth rate of broilers occurs in the range of 10–22°C. However, maximum feed efficiency is at about 27°C. In layers, he reported that in the temperature range 10–30°C the net energy available for egg production is almost constant, and feed costs per egg are minimal at 30°C.

As for the optimum temperature range, Charles (2002) reviewed the literature on the optimum temperature for performance and concluded that for growing broilers it is 18–22°C and for layers 19–22°C. We know, however, that what is ideal for growth is not ideal for feed efficiency, and what is ideal for feed efficiency is not ideal for egg weight. For example, we know that feed efficiency is always reduced at temperatures below 21°C. Egg production and growth rate are reduced at temperatures below 10°C. The overall optimum range is mainly dependent on the relative market value of the product produced, in proportion to feed cost. As the price ratio widens, the best temperature falls, and vice versa.

Another general consideration that is very relevant to this discussion is feed intake. There is no question that high and low environmental temperatures impose limitations on the performance of both broilers and laying hens that are unrelated to feed intake. Only part of the impairment in performance is due to reduced feed intake. The National Research Council (NRC, 1981) summarized several papers on laying hens, and concluded that the decrease in feed intake is about 1.5% per 1°C, over the range of 5–35°C, with a baseline of 20–21°C. Austic (1985) summarized work on growing chickens and concluded that it was 1.7% per 1°C over a baseline of 18–22°C. Response of birds subjected to cyclic temperatures is not too different from that due to constant temperature. This decrease in feed intake is not linear but becomes more severe as temperature rises. Table 6.1 shows the change in percentage decrease per 1°C rise in temperature as calculated and summarized from 12 references.

Some workers have tried to partition the detrimental effects on performance into those that are due to high temperature per se and those due to reduced feed intake, by conducting paired-feeding experiments. Smith and Oliver (1972), in work with laying hens subjected to 21°C and 38°C temperatures, showed that a 40–50% reduction in egg production and egg weight at 38°C is due to reduced feed intake, while the reductions in shell thickness and shell strength are mainly due to high temperature. Dale and Fuller (1979), in work with broilers, showed that 63% of the reduction in growth rate is due to reduced feed intake. In contrast to paired feeding, another interesting approach to this area of research that needs to be pursued is testing the effect of force-feeding of birds in a hot environment on different performance

Table 6.1. Effect of temperature on feed intake of laying hens (values summarized from various references).

Temperature (°C)	Decrease per 1°C rise
20	
25	1.4
30	1.6
35	2.3
40	4.8

parameters in layers and in broilers. The author conducted some paired-feeding and forced-feeding studies on broilers raised at high temperature and found that 67% of the reduction in growth rate in these birds is due to reduced feed intake (Daghir, N.J. and Hussein, A., 2007, unpublished results).

Energy Requirements

ME requirement decreases with increasing temperature above 21°C. This reduced requirement is mainly due to a reduction in energy requirements for maintenance, and the requirement for production is not influenced by environmental temperature. We have observed for many years that energy consumption during the summer drops significantly in contrast to winter or spring (Daghir, 1973). Energy intake during the summer was 10–15% lower than during the winter. Energy requirement for maintenance decreases with environmental temperature to reach a low at 27°C, followed by an increase up to 34°C. This has been demonstrated by Hurwitz et al. (1980) in work with broilers. Figure 6.1 shows the effects of environmental temperature on both feed efficiency and energy requirements for maintenance of male and female broilers. The use of high-energy rations for broilers has become quite common in warm regions. Some workers feel that this practice should be accompanied by raising the levels of the most critical amino acids. This is based on a few reports, one of which is McNaughton and Reece (1984), who found a significant weight response to additional energy only in the presence of relatively high levels of lysine (Fig. 6.2). They concluded that a diet energy response in warm weather is seen only when adequate amino acid levels are provided. This approach may increase performance, but it will also increase the heat load on the bird and its ability to survive.

The beneficial use of fats in hot-weather feeding programmes is well documented. We have known for some time now that the addition of fat stimulates feed and ME consumption at high temperatures. One of the very early reports on this was by Fuller and Rendon (1977), who showed that broilers fed a ration in which 33% of the ME was supplied by fat consumed 10% more ME and 10% more protein and gained 9% more weight than chicks fed a low-fat ration. Reid (1979) also observed improved performance in laying hens as a result of adding fat at high temperatures. We have observed that added fat at 31°C improved feed consumption in laying hens to a greater extent than at lower temperatures (Daghir, 1987). Table 6.2 shows that the addition of fat to laying rations increased feed intake by 17.2% at 31°C and by only 4.5% at lower temperatures (10–18°C).

The beneficial effects of fat at high temperature are several and well known. The higher fat content of the diet contributes to reduced heat production, since fat has a lower heat increment than either protein or carbohydrate. Energy intake is increased in both broiler and laying hens in a warm environment by the addition of fat. The addition of fat to the diet appears to increase the energy value of the other feed constituents (Mateos and Sell, 1981).

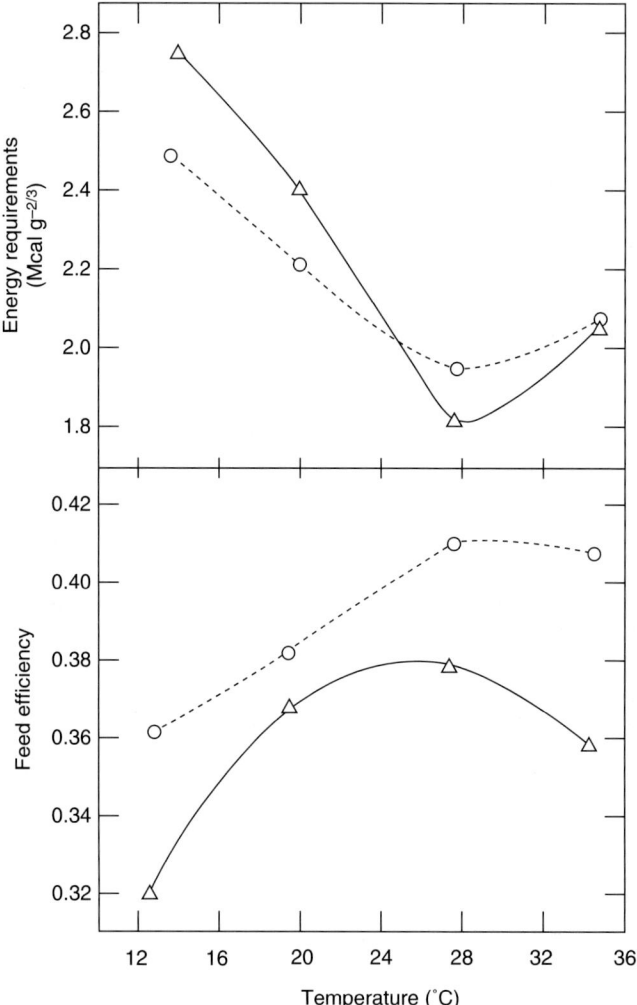

Fig. 6.1. Feed efficiency (lower graph) and energy requirements (upper graph) for maintenance of male (O) and female (△) broilers as functions of environmental temperature. (From Hurwitz et al., 1980.)

Fat has also been shown to decrease the rate of food passage in the gastrointestinal (GI) tract (Mateos et al., 1982) and thus increase nutrient utilization. This is interesting in light of an earlier observation by Wilson et al. (1980), who found that high environmental temperature caused an increase in food passage time in white Pekin ducks. Dietary fat can therefore help in counteracting this effect of high temperature.

Geraert et al. (1992) investigated the effect of high ambient temperature (32°C versus 22°C) on dietary ME value in genetically lean and fat, 8-week-old male broilers. Lean broilers exhibited higher apparent ME (AME) and true ME (TME) values than fat broilers. Hot climatic conditions significantly

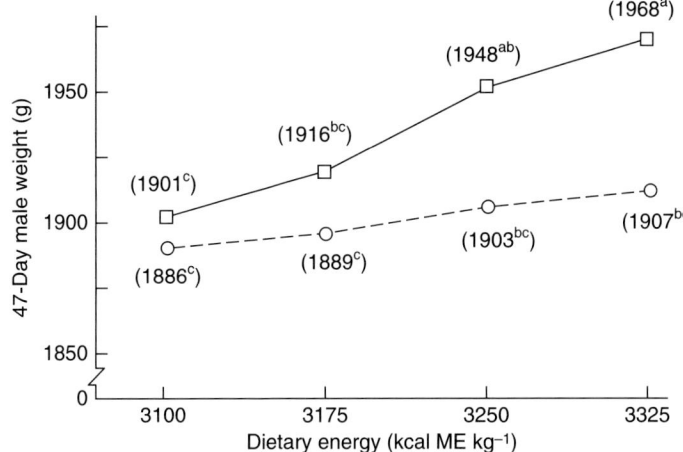

Fig. 6.2. Influence of dietary energy level on the lysine requirement of 23- to 47-day broiler males. Values in parentheses show mean of each treatment. Means followed by different small letters are significantly different ($P < 0.05$). Lysine at 0.308%/Mcal/kg (dashed line) or 0.322%/Mcal/kg (solid line). (From McNaughton and Reece, 1984).

Table 6.2. Interaction of temperature and added fat on feed consumption (g/hen/day) (from Daghir, 1987).

Temperature (°C)	Added fat %		% Increase
	0	5	
31	93	109	17.2
10–18	127	133	4.5

increased AME and TME values, particularly in leaner birds. Protein retention efficiency was enhanced by selection for leanness and increased with ambient temperature.

Protein and Amino Acid Requirements

The effects of high-temperature stress on protein requirements are not fully understood. Early work indicated that temperature changes neither increase nor decrease the protein requirement per unit gain. More recent work shows that there is decreased protein synthesis and increased breakdown under heat stress (Lin et al., 2006) and that this decreased protein synthesis cannot be restored by increasing the dietary protein level (Tenin et al., 2000). Cahaner et al. (1995) even suspected that high protein levels decrease broiler performance at high temperature and that this dietary protein effect is genotype related. Of course part of this decreased performance may be due to the increased heat increment, since protein has a high heat increment. It has been known for over a century that protein ingestion causes a larger increase in heat production than either carbohydrates or fat (Rubner, 1902). Therefore heat increment is lowered by decreasing dietary protein. Gonzalez-Esquerra and Leeson (2005) suggested that the length of exposure to heat stress may affect the response of birds to dietary protein. Short-term exposure has a different effect from long-term exposure. Therefore, reduction in crude protein levels in heat-stressed birds as a means to reduce heat production may not always be justified. Knowing the essential amino acid requirements for an ideal protein should lead to a minimum heat increment production. Musharaf and Latshaw (1999), however, in a review of heat increment, concluded that, to date, research has failed to document improved feed efficiency as a result of feeding an ideal protein.

Bray and Gesell (1961) reported that egg production could be maintained at 30°C provided a daily protein intake of about 15 g was ensured by appropriate dietary formulation. Figure 6.3 shows that temperature has no effect on egg numbers as long as protein intake is maintained. Egg mass, however, reaches a plateau at 30°C earlier than at 24.4 or 5.6°C. We continue to find responses from increasing protein intake to 18 g/day, and even 20 g/day in some strains, but this is independent of temperature.

Amino acid requirements as affected by temperature have been studied for many years. March and Biely (1972) demonstrated that high temperatures

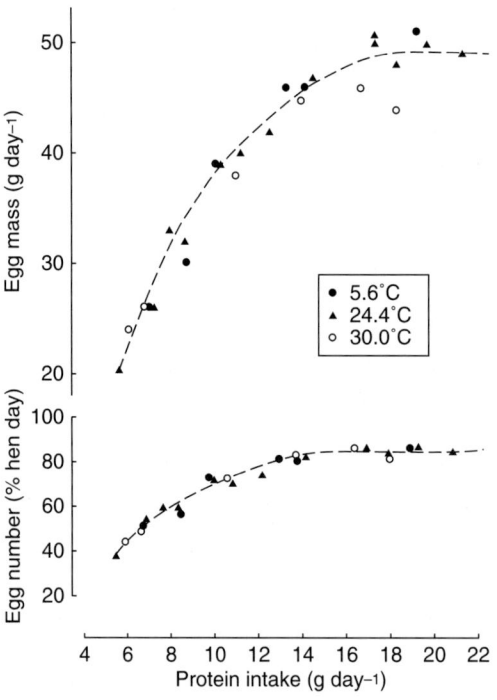

Fig. 6.3. Egg mass (upper curve) and number (lower curve) at different ambient temperatures, as affected by protein intake. (Adapted from Bray and Gesell, 1961.)

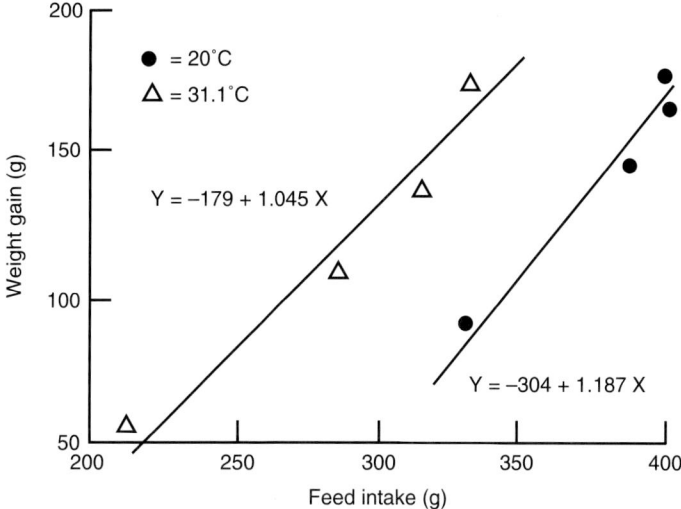

Fig. 6.4. Relationship between feed intake and weight gain of White Leghorn chicks fed for 15 days on diets with 0.73, 0.88, 1.03 or 1.33% lysine, at two ambient temperatures. (From March and Biely, 1972).

(31°C) do not affect the metabolic requirement for lysine. They carried out feeding experiments with chicks raised under two temperature conditions (20°C and 31°C). Chicks were fed diets with graded levels of lysine (0.73, 0.88, 1.03 and 1.33%). Two separate growth response lines were obtained when feed intake was related to body-weight gain at the two temperatures (Fig. 6.4). These results indicate that growth is reduced at the same rate as feed intake at the high temperature. March and Biely (1972) showed that, when these data are plotted on the basis of lysine intake versus body-weight gain, one response line can describe the relationship for both temperatures (Fig. 6.5). Balnave and Oliva (1990) reported that the methionine requirement of 3–6-week-old broilers kept at a constant 30°C or at cycling temperatures of 25–30°C was reduced compared with broilers kept at 21°C. Lin *et al.* (2006) reviewed the literature on amino acid balance and the enhanced protein breakdown at high temperature. These workers concluded that results from various studies are conflicting. Increasing lysine levels does not improve weight gain in heat-stressed broilers (Mendes *et al.*, 1997). Rose and Uddin (1997) observed that the growth rate response to lysine supplementation is decreased by high temperature in broilers. Brake *et al.* (1998), however, reported a significant response from increasing the arginine:lysine ratio on broiler performance at high temperature. Several factors may be involved in differing responses to amino acid supplementation at high temperature. Chen *et al.* (2005) reported that the response to crystalline amino acid supplementation is affected by dietary electrolytes such as sodium chloride. Brake *et al.* (1998) confirmed this and reported that increasing the Arg:Lys ratio in Australian diets fed to broilers raised at 31°C improved body-weight gain and feed conversions when the diets contained low levels of NaCl

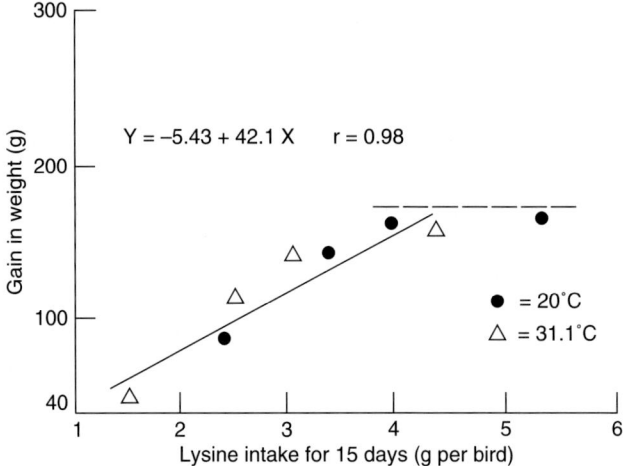

Fig. 6.5. Relationship between accumulative intake of lysine and accumulative growth of chicks fed for 15 days on diets with 0.73, 0.88, 1.03 or 1.33% lysine, at two ambient temperatures. (From March and Biely, 1972).

Table 6.3. Interaction means for body-weight gain and feed conversion of broilers subjected to chronic heat stress between 28 and 49 days of age and fed diets containing different Arg:Lys ratios and sodium chloride concentration (From Brake et al., 1998).

	NaCl (g/kg)			
	1.2	2.4	1.2	2.4
Arg:Lys (g:g)	BW gain (g)		Feed conversion (feed:gain)	
1.05	1.059^b	1.154^{ab}	2.38^c	2.19^b
1.20	1.060^b	1.125^{ab}	2.29^{bc}	2.26^{bc}
1.34	1.144^{ab}	1.142^{ab}	2.17^{ab}	2.16^{ab}
1.49	1.237^a	1.085^{ab}	2.20^a	2.21^b
SEM	53		0.05	

$^{a-c}$ Means with no common superscript are significantly different at $P < 0.05$.

(see Table 6.3). Balnave and Brake (2001) found that sodium bicarbonate improved broiler performance with high Arg:Lys ratio. Gonzalez-Esquerra and Leeson (2006) studied the effects of Arg:Lys ratios, methionine source and time of exposure to heat stress on performance. They concluded that the Arg:Lys ratio, methionine source and time of exposure to heat stress altered protein utilization in hyperthermic birds. Protein utilization was not affected by methionine source (L-Methionine (L-Met), 2-hydroxy-4-(methylthio) butanoic acid (HMB), or DL-Methionine (DLM)) when fed at high Arg:Lys ratios for birds under acute or chronic heat stress.

The above results indicate that the ideal amino acid balance for broilers raised at high ambient temperatures may vary with different dietary conditions.

The supplementation of essential amino acids to a diet with a poor protein quality or amino acid imbalance helps to improve performance by reducing heat increment and the harmful effects of high temperature. The industry therefore has followed the practice of adjusting the dietary levels of protein and amino acids in order to maintain a constant intake of these nutrients as house temperatures and thus feed intake levels vary. This is based on the assumption that temperature does not affect the efficiency with which amino acids are utilized for tissue growth or egg production. Hurwitz *et al.* (1980) suggested a method of estimating the protein and amino acid requirements which takes into account the reduced rate of production at high temperatures. They used a mathematical model which evaluates the amino acid requirements on the basis of the sum of the maintenance requirements per kcal or as percentage of diet increase as environmental temperature increases above the optimum range for growth. Then, as the temperature of 28–30°C approached, there was a decline in the amino acid requirements. This was demonstrated for arginine, leucine and the sulfur amino acid requirements. On this basis, Hurwitz *et al.* (1980) suggested that the use of a linear relationship between the amino acid requirements in feed formulation should be reconsidered. Austic (1985) recommended that we continue to increase amino acid levels as a percentage of the diet up to 30°C. Beyond that temperature, further increases are not justified because both growth and egg production will be depressed.

Sinurat and Balnave (1985) reported that the food intake and growth rate of broilers in a cycling temperature (25–35°C) were improved not only by increasing the dietary ME, but also by reducing the amino acid:ME ratio during the finishing period. The same workers (Sinurat and Balnave, 1986) observed that broilers at high temperatures (25–35°C) on a free-choice system selected similar amino acid:ME ratios, which were lower than the ratios contained in the complete diet. This is in agreement with work that goes back a number of years (Waldroup *et al.*, 1976), in which it was shown that growth rate and feed utilization of heat-stressed broilers was significantly improved when diets were formulated to minimize excess of amino acids. This is theoretically sound, since minimizing protein levels and improving the balance of amino acids should minimize the heat increment and thereby reduce the amount of heat which must be dissipated.

Zuprizal *et al.* (1993) studied the effects of high temperature on total digestible protein (TDP) and total digestible amino acids (TDAA) of rape-seed meals and soybean meals fed to broilers at 6 weeks of age. TDP and TDAA of both rape-seed and soybean meals decreased as the ambient temperature increased from 21 to 32°C. A 12 and 5% reduction in TDP value was observed with the rape-seed and soybean meals respectively. Wallis and Balnave (1984) studied the influence of environmental temperature, age and sex on the digestibility of amino acids in growing broiler chickens. Although sex had no major effect on amino acid digestibility at 30 or 50 days of age, the influence of environmental temperature was found to be sex-related. There was a decrease in amino acid digestibility at higher temperature in female but not in male birds. Scott and Balnave (1988) conducted studies to evaluate the use of diets varying in energy and nutrient density to overcome the nutritional stresses associated with the onset of lay and with periods of high temperature. Changes in dietary ME concentration had little influence on food and nutrient intake and egg mass output of hens in early lay kept at normal, cold or hot temperatures. At normal and cold temperatures, all dietary ME–nutrient density combinations allowed hens to meet the recommended daily protein intake, but only those fed the most concentrated diets were able to meet this recommendation at hot temperatures. Egg mass output of hens at hot temperature remained inferior even with the highest ME and protein intakes.

Vitamins

Vitamin C

Vitamin C is the most-studied vitamin in relation to ambient temperature, and yet its effects are still not fully defined. There is some evidence which indicates that under conditions of high environmental temperatures some mammals and birds are not able to synthesize sufficient ascorbic acid to replace the severe losses of this vitamin that occur during stress. As early as 1961, Thornton (1961) showed that blood ascorbic acid decreased with an increase in environmental temperature from 21 to 31°C. This action was postulated to be a result of both partial exhaustion of the endogenous stores and a reduction in the amount of vitamin being synthesized. Ahmad et al. (1967) also showed that ascorbic acid limited the increase of body temperature during heat stress up to 35°C. Supplemental ascorbic acid has also been reported to improve heat resistance and reduce mortality associated with elevated ambient temperatures (Pardue et al., 1984). Pardue et al. (1985a), in a study on the effects of high environmental temperatures on broilers, showed that ascorbic acid reduced mortality in both female and male broilers during a heating episode (38°C at bird level) in the production facility. Ascorbic acid supplementation stimulated growth in females during the early phase of growth but not males. Kafri and Cherry (1984) also showed improved growth rate at 32°C in males but not females. This suggests a need to investigate sex differences in ascorbic acid synthesis and/or metabolism.

Njoku (1984) showed improved growth performance of broilers reared in the tropics when their diet was supplemented with 200 mg/kg ascorbic acid. The same worker, in a second study, also showed improvement in feed conversion in broilers (Njoku, 1986).

A report by Thaxton (1986) showed that, following infection with infectious bursal disease (IBD) virus, vitamin C protected the immune biological tissues in growing birds and reduced their mortality to infection in a hot environment. Furthermore, Pardue et al. (1985b) showed that immunosuppression at high environmental temperatures could result from a reduction in thyroid activity. Therefore, it may be that the reduction of weight loss of an immunocompetent organ induced by feeding ascorbic acid may be associated with thyroid activity. Takahashi et al. (1991) studied the effects of supplemental ascorbic acid on broilers treated with propylthiouracil. Feeding ascorbic acid partly prevented the decreases in body-weight gain, feed conversion and weights of the bursa of Fabricius and thymus in chicks fed propylthiouracil. They suggested that ascorbic acid improves the performance of chicks with experimentally induced hypothyroidism. Hayashi et al. (2004) studied the effect of ascorbic acid on performance and antibody production in broilers vaccinated against infectious bursal disease under a hot environment. Ascorbic acid was fed at 158 ppm, starting at 20 days of age, and all birds were vaccinated against IBD at 15 and 20 days. They concluded that ascorbic acid can minimize the stress induced by the combined effects of vaccination and high temperature.

In work with laying hens, ascorbic acid supplementation was shown to improve egg weight, shell thickness and egg production (Perek and Kendler, 1962, 1963). Njoku and Nwazota (1989) found that the inclusion of ascorbic acid in the diet improved egg production, food intake and food utilization and decreased the cost of feed per kg egg. The addition of 400 mg ascorbic acid/kg diet gave the most efficient performance. Palm oil inclusion in the diet also reduced the effect of heat stress and increased egg production, egg weight, food intake and efficiency of utilization. Ascorbic acid and palm oil when fed alone or in combination reduced the incidence of cracked eggs. These authors concluded that the addition of both ascorbic acid and palm oil ameliorated the effects of heat stress in a hot tropical environment. Whitehead and Keller (2003), in a review on ascorbic acid, reported improvements in laying hen performance from diets containing 250–400 mg/kg. Anjun et al. (2002) reported that ascorbic acid reduces the incidence of blood spots in eggs.

With broiler breeders, dietary ascorbic acid supplementation has been shown to improve nutrient utilization as judged by the production of hatching eggs (Peebles and Brake, 1985). Wide variations are encountered in the recommendations for supplementing poultry diets with ascorbic acid. Maumlautnner et al. (1991) suggested that these variations might be due to high losses during storage at high temperature. They tested three forms of vitamin C (crystalline ascorbic acid, protected ascorbic acid and phosphate-ascorbic acid ester). These three were fed to hens from 21 to 30 weeks of age and the birds were kept at 20 or 34°C. Performance and eggshell quality of treated hens improved only at 34°C. The best results were with the protected ascorbic acid and the phosphate-ascorbic acid ester.

Vitamin A

Vitamin A requirements, as affected by temperature, have been investigated by several workers. As early as 1952, Heywang (1952) showed that hot weather caused a marked increase in vitamin A requirements of the laying and breeding chicken. Kurnick et al. (1964) reported more vitamin A storage in the liver in

Leghorn pullets during cool periods than in those fed during hot weather. Smith and Borchers (1972) suggested that environmental temperature is of minor importance as a factor influencing the conversion of β-carotene to vitamin A. Elevated body temperatures may interfere with absorption. A review by Scott (1976) documented a threefold increase in vitamin A requirement of breeder hens at 38°C compared with those at normal room temperature.

Moreng (1980) reviewed the early literature on the effects of high temperature on vitamin A requirements. His review shows that some evidence suggests improved bird performance in response to additional vitamin A. The differences in performance, however, were not always statistically significant. Lin et al. (2002) reported that performance of heat-stressed layers can be improved with high levels of vitamin A. Levels of 9000 IU/kg had a beneficial effect on feed intake and production compared to controls with 3000 IU/kg.

Vitamins E and D_3

It is known that vitamin E requirements increase with increased stresses, particularly those that are related to high temperature (Cheville, 1979). Generally, vitamin E serves as a physiological antioxidant through inactivation of free radicals and thus contributes to the integrity of the endothelial cells of the circulatory system. High environmental temperature may exert an effect on health and performance by modifying the cellular and hence dietary requirement of vitamin E (Heinzerling et al., 1974). Scott (1966) suggested that heat stress interferes with the conversion of vitamin D_3 to the active form, an important step for calcium metabolism. There is still no clear evidence that adding vitamin E or D_3 has any beneficial effect during heat stress.

Gheisari et al. (2004) studied the effect of dietary fat (0, 2.5, 5.0% sunflower oil), α-toropherol (288 ppm) and ascorbic acid (255ppm) on the performance and meat oxidative stability of heat-stressed broiler chicks. They showed that in heat-stress conditions supplementations with ascorbic acid and vitamin E reduce mortality and oxidative rancidity of dark and white meat during refrigerated storage. Puthgonsiripon et al. (2001), in work with laying hens, observed that supplemental vitamin E (65 IU/kg) did not affect egg production but improved egg mass during heat stress. Whitehead et al. (1998) found that very high levels of vitamin E helped to sustain egg production in layers under heat stress. These workers showed a 7% better egg production in heat-stressed layers fed a 500 IU vitamin E/kg diet. Plasma vitamin E levels were reported to be linearly related to dietary level. Bollengier-Lee et al. (1998) from the same research group found that layers at 32°C gave 20% higher egg production when fed a 500 versus 10 IU vitamin E/kg diet.

Thiamine

Mills et al. (1947) found that the requirement for this vitamin was significantly increased for chicks grown at 32.5°C, as compared with 21°C. They could not detect a change in requirement for pyridoxine, nicotinic acid, folic acid or choline.

Vitamin B_6

Celik et al. (2006) studied the effect of dietary vitamin B_6 and L-Carnitin on broilers reared under high temperature (34–36°C for 8 h and 20–22°C for 16 h/day). They observed that body-weight gain and feed intake were significantly improved at 42 days by feeding 3 mg B_6/kg feed and 60 mg L-Carnitin/l in the drinking water. When vitamin B_6 was fed alone, there was no effect, which indicated that there is an interaction with L-Carnitin.

Mineral Requirements

Calcium

Egg weight and eggshell strength decline at high environmental temperatures. This appears to be partly due to reduced calcium intake, but several physiological mechanisms are involved: (i) reduced blood flow through shell gland due to peripheral vasodilatation; (ii) respiratory alkalosis; (iii) reduced blood ionic calcium content; (iv) reduced carbonic anhydrase in shell gland and kidneys; and reduced Ca mobilization from bone stores. The methods that have been used to reduce environmental temperature effects on shell quality can be summarized as follows.

1. $NaHCO_3$ has been added to feed, but has not always given positive responses.
2. Carbon dioxide-enriched atmosphere has been very effective, but it is not practical if it involves reduced ventilation.
3. Introduction of oyster shell or hen-sized sources of Ca separately from other nutrients has been a very useful means of improving shell quality. The results of a study reported by Sauveur and Picard (1987) are shown in Table 6.4. Linchovnikova (2007) in a recent study showed that layer rations should contain two-thirds large particles of limestone or oyster shell during the third phase of production, and 4.1 g Ca/kg feed are needed during that phase.
4. Night cooling has been very effective as a tool in maintaining shell quality at high temperatures.
5. Use of carbonated drinking water during hot weather.

There is very little work on the combined effects of temperature and humidity on performance. It is somewhat accepted that high humidity aggravates the detrimental effects of high temperature. Picard et al. (1987) studied the effects of high temperature and relative humidity on egg composition. They found that, when high temperatures are combined with high

Table 6.4. Effects of temperature and Ca source on egg weight and shell quality (from Sauveur and Picard, 1987).

Diet		End of control period (20°C)	2nd day at 33°C	24–28th day at 33°C
Pulverized limestone	Egg wt (g)	61.1	60.0	57.7
	Eggshell wt (g)	5.84	4.47	5.22
	SWUSA*	8.04	6.26	7.48
Oyster shell	Egg wt (g)	62.2	60.0	58.9
	Eggshell wt (g)	5.82	4.96	5.42
	SWUSA*	7.93	6.90	7.67

*SWUSA – shell weight per unit surface area (g per 100 cm^2)

Table 6.5. Effects of high temperature and relative humidity on egg composition (from Picard et al., 1987).

Temperature/relative humidity	20°C/50%	33°C/30%	33°C/85%
Egg weight (g)	58.1	56.2	54.2
Eggshell weight (g)	5.72	5.47	4.89
Yolk weight (g)	15.0	14.3	13.9
Albumen weight (g)	37.5	36.4	35.3
Yolk dry matter (%)	52.7	51.0	50.9
Albumen dry matter (%)	12.8	12.4	12.0

Table 6.6. Effect of carbonated water on shell quality (from Odom et al., 1985).

		Temperature (°C)	
	Water treatment	23	35
% Eggshell	Tap	9.89	8.44
	Carbonated	9.84	8.88
Egg specific gravity	Tap	1.086	1.072
	Carbonated	1.088	1.074

humidity, there is a further decrease in egg weight and shell weight (Table 6.5). There is also a decrease in egg components.

Another dietary manipulation, tested by Odom et al. (1985), was to study the effect of drinking of carbonated water on production parameters of laying hens exposed to high ambient temperature (Table 6.6). Their data indicate that the use of carbonated drinking water during periods of hot weather can help relieve the associated problem of high environmental temperature-induced eggshell thinning. This has been confirmed by Koelkebeck et al. (1992), who reported that a carbonated drinking water system can be operated efficiently in a commercial cage layer facility and can help improve eggshell quality of flocks experiencing shell quality problems during the summer. The test was performed in a cage layer facility on both 46- and 86-week-old commercial layers during a 12-week period in the summer.

One of the problems of the fowl-processing industry is that processing yield from spent hens is low due to a high incidence of bone breakage. After 1 year of production, the hens' bones weaken, causing them to break easily, and this increases the incidence of bone fragments in the processed meat. Koelkebeck et al. (1993), in an experiment on heat-stressed laying hens, found that providing these birds with carbonated drinking water improves tibia breaking strength. They suggested that carbonated drinking water during heat stress may reduce bone breakage during the processing of spent hens. This has also been previously reported in male chickens by Kreider et al. (1990), who showed that carbonated drinking water improved tibia bone-breaking strength of cockerels exposed to a 37°C environment. This practice of supplying carbonated drinking water to laying hens in the summer is becoming quite popular on layer farms in the southern USA.

Phosphorus

Garlich and McCormick (1981) reported that Ca and P balance seems to have an effect on survival time during periods of acute heat stress. They showed a direct relationship between plasma phosphorus and survival time, and an inverse relationship with plasma calcium. Survival time in fasted chicks was greater when their previous diet contained low levels of Ca and high levels of P. This may be important to consider where feed withdrawal is practised to reduce mortality during high-temperature spells.

The detrimental effects of high P levels on shell quality during high temperature conditions have been well documented (Miles and Harms, 1982; Miles et al., 1983). We have shown (Daghir, 1987) that the decrease in shell thickness at high temperatures is greater with higher P levels than with low P levels (Table 6.7). Egg weight is affected similarly (Table 6.8). Usayran et al. (2001) reported that low dietary nonphytate P (0.15%) depressed egg production in post-peak hens, but improved shell thickness in hens subjected to

Table 6.7. Interaction of temperature and dietary phosphorus on shell thickness (mm) (from Daghir, 1987).

	Temperature (°C)		
Available phosphorus (%)	10–18	31	% Difference
0.45	0.362	0.336	−7.7
0.35	0.367	0.344	−6.7
0.25	0.351	0.348	−0.9

Table 6.8. Interaction of temperature and dietary phosphorus on egg weight (from Daghir, 1987).

	Temperature (°C)		
Available phosphorus	10–18	31	% Difference
0.45	60.7	47.4	−28.2
0.35	60.0	53.9	−11.4
0.25	56.9	52.1	−9.4

Table 6.9. Effect of NH_4Cl, $NaHCO_3$ and $CaCl_2$ on broilers (from Teeter *et al.*, 1985).

	Body-weight gain (g)	
Treatment	Thermoneutral*	Hot*
Basal	933[A]	442[C]
Basal + 0.3% NH_4Cl		484[BC]
Basal + 1% NH_4Cl		553[B]
Basal + 3% NH_4Cl		464[C]
Basal + 1% NH_4Cl + 0.5%$NaHCO_3$		594[B]
Basal + 0.5% $CaCl_2$		481[BC]
Basal + 1% $CaCl_2$		474[BC]

*Means with different superscripts are significantly different ($P < 0.05$).

a constant high temperature. Persia *et al.* (2002) did not observe any significant interactions between AP level and environmental temperature on egg production or mortality of laying hens. Attia *et al.* (2006) studied the response of broilers raised at a constant high temperature to a multi-enzyme or phytase supplementation. Phytase addition significantly improved phosphorus retention by 21.4% and reduced excrement phosphorus by 21.9%, which can reduce phosphorus pollution in hot regions.

Teeter *et al.* (1985) suggested that alkalosis and weight gain depression attributed to heat stress can be alleviated by dietary measures. They showed that including 0.5% $NaHCO_3$ in the diet enhanced body-weight gain by 9% (Table 6.9). Adding 0.3% or 1% NH_4C1 to diets increased body-weight gains by 9.5% and 25% respectively. Supplementing the 1% NH_4C1 diet with 0.5% $NaHCO_3$ increased weight gain an additional 9%. $CaCl_2$ addition had very little effect.

Potassium

It has been known for some time that the K requirement of growing chickens increases with increased temperature. Huston (1978) observed that blood K concentrations in growing chickens were reduced by high environmental temperature. The same has been reported for laying hens (Deetz and Ringrose, 1976). The K requirement was reported to increase from 0.4% of the diet at 25.7°C to 0.6% at 37.8°C.

Teeter and Smith (1986) and Smith and Teeter (1987) conducted several experiments to study the impact of ambient temperature and relative humidity (35°C and 70%) upon K excretion and the effects of KCl supplementation of broilers exposed to chronic heat and cycling temperature stress. They concluded that dietary K levels should be increased for birds reared in heat-stressed environments. A level of 1.5–2% total or 1.8–2.3 g K daily is needed to maximize gain in 5–8-week-old broilers. They also suggested that this could be added to drinking water at 0.24–0.3% K in the form of KCl. This may be preferred since birds are more likely to drink water than consume feed under heat stress.

Zinc

Klasing (1984) reported a redistribution of zinc during immunological stress with a reduction in plasma zinc levels and an increase in hepatic zinc. The requirement for zinc, therefore, may be increased during exposure to heat stress. Sahin and Kucuk (2003) reported that supplementation of quail diets with 60 mg zinc/kg reduced the negative effect of heat stress on performance and egg quality.

Bartlett and Smith (2003) evaluated the effect of zinc on performance and immune competence of broilers subjected to high temperature. Broilers were raised at either 24°C or 24 to 35°C cycling and fed a low zinc diet (34 mg/kg), adequate diet (68 mg/kg) or a supplemental zinc diet (181 mg/kg). Their results showed that the immune response of broilers can be influenced by the level of zinc in the diet and by environmental temperature.

Sahin et al. (2005) evaluated two sources of zinc ($ZnSO_4.H_2O$ and Zn picolinate) for their effects on performance of Japanese quail exposed to high ambient temperature (34°C). Supplementation with zinc improved carcass weight and antioxidant status of birds and the effects of Zn picolinate were relatively greater than $ZnSO_4.H_2O$.

Dietary Electrolyte Balance (DEB)

It has been known for some time that there is need for a balance between cations and anions in the body and not just for absolute amounts. Electrolyte balance is described by the formula of Na + K – Cl expressed as mEq/kg of diet. This electrolyte balance, which is also known as acid–base balance, is not only affected by the balance and proportion of electrolytes in the diet, but also by endogenous acid production and rate of renal clearance. Mongin (1980) proposed that an overall diet balance of 250 mEq/kg is optimum for normal physiological functions.

Different workers reported different results on the most appropriate DEB for birds under high temperature conditions. Ahmad and Sarwar (2006) reviewed the literature on this subject and concluded that these differences in response to DEB depend on ambient temperature, age of bird and length of exposure to high temperature. Very high (360 mEq/kg) and very low (0 mEq/kg) DEB can result in metabolic alkalosis and acidosis, respectively. Therefore, in diet formulation, one should avoid either high or low DEB. At the same time, excesses or deficiencies of any specific mineral should be avoided. These workers (Ahmad and Sarwar, 2006) also concluded that birds under heat stress perform best at a DEB of 250 mEq/kg and maintain their blood physiological parameters and blood nutrients well.

In a study on heat-stressed broilers kept at a daily average of 23–31°C, Borges et al. (2003a) reported that the optimum DEB was 190 mEq/kg for the starter period and 220 mEq/kg for grower and finisher. In a later study, the same authors (Borges et al., 2003b) could not find any significant effect of DEB over the range 40–340 mEq/kg on the performance of heat-stressed broilers fed nutritionally adequate diets.

Mushtag et al. (2005) evaluated the effect and interactions of different levels of Na^+ and Cl^- on broilers at cyclic temperature of 32 to 39°C while maintaining the DEB at 250 mEq/kg. These authors concluded that a minimum of 0.25 %Na^+ and 0.30% Cl^- are necessary for adequate performance under heat stress (32 to 39° cyclic) if the DEB is maintained at 250 mEq/kg. In a recent study the same authors (Mushlag et al., 2007) reported that the dietary requirement of Na^+ to be 0.20–0.25% with that of Cl^- to be 0.30% during the finishing phase (29–42 days) at temperature ranging from 32 to 40°C.

Non-nutrient Feed Additives

Several non-nutrient feed additives have been tested to reduce the harmful effects of high-temperature stress. Examples of these follow.

Antibiotics

The incorporation of antibiotics in so-called 'stress feeds' has been widely practised all over the world. Extremes in temperature are among the stresses that have at times been handled by antibiotic feeding. The literature on this subject is very skimpy and reports are often contradictory (Freeman et al., 1975). Because of the restrictions imposed on the use of antibiotics in poultry feeds in some parts of the world and the possible expansion of these restrictions to other parts, this area of research will probably not receive much attention in the future.

Aspirin

Aspirin has been used as a tranquillizing drug in order to help birds subjected to stress, particularly heat stress. The effects of aspirin on growth and egg production have been studied by few researchers. The results of feeding this drug to broilers during heat stress have been variable. An early report by Glick (1963) showed that supplementation of acetylsalicylic acid (ASA) at 0.3% of the diet significantly improved growth. Later studies, however, using ASA levels ranging from 0.005% to 0.9%, found no improvement or even negative effects on growth rate (Reid et al., 1964; Nakaue et al., 1967; Adams and Rogler, 1968).

Few reports are available on the effects of dietary aspirin on laying chickens. Balog and Hester (1989) fed aged layer breeders 0.05% ASA for a period of 4 weeks. Aspirin reduced production of shell-less eggs in these birds, but had no effect on soft-shelled eggs. Recently Abou El-Soud et al. (2006) studied the effects of ASA on Japanese quail subjected to high temperature (33–36°C and RH 60–70%). Feeding 0.05 or 0.1% ASA from 8 to 16 weeks of age decreased rectal temperature, increased hen-day egg production and improved shell thickness. These workers concluded that ASA was beneficial at high temperature because it reduced body temperature and oxidative stress. Hassan et al. (2003) had also previously studied the effect of ASA administered in drinking water on Japanese quail and found that adding 1.5 g/l improved body weight, fertility, hatchability, egg production, egg specific gravity and sperm storage capacity of these birds under heat stress.

The base of heat-stress reduction by ASA is not known. It may be through cyclooxygenase inhibition, since Edens and Campbell (1985) reported reduced heat stress in broilers given flunixin, a non-steroidal cyclooxygenase inhibitor.

Coccidiostats

McDougal and McQuistion (1980) studied the effects of anticoccidial drugs on heat-stress mortality in broilers. Overall mortality during the 8-week study averaged 6% in unmedicated or monensin-medicated birds, 10% in arprinocid-medicated birds and 36% in nicarbazin-medicated birds. Hooge et al. (1999) studied the effect of dietary sodium bicarbonate (0 or 0.2%) with each of these ionophore coccidiostats in two diets (maize–soy versus maize–soy–meat) on broilers raised on built-up litter or battery brooders. This was done because maize–soy–meat diets contain lower potassium and higher chloride than maize–soy diets. Treatment with sodium bicarbonate significantly improved coccidial lesion score, body weight and feed efficiency compared with monensin alone. The effects were similar in maize–soy and maize–soy–meat diets.

Reserpine

Several reports have shown improved heat tolerance occurring following administration of reserpine, an alkaloid that is extracted from the *Rauwolfia* plant. Edens and Siegel (1974) evaluated the role of reserpine in three experiments with young chicks. They showed that pre-treatment with the alkaloid prevented the rapid loss of CO_2, which normally occurs when birds are subjected to acute high temperature, and thereby stabilized the acid–base blood status.

Flunixin

This anti-inflammatory analgesic has been studied by Edens (1986), who added it to the drinking water at levels ranging from 0.28 to 2.20 mg/kg body weight/day. It was given to 5-week-old broilers that were exposed to 37°C for 5 h per day. Flunixin treatment increased water consumption by 100 ml before heat treatment and during heat stress. Water consumption was increased by 150–300 ml per bird per day.

Genestein

Genestein is a strong antioxidant and acts by inhibiting formation of oxygen free radicals, reducing lipid oxidation and stimulating antioxidant enzymes. Sahin *et al.* (2004) fed genestein at 200, 400 and 800 mg/kg diet to Japanese quail reared at 34 versus 22°C. These workers observed that genestein improved digestibility of CP, DM and ash, and improved bone mineralization in birds reared under heat-stress conditions.

Conclusions and Recommendations

1. Nutritional manipulations can reduce the detrimental effects of high environmental temperatures but cannot fully correct them, for only part of the impairment in performance is due to poor nutrition.
2. The addition of fat stimulates feed and ME consumption at high temperature. This beneficial effect of fat is well established in broiler feeding programmes and has been shown to be of value in pullet-rearing and laying rations in hot climates.
3. Temperature changes neither decrease nor increase the requirements for proteins and amino acids per unit gain. The major adjustments needed for temperature changes are those that ensure adequate daily intakes for the various amino acids. Avoiding excesses of amino acids and giving more attention to amino acid balance is important in helping to overcome some of the adverse effects associated with high-temperature stress because nutritional imbalances are more detrimental in hot than in temperate climates.
4. Research conducted so far on vitamin requirements as affected by temperature changes does not indicate any change in the absolute requirements. There is some evidence in the literature to suggest improved bird performance

at high temperatures in response to additional vitamins A, E and C. Vitamin D_3, thiamine and vitamin B_6 need to be in investigated further.

5. Besides the effect on calcium and electrolyte balance, there is little work on the effect of high temperature on mineral metabolism. Calcium and phosphorus balance has some effect on survival time during acute heat stress. The practice of separate feeding of oyster shell or hen-sized sources of calcium continues to be the most effective dietary approach to reducing the effect of environmental temperature on shell quality.

6. Several reports indicate the need for increased dietary K in heat-stressed conditions. In doing this, care should be taken to prevent the negative effect that this may have on the acid–base balance.

7. In diet formulations for hot climates, we should avoid either high or low DEB. In most studies, birds under heat stress performed best at a DEB of 250 mEq/kg.

8. Several non-nutrient feed additives have been tested to reduce the harmful effects of high-temperature stress. Examples of these are antibiotics, aspirin, reserpine, flunixin, genestein, etc.

References

Abou El-Soud, S.B., Ebeid, T.A. and Eid, Y.Z. (2006) Physiological and antioxidative effects of dietary acetylsalicylic acid in laying Japanese quail under high temperature. *The Journal of Poultry Science* 43, 255–265.

Adams, R.L. and Rogler, J.C. (1968) The effects of dietary aspirin and humidity on the performance of light and heavy breed chicks. *Poultry Science* 47, 1344–1348.

Ahmad, M.M., Moreng, R.E. and Muller, H.D. (1967) Breed responses in body temperature to elevated environmental temperature and ascorbic acid. *Poultry Science* 46, 6–15.

Ahmad, T. and Sarwar, M. (2006) Dietary electrolyte balance: implications in heat stressed broilers. *World's Poultry Science Journal* 62, 638–653.

Anjun, M.S., Ziar Rahman, A., Ali, S. and Sandhu, M.A. (2002) Egg quality characteristics influenced by various heat combating systems during summer. *Archiv Fur Geflukellkunde* 66, 136.

Attia, Y.A., Bohmer, B.M. and Rath-Maier, D.A. (2006) Responses of broiler chicks raised under constant relatively high ambient temperature to enzymes, amino acid supplementations, or a high-nutrient diet. *Archiv Fur Geflukellkunde* 70, 80–91.

Austic, R.E. (1985) Feeding poultry in hot and cold climates. In: Youssef, M. (ed.) *Stress Physiology In Livestock,* Vol. 3, *Poultry,* CRC Press, Boca Raton, Florida, pp.123–136.

Balnave, D. (2004) Challenges of accurately defining the nutrient requirements of heat-stressed poultry. *Poultry Science* 83, 5–14.

Balnave, D. and Brake, J. (2001) Different responses of broilers at low, high or cyclic moderate–high temperature to dietary sodium bicarbonate supplementation due to differences in dietary formulation. *Australian Journal of Agricultural Research* 52, 609–613.

Balnave, D. and Brake, J. (2005) Nutrition and management of heat-stressed pullets and laying hens. *World's Poultry Science Journal* 61, 399–406.

Balnave, D. and Oliva, A. (1990) Responses of finishing broilers at high temperature to dietary methionine source and supplementation levels. *Australian Journal of Agricultural Research* 41, 557–564.

Balog, J.M. and Hester, P.Y. (1989) The effect of dietary ASA on egg shell quality. *Poultry Science* 68 (Suppl. 4), 9 (Abstract).

Bartlett, J.R. and Smith, M.O. (2003) Effect of different levels of zinc on the performance and immunocompetence of broilers under heat stress. *Poultry Science* 82, 1580–1588.

Bollengier-Lee, S., Mitchell, M.A., Utomo, D.B., Williams, P.E. and Whitehead, C.C. (1998) Influence of high dietary vitamin E supplementation on egg production and plasma characteristics in hens subjected to heat stress. *British Poultry Science* 39, 106–112.

Borges, S.A., Fischer Da Silva, A.V., Ariki, J., Hooge, D.M. and Cummings, R.R. (2003a) Dietary electrolyte balance for broiler chickens under moderately high ambient temperature and relative humidities. *Poultry Science* 82, 301–308.

Borges, S.A., Fischer Da Silva, A.V., Ariki, J., Hooge, D.M. and Cummings, R.R. (2003b) Dietary electrolyte balance for broiler chickens exposed to thermoneutral or heat-stress environments. *Poultry Science* 82, 428–435.

Brake, J., Balnave, D. and Dibner, J.J. (1998) Optimum dietary arginine:lysine ratio for broiler chickens is altered during heat stress in association with changes in intestinal uptake and dietary sodium chloride. *British Poultry Science* 39, 639–647.

Bray, D.J. and Gesell, J.A. (1961) Studies with corn–soya laying diets. IV. Environmental temperature. *Poultry Science* 40, 1328–1335.

Cahaner, A., Pinchasov, Y. and Nir, I. (1995) Effects of dietary protein under high ambient temperature on body weight, breast meat yield, and abdominal fat deposition of broiler stocks differing in growth rate and fatness. *Poultry Science* 74, 968–975.

Celik, L., Tekeli, A., Kutlu, H.R. and Corgulu, M. (2006) Effects of dietary vitamin B_6 and L-Carnitin on growth performance and carcass weight of broilers reared under high temperature regime. *World's Poultry Science Journal* XII, *European Poultry Conference Proceedings*, pp. 345–346.

Charles, D.R. (2002) Responses to the thermal environment. In: Charles, D.R. and Walker, A.W. (eds) *Poultry Environment Problems, a guide to solutions*. Nottingham University Press, UK, pp.1–16.

Chen, J., Li, X., Balnave, D. and Brake, J. (2005) The influence of dietary sodium chloride, arginine:lysine ratio, and methionine source on apparent ileal digestibility of arginine and lysine in acutely heat-stressed broilers. *Poultry Science* 84, 294–297.

Cheville, N.F. (1979) Environmental factors affecting the immune response of birds – a review. *Avian Diseases* 23, 166–170.

Daghir, N.J. (1973) Energy requirements of laying hens in a semi-arid continental climate. *British Poultry Science* 14, 451–461.

Daghir, N.J. (1987) Nutrient requirements of laying hens under high temperature conditions. *Zootecnica International* 5, 36–39.

Daghir, N.J., Farran, M.T. and Kaysi, S.A. (1985) Phosphorus requirements of laying hens in a semi-arid continental climate. *Poultry Science* 64, 1382–1384.

Dale, N.M. and Fuller, H.L. (1979) Effects of diet composition on feed intake and growth of chicks under heat stress. I. Dietary fat levels. *Poultry Science* 58, 1529–1534.

Deetz, L.E. and Ringrose, R.C. (1976) Effect of heat stress on the potassium requirement of the hen. *Poultry Science* 55, 1765–1769.

Edens, F.W. (1986) Flunixin-induced increased water consumption in broiler chickens before and during heat stress. *Poultry Science* 65, 166–168.

Edens, F.W. and Campbell, D.G. (1985) Reduced heat stress in broilers given flunixin, a non-steroidal cyclooxygenase inhibitor. *Poultry Science* 64 (Suppl. 4), 93 (Abstract).

Edens, F.W. and Siegel, H.S. (1974) Reserpine modification of the blood pH, pCO_2 and pO_2 of chickens in high environmental temperature. *Poultry Science* 53, 279–284.

Freeman, B.M., Manning, A.C.C., Harrison, G.F. and Coates, M.E. (1975) Dietary aureomycin and the responses of the fowl to stressors. *British Poultry Science* 16, 395–404.

Fuller, H.L. and Rendon, M. (1977) Energetic efficiency of different dietary fats for growth of young chicks. *Poultry Science* 56, 549–557.

Garlich, J.D. and McCormick, J.D. (1981) Interrelationships between environmental temperature and nutritional status of chicks. *Federation Proceedings* 40, 73–76.

Geraert, P.A., Guillaumin, S. and Leclercq, B. (1992) Effect of high ambient temperature on growth, body composition and energy metabolism of genetically lean and fat male chickens. *Proceedings of the 19th World's Poultry Congress* 2, 109–110.

Gheisari, A., Sarie, A., Pourreza, J., Khoddami, A. and Gheisari, M. (2004) Effects of dietary fat, α-Tocopherol and ascorbic acid supplementation on the performance and meat oxidative stability of heat stressed broiler chicks. *Proceedings of the 22nd World's Poultry Congress*, Istanbul, Turkey, p. 404.

Glick, B. (1963) Research reports. *Feedstuffs* 35, 14.

Gonzalez-Esquerra, R. and Leeson, S. (2005) Effects of acute versus chronic heat stress on broiler response to dietary protein. *Poultry Science* 84, 1562–1569.

Gonzalez-Esquerra, R. and Leeson, S. (2006) Effect of arginine:lysine ratio and source of methionine on growth and body protein accretion in acutely and chronically heat-stressed broilers. *Poultry Science* 85, 1594–1602.

Gous, R.M. and Morris, T.R. (2005) Nutritional interventions in alleviating the effects of high temperature in broiler production. *World's Poultry Science Journal* 61, 463–475.

Hassan, S.M., Mady, M.E., Cartwright, A.L., Sabri, H.M. and Mobarak, M.S. (2003) Effect of acetyl salicylic acid in drinking water on reproductive performance of Japanese quail (*Cotumix coturnix japonica*). *Poultry Science* 82, 1174–1180.

Hayashi, K., Yoshizaki, R., Ohtsuka, A., Toroda, T. and Tuduki, T. (2004) Effect of ascorbic acid on performance and antibody production in broilers vaccinated against IBD under a hot environment. *Proceedings of the 22nd World's Poultry Congress*, Istanbul, Turkey, p. 550.

Heinzerling, R.H., Nockls, C.F., Quarles, C.L. and Tengardy, R.P. (1974) Protection of chicks against *E. coli* infection by dietary supplementation with vitamin E. *Proceedings of the Society of Experimental Biology and Medicine* 146, 279–282.

Heywang, B.W. (1952) The level of vitamin A in the diet of laying and breeding chickens during hot weather. *Poultry Science* 31, 294–301.

Hooge, D.M., Cummings, K.R. and McNarylton, J.L. (1999) Evaluation of sodium bicarbonate, chloride or sulphate with a coccidostat in corn-soy or corn-soy-meat diets for broiler chickens. *Poultry Science* 78, 1300–1306.

Hurwitz, S., Weiselberg, M., Eisner, U., Bartov, I., Riesenfield, G., Shareit, M., Nir, A. and Bornstein, S. (1980) The energy requirements and performance of growing chickens and turkeys as affected by environmental temperature. *Poultry Science* 59, 2290–2299.

Huston, T.M. (1978) The effect of environmental temperature on potassium concentrations in the blood of the domestic fowl. *Poultry Science* 57, 54–56.

Kafri, I. and Cherry, J.A. (1984) Supplemental ascorbic acid and heat stress in broiler chicks. *Poultry Science* 63, 125–126.

Kampen, M.V. (1984) Physiological responses of poultry to ambient temperature. *Archiv fur Experimented Veterinar Medizin* 38, 384–391.

Klasing, K.C. (1984) Effects of inflammatory agent and interleukin on iron and zinc metabolism. *American Journal of Physiology* 247, R901–R904.

Koelkebeck, K.W., Harrison, P.C. and Parsons, C.M. (1992) Carbonated drinking water for improvement of egg shell quality of laying hens during summer time months. *Journal of Applied Poultry Research* 1, 194–199.

Koelkebeck, K.W., Harrison, P.C. and Mandindou, T. (1993) Effect of carbonated drinking water on production performance and bone characteristics of laying hens exposed to high environmental temperature. *Poultry Science* 72, 1800–1803.

Kreider, E.M., Nelson, S.M. and Harrison, P.C. (1990) Influence of carbonated drinking water on tibia strength of domestic cockerels reared in hot environments. *American Journal of Veterinary Research* 51, 1948–1949.

Kurnick, A.A., Heywang, B.W., Hulett, B.J., Vavich, M.G. and Reid, B.L. (1964) The effect of dietary vitamin A, ambient temperature and rearing location on growth, feed conversion and vitamin liver storage of white Leghorn pullets. *Poultry Science* 43, 1582–1586.

Leeson, S. (1986) Nutritional considerations of poultry during heat stress. *World Poultry Science Journal* 42, 69–81.

Lin, H., Wang, L.F., Sang, J.L., Yie, Y.M. and Yany, R.M. (2002) Effect of dietary supplemental levels of vitamin A on the egg production and immune response of heat-stressed laying hens. *Poultry Science* 81, 458–465.

Lin, H., Jiao, H.C., Buyse, J. and Decuypere, E. (2006) Strategies for preventing heat stress in poultry. *World's Poultry Science Journal* 62, 71–85.

Linchovnikova, M. (2007) The effect of dietary calcium source, concentration of particle size on calcium retention, egg shell quality and overall calcium requirement in laying hens. *British Poultry Science* 48, 71–75.

McDougal, L.R. and McQuiston, T.E. (1980) Mortality from heat stress in broiler chickens influenced by anticoccidial drugs. *Poultry Science* 59, 2421–2423.

McNaughton, J.L. and Reece, F.N. (1984) Response of broiler chickens to dietary energy and lysine levels in a warm environment. *Poultry Science* 63, 1170–1174.

March, B.E. and Biely, J. (1972) The effect of energy supplied from the diet and from environment heat on the response of chicks to different levels of dietary lysine. *Poultry Science* 51, 665–668.

Mardsen, A. and Morris, T.R. (1987) Quantitative review of the effects of environmental temperature on food intake, egg output, and energy balance in laying pullets. *British Poultry Science* 28, 693–704.

Mateos, G.G. and Sell, J.L. (1981) Influence of fat and carbohydrate source on rate of food passage of semi-purified diets for laying hens. *Poultry Science* 60, 2114–2119.

Mateos, G.G., Sell, J.L. and Eastwood, J.A. (1982) Rate of food passage as influenced by level of supplemental fat. *Poultry Science* 61, 94–100.

Maumlautnner, K., Singh, R.A. and Kamphues, J. (1991) Influence of varying vitamin C sources on performance and egg shell quality of layers at varying environmental temperature. *Vitamine und Weitere Zusatzstofft bei Mensch und Tin. Symposium Proceedings*, pp. 266–269.

Mendes, A.A., Watkins, S.E., England, J.K., Saleh, E.A., Waldroup, A.L. and Waldroup, P.W. (1997) Influence of dietary lysine levels and arginine:lysine ratio on performance of broilers exposed to heat or cold stress during the period of three to six weeks of age. *Poultry Science* 76, 472–481.

Miles, R.D. and Harms, R.H. (1982) Relationship between egg specific gravity and plasma phosphorus from hens fed different dietary calcium, phosphorus and sodium levels. *Poultry Science* 61, 175–177.

Miles, R.D., Costa, P.T. and Harms, R.H. (1983) The influence of dietary phosphorus level on laying hen performance, egg shell quality and various blood parameters. *Poultry Science* 62, 1033–1037.

Milligan, J.L. and Winn, P.N. (1964) The influence of temperature and humidity on broiler performance in environmental chambers. *Poultry Science* 43, 817–824.

Mills, C.A., Cottingham, E. and Taylor, E. (1947) The influence of environmental temperature on dietary requirement for thiamine, pyridoxine, nicotinic acid, folic acid and choline in chicks. *American Journal of Physiology* 149, 376–379.

Mongin, B. (1980) Electrolytes in nutrition. *3rd International Minerals Conference*, Orlando, Florida, pp. 1–16.

Moreng, R.E. (1980) Temperature and vitamin requirements of the domestic fowl. *Poultry Science* 59, 782–785.

Musharaf, N.A. and Latshaw, J.D. (1999) Heat increment as affected by protein and amino acid nutrition. *World's Poultry Science Journal* 55, 233–240.

Mushlag, J., Mirza, M.A., Athor, M., Hooge, D.M., Ahmad, T., Ahmad, G., Mushttag, M.M and Noreen, U. (2007) Dietary sodium and chloride for 29 to 42 d old broilers at constant electrolyte balance under subtropical conditions. *Journal of Applied Poultry Research* 16, 161–170.

Mushtag, J., Sarwar, M., Nawaz, H., Mirza, M.A. and Ahmad, T. (2005) Effects of interactions of dietary sodium and chloride on broiler starter performance under summer conditions. *Poultry Science* 84, 1716–1722.

Nakaue, H.S., Weber, C.W. and Reid, B.L. (1967) The influence of ASA on growth and some reparatory enzymes in broiler chicks. *Proceedings of the Society of Experimental Biology and Medicine* 125, 663–664.

National Research Council (NRC) (1981) *Effect of Environment on Nutrient Requirements of Domestic Animals*. National Academy Press, Washington, DC, pp. 109–133.

Njoku, P.C. (1984) The effect of ascorbic acid supplementation on broiler performance in a tropical environment. *Poultry Science* 63 (Suppl.), 156.

Njoku, P.C. (1986) Effect of dietary ascorbic acid supplementation on broiler chickens in a tropical environment. *Animal Feed Science and Technology* 16, 17–24.

Njoku, P.C. and Nwazota, A.O.U. (1989) Effect of dietary inclusion of ascorbic acid and palm oil on the performance of laying hens in a hot tropical environment. *British Poultry Science* 30, 831–840.

Odom, T.W., Harrrison, P.C. and Darre, M.J. (1985) The effects of drinking carbonated water on the egg shell quality of single comb white Leghorn hens exposed to high environmental temperature. *Poultry Science* 64, 594–596.

Pardue, S.L., Thaxton, J.P. and Brake, J. (1984) Effects of dietary ascorbic acid in chicks exposed to high environmental temperature. *Journal of Applied Physiology* 58, 1511–1516.

Pardue, S.L., Thaxton, J.P. and Brake, J. (1985a) Influence of supplemental ascorbic acid on broiler performance following exposure to high environmental temperature. *Poultry Science* 64, 1334–1338.

Pardue, S.L., Thaxton, J.P. and Brake, J. (1985b) Role of ascorbic acid in chicks exposed to high environmental temperature. *Journal of Applied Physiology* 58, 1511–1520.

Peebles, E.D. and Brake, J. (1985) Relationships of dietary ascorbic acid to broiler breeder performance. *Poultry Science* 64, 2041–2048.

Perek, M. and Kendler, J. (1962) Vitamin C supplementation to hen's diet in a hot climate. *Poultry Science* 41, 677–678.

Perek, M. and Kendler, J. (1963) Ascorbic acid as a dietary supplement for white Leghorn hens under conditions of climatic stress. *British Poultry Science* 4, 196–200.

Persia, M.E., Biggs, P.E., Koelkebeck, K.W. and Parsons, C.M. (2002) Effects of heat stress and phosphorus deficiency on egg production and mortality of laying hens. *Poultry Science Poscal 80* (Supplement 1), 118.

Picard, M., Angulo, I., Antoine, H., Bouchot, C. and Sauveur, B. (1987) Some feeding strategies for poultry in hot and humid environments. *Proceedings of the 10th Annual Conference of the Malaysian Society of Animal Production*, pp. 110–116.

Puthgonsiripon, U., Sheideler, S.E., Sell, J.L. and Beck, M.M. (2001) Effects of vitamin E and C supplementation on performance, in-vitro lymphocyte proliferation and antioxidant status of laying hens during heat stress. *Poultry Science* 80, 1190–1200.

Reid, B.L. (1979) Nutrition of laying hens. *Proceedings, Georgia Nutrition Conference*, University of Georgia, Athens, Georgia, pp. 15–18.

Reid, B.L., Kurnick, A.A., Thomas, J.M. and Hulett, B.J. (1964) Effect of ASA and oxytetracycline on the performance of white Leghorn breeders and broiler chicks. *Poultry Science* 43, 880–884.

Rose, S.P. and Uddin, M.S. (1997) Effect of temperature on the response of broiler chickens to dietary lysine balance. *British Poultry Science* 38, 536–537.

Rubner, M. (1902) *The laws of energy consumption in nutrition*. Translated and reprinted in 1968. Clearing House of Federal, Scientific and Technical Information, Springfield, Virginia.

Sahin, K. and Kucuk, O. (2003) Zinc supplementation alleviates heat stress in laying Japanese quail. *Journal of Nutrition* 133, 2808–2811.

Sahin, K., Sahin, N., Onderci, M., Sarkar, F.H., Doerga, D., Prasad, A. and Kucuk, O. (2004) Genistein alleviates heat stress in Japanese quail. *Proceedings of the 22nd World's Poultry Congress*, Istanbul, Turkey, p. 546.

Sahin, K., Smith, M.O., Onderci, M., Sahin, N., Gurser, M.F. and Kucuk, O. (2005) Supplementation of zinc from organic or inorganic source improves performance and antioxidant states of heat-distressed quail. *Poultry Science* 84, 882–887.

Sauveur, B. and Picard, M. (1987) Environmental effects on egg quality. In: Wells, R.G. and Belyavin, C.G. (eds) *Egg Quality: Current Problems and Recent Advances*. Butterworths, London, pp. 219–234.

Scott, M.L. (1966) Factors in modifying the practical vitamin requirements of poultry. *Proceedings, Cornell Nutrition Conference*, pp. 34–35.

Scott, M.L. (1976) Effects of heat on vitamin metabolism. In: Tromps, S.W. (ed.) *Progress in Biometeorology*. Swets and Zeitinger, Amsterdam, pp. 275–282.

Scott, T.A. and Balnave, D. (1988) Influence of dietary energy, nutrient density and environmental temperature on pullet performance in early lay. *British Poultry Science* 29, 155–165.

Shane, S.M. (1988) Factors influencing health and performance of poultry in hot climates. *Critical Reviews in Poultry Biology* 1, 247–267.

Sinurat, A.P. and Balnave, D. (1985) Effect of dietary amino acids and ME on the performance of broilers kept at high temperatures. *British Poultry Science* 26, 117–128.

Sinurat, A.P. and Balnave, D. (1986) Free-choice feeding of broilers at high temperatures. *British Poultry Science* 27, 577–584.

Smith, A.J. and Oliver, J. (1972) Some nutritional problems associated with egg production at high environmental temperatures: the effect of environmental temperature and rationing treatments on the productivity of pullets fed on diets of different energy content. *Rhodesian Journal of Agricultural Research* 10, 3–21.

Smith, J.E. and Borchers, R. (1972) Environmental temperature and the utilization of β-carotene by the rat. *Journal of Nutrition* 102, 1017–1024.

Smith, M.O. and Teeter, R.G. (1987) Potassium balance of the 5–8-week-old broiler exposed to constant heat or cycling high temperature stress and the effects of supplemental potassium chloride on body weight gain and feed efficiency. *Poultry Science* 66, 487–492.

Takahashi, K., Akiba, Y. and Horiguctti, M. (1991) Effects of supplemental ascorbic acid on performance, organ weight and plasma cholesterol concentration in broilers treated with propylthiouracil. *British Poultry Science* 32, 545–554.

Teeter, R.G. and Smith, M.O. (1986) High chronic ambient temperature stress effects on broiler acid-base balance and chloride and potassium carbonate. *Poultry Science* 65, 1777–1781.

Teeter, R.G., Smith, M.O., Owens, F.N. and Arp, S.C. (1985) Chronic heat stress and respiratory alkalosis: occurrence and treatment in broiler chicks. *Poultry Science* 64, 1060–1064.

Tenin, S., Chagneau, A.M., Peresson, R. and Tesseraud, S. (2000) Chronic heat exposure alters protein turnover of three different skeletal muscles in finishing broiler chickens fed 20 or 25% protein diets. *Journal of Nutrition* 130, 813–819.

Thaxton, J.P. (1986) Role of ascorbic acid for relief of stress effects. *Proceedings, Colorado State University 2nd Poultry Symposium on the Impact of Stress*, pp. 53–65.

Thornton, P.A. (1961) Increased environmental temperature influences on ascorbic acid activity in the domestic fowl. *Federation Proceedings* 20, 210A.

Usayran, N., Farran, M.T., Awadallah, H., Al-Hawi, I., Asmar, R. and Ashkarian, K. (2001) Effects of added dietary fat and phosphorus on the performance and egg quality of laying hens subjected to a constant high environmental temperature. *Poultry Science* 80, 1695–1701.

Waldroup, P.W., Mitchell, R.J., Payne, J.R. and Hazen, K.R. (1976) Performance of chicks fed diets formulated to minimize excess levels of essential amino acids. *Poultry Science* 55, 243–253.

Wallis, I.R. and Balnave, D. (1984) The influence of environmental temperature, age, and sex on the digestibility of amino acids in growing broiler chickens. *British Poultry Science* 25, 401–407.

Whitehead, C.C. and Keller, T. (2003) An update on ascorbic acid in poultry. *World's Poultry Science Journal* 59, 161–184.

Whitehead, C.C., Bollengier-Lee, S., Mitchell, M.A. and Williams, P.E. (1998) Vitamin E can alleviate the depressed egg production of heat-stressed laying hens. *British Poultry Science* 39, 544–546.

Wilson, E.K., Pierson, F.W., Hester, P.V., Adams, R.L. and Stadelman, W.J. (1980) The effects of high environmental temperature on feed passage time and performance traits of white Pekin ducks. *Poultry Science* 59, 2322–2330.

Zuprizal, Larbier, M., Chagneau, A.M. and Geraert, P.A. (1993) Influence of ambient temperature on true digestibility of protein and amino acids of rapeseed and soybean meals in broilers. *Poultry Science* 72, 289–295.

7 Feedstuffs Used in Hot Regions

N.J. DAGHIR

Faculty of Agricultural and Food Sciences, American University of Beirut, Lebanon

Introduction	161
Cereals and their by-products	161
Barley (*Hordeum sativum*)	161
Millet (*Setoris* spp.)	164
Rice by-products (*Oryza sativa*)	164
Paddy rice	166
Sorghum (*Sorghum bicolor*)	166
Triticale	168
Protein supplements	169
Coconut meal (Copra)	169
Cottonseed meal (*Gossypium* spp.)	170
Groundnut meal	172
Sunflower seed and meal (*Helianthus annuus*)	173
Safflower meal (*Carthamus tinctorius*)	175
Sesame meal (*Sesamum indicum*)	175
Linseed meal	176
Mustard seed meal	176
Single-cell protein	177
Novel feedstuffs	177
Cassava root meal	177
Dates and date by-products	178
Palm kernel meal	179
Mungbean (*Vigna radiata*)	180
Bread fruit meal	180
Ipil-ipil leaf meal (*Leucaena leucocephala*)	180
Salseed (*Shorea robusta*)	181
Dried poultry waste	181

Buffalo gourd meal (*Cucurbita foetidissima*)	182
Guar meal (*Cyamopsis tetragonaloba*)	182
Bambara groundnut meal (*Voandzeia subterranea*)	183
Jojoba meal (*Simmondsia chinensis*)	183
Velvet beans (*Mucuna* spp.)	184
Conclusions	184
References	186

Introduction

The future development of the poultry industry in many regions of the world depends to a large extent on the availability of feedstuffs in those areas that are suitable or can be made suitable for use in poultry feeds. This is because feed costs constitute about 50–70% of the total cost of producing eggs and poultry meat, and the less a country can depend on imported feeds, the lower feed costs can be. In North America, yellow maize and soybean meal are the two major ingredients used in poultry feed. With the exception of some Latin American countries, such as Brazil and Argentina, these two feed ingredients are not plentiful in most of the hot regions of the world. This has stimulated poultry nutritionists in the hot regions to search for cheaper locally available feedstuffs and to investigate the composition and nutritional value of these feedstuffs. Furthermore, the use of cereal grains for poultry feeds is questionable because cereal grains still form the staple diet of the people in those areas. In many of these countries, the feed industry is owned and directed by multinational corporations, who prefer to rely on known technology rather than investing in the development of new technology appropriate to a developing region. The value of this chapter, therefore, is to point out the ingredients that can be used and are being used for poultry feeding in tropical and hot regions and present research findings on chemical composition and nutritional value of these ingredients. The major sources of nutrients in poultry feeds are the grains and their by-products, oilseed meals, meat-packing by-products, fish meal, several other agricultural and industrial by-products and a host of feed additives. This chapter covers three groups of feedstuffs: the cereals and their by-products, protein supplements and novel feedstuffs. The first two groups are those that are commonly used in hot regions, and the novel feedstuffs group consists of those that are less commonly used but have potential for greater use in these regions. It is hoped that collecting these data on feedstuffs' composition and nutritional value will help countries in these regions to take full advantage of available knowledge and will stimulate further research in this field.

Cereals and their By-products

Barley (*Hordeum sativum*)

Barley is one of the most extensively cultivated cereal grains in the world. The annual world production is over 135 million tonnes (FAOSTAT, 2005).

It is of significance in hot, dry regions where annual precipitation is not sufficient for the production of other grains, such as maize and wheat. It is the most commonly raised feed grain in Europe, where it is grown on relatively poorer soil than that used for other grains. The annual production of barley in Europe is over 82 thousand tonnes (FAOSTAT, 2005). Historically, barley has been cultivated in the eastern Mediterranean regions for over 12,000 years and has been used for both human and animal consumption. New varieties of barley that are adapted to local cultivation practices and environmental conditions are being constantly selected (ICARDA, 2004). Barley types include spring and winter varieties, with two groups predominating: the six-rowed and the two-rowed barleys. The mature grain contains a hull which encloses the kernel. In most varieties, the hulls are cemented to the kernel and are carried through the threshing process. In the naked varieties, the kernel threshes free. The hull forms about 10–14% of the weight of the grain.

Barley is considered a low-energy feed grain due to its high fibre content, and the metabolizable energy (ME) value is about 12.5 MJ per kg dry matter. Therefore, in diets requiring high energy levels, such as those for broilers, the amount of barley used is minimal. The crude protein content of barley grain ranges from about 6 to 16% on a dry matter basis. Therefore many varieties of barley are higher in protein and amino acid than maize (National Research Council, 1994). The lipid content of this grain is low, usually less than 2.5%. Crude fibre content of barley is about 5–6%, with higher levels occurring if grown in arid regions. Barley contains anti-nutrient compounds called β-glucans. These are non-starch polysaccharides made up of glucose units and found in the endosperm cell wall. They are considered to be responsible for the sticky droppings and reduced feed utilization and growth rate in growing chickens fed high levels of barley (Fuente *et al.*, 1995). Acen and Varum (1987) compared the β-glucan levels in ten Scandinavian varieties of barley and found a variation in mean values for 2 years from 3.08 to 4.83%. There were more β-glucans in two-rowed barley than in six-rowed barley. The β-glucan content of barley increases with stage of ripeness (Hasselman *et al.*, 1981).

The use of enzyme mixtures containing mainly β-glucanase is now common in barley-based poultry rations, especially broiler rations, resulting in the breakdown of the cell-wall-soluble β-glucans, decreasing digesta viscosity and improving nutrient digestibility and thus performance in birds (Bedford, 1996; Esteve-Garcia *et al.*, 1997; Scott *et al.*, 1999).

Research workers have for some time been investigating not only factors that influence the value of this grain as a feed but also the possible methods of improving its nutritional value. Jeroch and Danicke (1995) reviewed work on barley including its chemical composition, feeding value for broilers and layers, and treatments for improving its nutritional value. Both water treatment and enzyme supplementation have been shown to improve the nutritional value of barley. The effectiveness of such treatment was shown to be dependent on geophysical area and weather conditions in which barley was grown. Daghir and Rottensten (1966) reported on experiments conducted to study the influence of variety and enzyme supplementation on the nutritive value of barleys grown in the same location and receiving similar cultural treatments. Enzyme supplementation significantly improved growth of chicks in all varieties tested and had a more marked effect on growth than the variety used. In general, barley grown under arid conditions responds more positively to the enzyme and water treatments than that grown in more humid areas. Figures of ME content given by Allen (1990) show that barley from the dry Pacific coast of the USA is lower than that of Midwestern barley. Herstad (1987) in a study on broilers and laying hens showed that supplementation of broiler diets with β-glucanase improved litter conditions and overall performance. Barbour et al. (2006) showed that enzyme supplementation to diets containing different barley cultivars resulted in an average AME_n improvement of 11%. In laying hens, using barley at 73% of the diet did not reduce egg production or egg weight but had a slight negative effect on feed efficiency. Coon et al. (1988) fed two barley cultivars (Morex and Glenn) and a high-protein, feed-grade barley (12.6% crude protein (CP)) to laying hens at different dietary levels. They concluded that the use of barley can be beneficial for regulating egg size and minimizing body-weight gains in post-peak layers if barley is priced low enough to offset resulting increased feed consumption and lower feed utilization.

Barley varieties vary considerably in proportion of hull. Hull-less barley has an energy value similar to that of wheat, but yield of this variety is much lower per acre than the hulled varieties. Some researchers reported no difference in performance of chicks fed hull-less or hulled barley (Anderson et al., 1961; Coon et al., 1979). Newman and Newman (1988) evaluated the hull-less barley mutant Franubet in broiler feeding trials and found that this variety supported chick growth equal to that of maize and without the problem of wet, sticky droppings. The authors suggested that the superior value of this barley cultivar may be due to absence of anti-nutritional factors, the presence of a different form of β-glucans and the character of its unique starch, which could affect energy availability.

Rotter et al. (1990) studied the influence of enzyme supplementation on the bioavailable energy of a hull-less (Scout) barley. Barley replaced up to 75% of the wheat in the diet. The available energy for young chicks increased significantly owing to enzyme supplementation as the barley component of the diet increased. No significant increase in ME was observed when the same diets were fed to adult roosters. They suggested that the mature bird has a more developed gastrointestinal tract, which can neutralize the negative effects of the β-glucans.

Classen et al. (1991), in a review on the use of feed enzymes, concluded that the efficacy of these enzyme products has been well demonstrated in both university and on-farm trials. These workers believed that enzyme supplementation would become an invaluable tool within the feed industry. This prediction has come about and enzymes are now common supplements in poultry feeds.

Barley has to be checked for moisture, which should not exceed 13%, and crude protein, which can vary from 7 to 12%. Crude fibre in barley analyses at about 5 to 6%. The texture of a mash feed containing more than 20% barley may adversely affect palatability and feed consumption. Pelleting will improve palatability and will enable use of higher levels of this grain.

Millet (*Setoris* spp.)

The name millet is applied to several species of cereals which are widely cultivated in the tropics and warm temperate regions of the world. India, Nigeria, China and Burkina Faso are the top producers of millet in the world. These four countries combined produce is more than 24 million tonnes per year. FAOSTAT (2005) show that 50% of the total millet grain production is pearl millet, 30% proso/golden and foxtail millet, and 10% finger millet. The remaining 10% are of different minor species. Millet has a chemical composition and ME content very similar to those of sorghum. Its feeding value, however, is more like that of barley and oats.

The composition of millet is very variable: the protein content is generally within the range of 10–12%, the ether extract 2–5% and the crude fibre 2–9%. Few studies have been published on the feeding value of millet for chickens. Sharma *et al.* (1979) found that broilers fed on millet competed favourably with those on maize, wheat and sorghum. Abate and Gomez (1983) also reported similar results. Karmajeewa and Than (1984) fed pullets a protein concentrate plus wheat, millet or paddy rice as crushed or whole grain from 8 to 20 weeks. Pullets fed on millet had higher linoleic acid content in their livers and laid larger eggs than those reared on wheat. They concluded that replacement pullets can be grown successfully on whole wheat or millet and a protein concentrate offered on a free-choice basis with no adverse effects on their subsequent laying performance. Millet has higher linoleic acid content than wheat and when included in growing diets would improve egg weight in the subsequent laying phase.

Pearl millet (*Pennisetum typhoides*), also known in India and Pakistan as bajra, has been investigated for use in laying-hen diets by several workers. Chawla *et al.* (1987) reported that it can be used at 20% of the diet, although Reddy and Reddy (1970) were able to incorporate as much as 32% into the diet without detrimental effects on production. Mohan Ravi Kumar *et al.* (1991) found that the inclusion of pearl millet at 60% in laying-hen diets made isocaloric and isonitrogenous with maize-based diets did not influence hen-day egg production, feed intake, feed efficiency or body weight. Such high levels of pearl millet in laying-hen diets were possible because these workers had incorporated fairly high levels of fish meal in those diets.

Rice by-products (*Oryza sativa*)

Rice is second to wheat in worldwide production. The top three producers of rice in the world are China, India and Indonesia. World production in 2005 was nearly 610 million tonnes, 90% of which is produced and consumed in Asia. Rice is harvested from the field in the form of paddy, which is the complete rice seeds. Each grain of paddy contains one rice kernel and many other layers. The outer layer is the husk, which consists mostly of silica and cellulose. The next layer is a thin film of bran. This consists of fibre, B-vitamins, protein and fat, and is the most nutritional part of the rice. At the base of the grain is the embryo. The remaining large layer consists mainly of starch. Only when it is produced in abundance is rice incorporated in sizeable amounts in poultry feed. In the processing of whole rice, the first step is removal of the husks. Once the husks have been removed, the rice becomes 'brown rice'. The second step is removal of the bran, which yields white rice. White rice is usually further processed or polished and the residues are called rice polishings. Basically, there are two rice by-products used in poultry feeding. These are rice bran and rice polishings. Because many of the hot and tropical regions of the world are rice-producing areas, huge quantities of these two by-products, as well as other rice by-products, are produced in those regions. Over 50 million tonnes of rice bran are produced each year. Indonesia is one of the biggest users of rice bran in poultry feeds. Broiler feeds range from 0 to 10% in rice bran, while layer and breeder feeds may contain 10–30%. Rice bran is recognized as being variable in its composition, depending on the severity with which the rice is threshed and the extent to which the oil is extracted. McNab (1987) looked at 14 samples of rice bran and found them to vary from 2 to 20% in oil and from 7.17 to 12.91 kJ/g in energy. Protein content did not vary much and was in the range of 13–14%. Panda (1970) reported on rice bran produced in India and varying from 1 to 14% in oil content and 13 to 14% in crude protein. The oil is highly unsaturated and may become rancid very quickly. It is therefore removed to produce a product with better keeping quality. Torki and Falahati (2006) fed rice bran from Iran in wheat–soybean diets to broilers with and without phytase. They found that levels of rice bran up to 22.5% have no adverse effects, and phytase improved feed conversion. Farrell (1994), using Australian rice bran, observed that this ingredient can be used at 10, 20 and 25% in broiler starter, finisher, and layer diets respectively. Attia et al. (2003) studied the feeding value of rice bran and found that it can be improved by phytase and other enzyme supplementation.

Feltwell and Fox (1978) reported that the feeding value of rice polishings depends upon the degree of polishing to which the grain has been subjected. It is unlikely to contain much of the rice flour. Panda (1970) gives a figure of 11–12% crude protein and 12–13% oil for rice polishings produced in India, which is close to Scott et al. (1982), who list this product as having 12% crude protein and 12% oil. Broken rice and rice polishings are produced in fairly large amounts in Asian countries. Pakistan, for example, produces about 800,000 tonnes of broken rice and 300,000 tonnes of rice polishings per year. About 25% of the broken rice and 80% of the rice polishings are used by the poultry-feed industry (Razool and Athar, 2007). Ali and Leeson (1995) reported that the AME_n of rice polishing was 13.96 MJ/kg and that levels of 20% in a maize–soybean ration reduce body-weight gain.

Both rice bran and rice polishings can be used in poultry rations at fairly high levels if they are low in rice hulls and if the high oil level in them can be stabilized by an antioxidant so that much of the energy value will not be lost through oxidative degradation. Hussein and Kratzer (1982) demonstrated the detrimental effects of rancid rice bran on the performance of chickens. Cabel and Waldroup (1989) tested the effects of adding an antioxidant or a metal chelator or both on the nutritive quality of rice bran stored at high temperature (35–38°C) and high humidity (80–90%). The addition of 250 p.p.m. ethoxyquin proved effective in significantly reducing the initial peroxide value, and 1000 p.p.m. ethylenediamine tetra-acetic acid (EDTA) significantly lowered the 20 h active oxygen method value. They therefore concluded that rancidity development in stored rice bran can be slowed by the addition of ethoxyquin or EDTA.

Because the production of edible oil from rice bran can provide a much-needed source of energy for people in rice-producing areas, Randall et al. (1985) developed a process to stabilize the bran and prevent enzymatic hydrolysis of its oil. This stabilization process heats the freshly milled bran in an extrusion cooker to 130°C. The hot extruded bran is maintained at near 100°C on an insulated holding belt for 3 min prior to cooling in an ambient airstream. The processed rice bran is in the physical form of small, free-flowing flakes with a low microbiological load. Lipolytic enzymes are permanently inactivated and the extracted bran residue retains the flake form, thereby reducing dust and fines. Sayre et al. (1987) conducted feeding trials with broilers to compare this stabilized rice bran with the raw bran. Free fatty acid (FFA) content in oil from raw rice bran stored at elevated temperatures (32°C) reached 81% whereas FFA in stabilized bran oil remained at about 3%. Chickens fed stabilized rice bran made significantly greater gains than those fed raw bran diets. Farrell and Martin (1998) reported that adding food phytase to rice-bran diets improved their nutritional value even though these diets were adequate in available phosphorus, which may indicate that phytase helped in improving availability of other minerals. It should be noted here that rice bran contains 1.6–1.8% phosphorus, all as phytic phosphorus.

Paddy rice

In the process of dehulling and milling of paddy rice, a huge quantity of broken grains, popularly known as rice kani, is obtained as a by-product. Verma et al. (1992) examined the usefulness of this by-product as a substitute for maize. Graded amounts of rice kani were substituted for yellow maize up to 40% in isonitrogenous diets for broilers. No significant differences were observed in body-weight gain, feed intake or feed efficiency. The ME value obtained for rice kani was reported by these workers to be 13.3 MJ/kg. The protein content of broken rice is relatively low, about 7%. The authors concluded that rice kani is a good substitute for maize in broiler diets.

Sorghum (*Sorghum bicolor*)

Sorghum is one of the most important feed and food crops in the arid and semi-arid tropics (Hulse et al., 1980). The leading countries in the production of sorghum are India, Nigeria, Mexico, Sudan and the USA. These five countries alone produce more than 38 million tonnes per year. It is the main food grain in Africa and parts of India and China. Sorghum is extensively used in poultry feeding in many countries of the world. One of the major limitations for its use in poultry feeds is its relatively high tannin content. The detrimental effects of high-tannin sorghums on the growth and feed efficiency of the growing chicken are well documented (Chang and Fuller, 1964; Armstrong *et al.*, 1973). Vohra *et al.* (1966) showed that as little as 0.5% tannic acid will depress growth. The question has always been raised as to whether naturally occurring sorghum tannins are as toxic as tannic acid. Dale *et al.* (1980) observed a growth depression regardless of the source of the tannin, but a higher sorghum tannin content was necessary to cause growth depression equivalent to commercial tannic acid. Supplementation of high-tannin sorghum diets with methionine or choline and methionine alleviated the growth depression. Elkin *et al.* (1978) found that addition of 0.15% methionine to a high-tannin sorghum–soybean meal diet brought the growth of broilers up to that obtained with a similar diet containing a low-tannin sorghum grain. Douglas *et al.* (1990a) studied the composition of low- and high-tannin sorghums and found that they only had minor differences in their protein and amino acid contents. Other methods for improving the nutritional value of high-tannin sorghum are the addition of fats and the adequate grinding of the grain (Douglas *et al.*, 1990b). Nir *et al.* (1990) concluded that the main effect of grinding is to improve feed utilization, which is accomplished by increasing the surface area of the grain relative to its reduced particle size.

There are several causes for the growth-depressing and toxic effects of tannins. Tannins affect the palatability of diets and thus reduce feed intake, but this is not a major factor in growth depression caused by high-tannin diets (Vohra *et al.*, 1966). Nyachoti *et al.* (1997) reviewed the literature on sorghum tannins and methods of detoxification. These workers believe that the poor performance observed from feeding sorghum may not all be due to tannin level but also to other factors. Garcia *et al.* (2004) observed that the digestibility of rations containing sorghum, with or without tannins, differs with different temperatures that broilers are subjected to. Digestibility coefficients were higher at 32°C versus 25°C. Nelson *et al.* (1975) reported a high variability in the metabolizable energy values of sorghum grains. These workers observed that both metabolizable energy and amino acid availability increased as the tannic acid content of the sorghum grains decreased. Douglas *et al.* (1990a) also showed that the ME content of high-tannin sorghum was lower than that of the low-tannin sorghum varieties. Furthermore, the high-tannin variety contained higher levels of both acid detergent fibre and neutral detergent fibre than the low-tannin variety.

Besides the negative effects of tannins in sorghum grains, the presence of phytates has also been shown to reduce growth and increase incidence of locomotor disorders (Luis *et al.*, 1982; El-Alaily *et al.*, 1985; Mohammadain *et al.*, 1986). Ibrahim *et al.* (1988) were able to improve the nutritional quality of Egyptian and Sudanese sorghum grains by the addition of phosphates. These authors suggested that treatment of sorghum by dry-mixing with dicalcium phosphate could extend the use of high-tannin sorghum in poultry feeds.

Few studies have been conducted to determine the effects of sorghum tannin on the performance of the laying hen. Supplementation of tannic acid to a laying-hen diet reduced feed consumption, egg production and egg weight and resulted in abnormal colouring and mottling of the yolk (Armanious et al., 1973). Bonino et al. (1980) conducted a 32-week study in which high-tannin sorghum was found to cause a small but significant decrease in egg production in comparison with low-tannin sorghum-fed controls. Sell et al. (1983) observed that tannin significantly reduced egg production and feed efficiency but had no effect on egg weight or body weight. At the end of the experiment, all hens were placed on a commercial laying ration for a 31-day period and recovery was complete by the end of this period.

In general, low-tannin sorghums are nearly equal to maize when fed to broilers. High-tannin sorghums, however, are lower in energy than maize, and therefore when sorghum grains are used in place of maize in broiler and layer diets xanthophyll supplement and an additional amount of fat are needed in the ration. Finally, sorghum grains must be adequately ground to ensure maximum utilization.

Triticale

Triticale is a synthetic, small grain crop resulting from the intergeneric cross between durum wheat and rye. Its name is derived from a combination of the two generic terms of the parent cereals (*Triticum* and *Secale*). The objectives of crossing these two cereals were to produce a grain having the high quality and productivity of wheat and the hardiness of rye. Triticale is now grown commercially in many countries of the world. Its composition is variable, with crude protein ranging from 11 to 18%. It has been grown in several hot regions of the world and found to be as high-yielding as wheat.

The major triticale-producing countries are China, Poland, Germany, France, Australia, Belgium and Hungary. The FAO estimate of the area grown all over the world is in excess of 3,000,000 ha, 90% of which are in Europe. FAOSTAT (2005) give a total world production of 13.5 million tonnes.

The use of triticale for broilers has been tested by several workers. Sell et al. (1962) fed chicks increasing levels of triticale at the expense of wheat in a basal diet and concluded that methionine and lysine were limiting for chick growth when the diets containing the triticale were isonitrogenous with the basal wheat diet. Fernandez and McGinnis (1974) compared triticale with maize and provided evidence that chicks grown on diets with 53 or 73% triticale were lighter in weight than those on a maize diet.

Wilson and McNab (1975) also found that broiler weights at 56 days were depressed when triticale was substituted for maize at a 50% level. Proudfoot and Hulan (1988) concluded that triticale should not be used over 15% in broiler diets. Bragg and Sharby (1970), on the other hand, found that triticale could partially or totally replace wheat in broiler diets without adverse effects on growth or feed efficiency. Rao *et al.* (1976) were also able to replace 75% of the maize in broiler diets with no adverse effect on weight gain or feed efficiency. If included in the diet in finely ground form, triticale impairs feed intake. Therefore, part of the negative effect observed by the above researchers may have been due to differences in the grinding of the grain.

Ruiz *et al.* (1987), in studying the nutritive value of triticale (Beagle 82) as compared with wheat, found that broiler growth response was inconsistent with triticale fed at levels of 25 and 50% of the diet. These workers indicated that variations in chemical composition of different triticale cultivars may account for the variation in results from studies conducted to evaluate the nutritive value of triticale. Daghir and Nathanael (1974) studied the composition of triticale (Armadillo 107) as compared with wheat (Mexipak) and found that the proximate analysis of these two grains was practically identical. In mineral composition, they were also similar, except in iron, which was lower in triticale than in wheat. ME values as determined by these authors with both chicks and laying hens were found to be about 5–6% lower in triticale than in wheat.

In studies with broilers and laying hens, Jerock (1987) concluded that, because of the great variability among varieties of triticale, it is necessary to limit the level of usage to 30 and 50% in the rations of broilers and laying hens, respectively. Flores *et al.* (1992) confirmed these varietal differences in triticale and added that enzyme supplementation could be a tool to improve some of these poor varieties. More recent studies indicate that triticale compares favourably with wheat in terms of feeding value (Briggs, 2001). Leeson and Summers (1987) observed a decrease in laying-hen performance when triticale was used at 70% level in isocaloric diets, compared with maize. Cuca and Avila (1973) were able to use 50% triticale in a milo-based diet with no detrimental effect on laying-hen performance. Luckbert *et al.* (1988) showed that triticale can be used up to 40% in diets for laying hens.

Castanon *et al.* (1990) studied the effect of high inclusion levels of triticale in diets for laying hens containing 30% field beans. Partial replacement of maize by triticale did not affect feed consumption, laying rate or feed-to-gain ratio. Mean egg weight increased slightly with increasing levels of triticale, while yolk colour decreased with increases in triticale.

The above studies indicate that triticale can partially replace maize and/or wheat in broiler and layer rations. The level of replacement depends on the cultivar being used and the chemical composition of that cultivar.

Furthermore, higher levels can be used in laying than in broiler rations. When used at high levels, it may be advisable to supplement the ration with synthetic lysine.

Protein Supplements

Coconut meal (Copra)

Coconut meal is produced from copra, which comes from dried coconuts. It is also known as copra meal. The use of coconut meal in poultry diets is somewhat limited because of its high fibre content, averaging about 12–14%. The average protein content of this meal (solvent processed) is about 20–22%. Light-coloured meals are usually better in quality than dark-coloured ones. The meal protein is low in the amino acids lysine and histidine (McDonald *et al.*, 1995). The oil content of this meal varies from 2.5 to 7.5% and, in areas where energy sources are scarce, it can be useful for the preparation of high-energy diets. Such diets, however, have the disadvantage of being susceptible to becoming rancid during storage unless they are properly treated with an antioxidant. Panigrahi *et al.* (1987) reported that coconut meals from small-scale screw-press expellers generally have a high lipid content and produce good growth in broiler chicks when fed at high levels of the ration. These workers, however, found that these high-coconut meal diets induced inquisitive and excited behaviour and increased feed spillage and water intake. Increasing the energy content of the diet by the addition of maize oil improved efficiency of food utilization, food intake and growth rate and reduced abnormal behaviour (Panigrahi, 1992). The maximum level of coconut meal used in poultry diets should not exceed 10% of the diet. When used at such levels, amino acid supplementation may be necessary.

Cottonseed meal (*Gossypium* spp.)

Early work on cottonseed meal (CSM) composition and nutritive value has been reviewed by Phelps (1966). Since that date, a great deal has been written on CSM and on methods of improving its nutritional value for poultry. Decorticated good-quality CSM has become a very useful source of supplementary protein for poultry in many regions of the world, particularly the cotton-producing areas, and the most recent review has been published by Nagalakshmi *et al.* (2007).

Although CSM is a rich source of protein, it is low in lysine and methionine compared with soybean meal. Its use in poultry rations has been limited because of certain undesirable characteristics. It contains gossypol, a polyphenolic pigment found in the pigment glands of most varieties of CSM. Gossypol usually exists as a mixture of two enantiomers that exhibit different biological activities. These exist in positive (+) or negative (–) forms and their levels and ratio could be altered by developing new genetic strains of cotton. Lordelo *et al.* (2005) reported that both gossypol enantiomers are toxic to broilers but that (–) – gossypol was more harmful than (+) – gossypol. Gossypol has been known for many years to depress growth, feed intake, feed efficiency and hatchability. Additional deleterious effects that have been reported include enlarged gall-bladders, and blood and bone marrow changes, including a reduction of haemoglobin, red cell count and serum proteins. Other symptoms reported were oedema in body cavities, degenerative changes in liver and spleen, haemorrhages of liver, hypoprothrombinaemia and diarrhoea. The anaemia resulting from gossypol toxicity is due to its property of binding iron. Gossypol is present in either the bound or free state. Our concern is mainly with free gossypol because the bound form is non-toxic to animals. The amount of free gossypol in CSM depends to a large extent on the type of processing, i.e. screw-press, prepress or solvent extraction (Jones, 1981). Although screw-pressing produces a meal that is lowest in free gossypol, prepress solvent extraction is the method in common use. Waldroup (1981) suggested that 100 p.p.m. of free gossypol in broiler diets is considered acceptable on the basis of growth and feed efficiency. The presence of gossypol in poultry diets may be counteracted by the addition of iron salts, which bind the gossypol. Scott et al. (1982) suggested that gossypol was inactivated or chelated by iron in a 1:1 molar ratio. This has been confirmed by El-Boushy and Raternick (1989), who indicated that free gossypol can be chelated by added iron or iron available naturally in the feed. These workers also cautioned against using high levels of iron because this can have a negative effect on body-weight gains of broilers.

Another method of improving CSM was reported by Rojas and Scott (1969) and consists of treatment of the meal with phytase produced by *Aspergillus ficcum*. The hydrolysis of phytin in the CSM released the phosphorus for utilization and may have freed some proteins from protein–phytate complexes according to these workers.

The hull-less 50% protein CSM available in some areas of the world has given excellent results when fed to broilers. This is particularly true if the gossypol content is minimized. This type of CSM has an energy value close to that of soybean meal. Plant breeders have developed cultivars that are practically free of gossypol. The development of these glandless varieties containing low gossypol levels has increased the use of this meal in poultry rations. Unfortunately, however, these cultivars are more susceptible to insect infestation and do not yield as much as the gossypol-rich varieties.

Another problem with CSM is the cyclopropene fatty acids. The cyclopropene fatty acids, malvalic and sterculic acids, existing in cottonseed oils have been shown to cause a pink discoloration of egg white (Phelps et al., 1965). The level of these acids in the crude oil ranges from 0.6 to 1.2% (Bailey et al., 1966). The level of these acids in the meal depends upon the amount of the residual oil but usually averages about 0.01% in meals obtained from commercial processors (Levi et al., 1967). These fatty acids also cause a greater deposition of stearic and palmitic acid in depot fat. Thus egg and body fat of hens consuming cottonseed oil has a higher proportion of stearic acid than that found when other fats are fed. Besides the effects of the cyclopropene fatty acids on pink discoloration of egg whites, free gossypol also causes yolk discoloration. Heywang et al. (1955) reported that as little as 10 p.p.m. free gossypol in the laying-hen diet produced discoloured yolks. Panigrahi and Hammonds (1990) reviewed the literature on the effects of screw-press CSM on eggs produced. Recently, Lordelo et al. (2007) reported that the ingestion of (+) – gossypol has a greater effect on egg yolk discoloration than the consumption of (–) – gossypol.

The appearance of aflatoxin in CSM has prompted some workers to use ammonia treatment, which effectively alters the aflatoxins without affecting the performance of laying hens (Vohra et al., 1975). Isopropanol extraction, which has been used for aflatoxin removal, also reduces the free gossypol and residual oil levels in the extracted CSM. Reid et al. (1987) reported that isopropanol-treated CSM can be safely utilized in laying-hen diets at levels up to 15% without any detrimental effects on either performance or yolk colour and quality.

Panigrahi et al. (1989) studied the effect of feeding a screw-press-expelled CSM to laying hens at dietary concentrations of up to 30% of the diet. The overall performance of hens fed 7–5% CSM was not significantly different from that of controls, but a 30% CSM diet, containing 255 g free gossypol/kg and 87 mg cyclopropenoid fatty acids, significantly reduced feed intake and egg production. The 15% CSM diet did not produce adverse effects initially, but egg production was depressed towards the end of the production cycle. Treatment of 30% CSM diet with a solution of ferrous sulfate heptahydrate (100 mg supplemented dietary iron per kg) further reduced feed intake and egg production. Storage of eggs at room temperature for up to 1 month did not lead to discoloration of any kind in the CSM diet groups but resulted in yolk mottling. Storage of eggs at cold temperatures for 3 months resulted in brown yolk discoloration and the initial stages of pink albumen discoloration when the 30% CSM diet was fed. The brown yolk discoloration was reduced by treatment of the CSM with iron. Panigrahi and Plumb (1996), however, stated that iron supplementation is costly and contributes to heavy metal content of faeces. It can also depress bird performance by reducing dietary P availability.

Panigrahi et al. (1989) showed variability in the responses of different flocks of hens to dietary CSM. In a later study, Panigrahi and Morris (1991) examined the effects of genotype of the hen on the depression of egg production and discolorations in eggs resulting from dietary screw-pressed CSM. They reported that the strongest interaction between breed and diet occurred with food intake. The susceptibility of eggs to the cyclopropenoid fatty acid-related cold storage effects also depended on the genotype of the hen.

Ravindran and Blair (1992) believe that cottonseed meal if supplemented with lysine can replace 40% of the protein from soybean meal in broiler diets, but its use for layers is very limited because of the cyclopropenoid fatty acids. Perez-Maldonado et al. (2003) reported that 20% Australian solvent-extracted cottonseed meal can be used in broiler starter rations and 30% in finisher rations with no effect on performance when diets are formulated on a digestible amino acid basis.

Before use in poultry feeds, CSM should be checked for gossypol content, cyclopropenoid fatty acids, fibre and protein.

Groundnut meal

Groundnut meal is also known as peanut meal, *Arachis* nut, earth-nut and monkey-nut meal. It is used extensively in poultry feeds in many hot and tropical regions. Generally, groundnut meal should not be used as the major source of protein unless the diet is supplemented with the essential amino acids, lysine and methionine. Furthermore, it is one of those feedstuffs that is very susceptible to contamination by *Aspergillus flavus*, which produces the group of toxins known as aflatoxins.

The early literature on the use of groundnut meal as a feedstuff for poultry was reviewed by Rosen (1958). Daghir *et al.* (1969) evaluated groundnut meals produced in the Middle East by using animal assays as well as several types of chemical assays. In all meals, methionine and lysine were the first and second limiting amino acids, respectively. In an earlier study, Daghir *et al.* (1966) reported that in all-plant diets, in which groundnut meal served as the principal source of protein, lysine was more limiting than methionine. This amino acid limitation of groundnut meal has been confirmed by several workers (Anderson and Warnick, 1965; Mezoui and Bird, 1984; El-Boushy and Raternick, 1989). Groundnut meal can be used in broiler diets if adequate levels of dietary lysine and methionine are present (Zhang and Parsons, 1996). Costa *et al.* (2001) compared the performance of broilers fed groundnut meal or soybean meal as protein sources. They observed that maize–groundnut meal diets supplemented with lysine, methionine and threonine could be used in broiler diets up to 20%. Pesti *et al.* (2003) compared maize–groundnut meal diets supplemented with threonine to maize–soybean diets for laying hens at protein levels of 16, 18.5 and 21%. They concluded groundnut meal was as good as soybean at all protein levels. Groundnut meal quality has also been shown to be improved by the addition of phytase to the diet. This enzyme increased the AME_n of groundnut meal by 9% (Driver *et al.*, 2006).

Sunflower seed and meal (*Helianthus annuus*)

Sunflower seed production for the primary purpose of producing oil is significant in many countries in the hot regions of the world as well as in temperate regions. In fact, sunflower seed production in the USA is second only to soybeans among the oilseeds. Sunflower meal (SFM) is, therefore, available in large quantities for use in animal feeds. The CP and fibre contents of this meal vary from 29 to 45% and from 14 to 32% respectively, depending on processing (Villamide and San Juan, 1998).

Morrison *et al.* (1953) showed that one reason for the variability in nutrient value of SFM found by earlier workers was the relatively high temperature used in their processing. An excellent review of the value of SFM for poultry was published by McNaughton and Deaton (1981). Klain *et al.* (1956) were the first to report that the protein of SFM was deficient in lysine for chicks. Rad and Keshavarz (1976) and Valdivie *et al.* (1982) used isocaloric diets and showed that SFM could be used in lysine-supplemented diets at about 15–20% without adversely affecting chick performance. Gippert *et al.* (1988) reported that Hungarian SFM can replace 25, 50 and 75% of the soybean meal in a starter, grower and finisher broiler ration, respectively, without any detrimental effect on body weight or feed conversion. Zatari and Sell (1990) reported that up to 10% SFM could be used in diets adequate in lysine and containing 6% animal–vegetable fat without adversely affecting growth or feed efficiency of broilers to 7 weeks of age. Senkoylu and Dale (1999) reviewed the literature on sunflower meal and described the nutritional characteristics of the meal in comparison with other meals, pointing out that one of its good traits is that it does not have the anti-nutritional factors that are found in soybean, cottonseed or rapeseed. Mushtaq et al. (2006) studied the effect of adding a multi-enzyme preparation on sunflower meal (20 and 30% of diet) and three levels of digestible lysine (0.8, 0.9, 1.0%). A level of 1.0% digestible lysine was best for body-weight gain and feed/gain. Enzyme supplementation had no significant effect.

There is very little information on the feeding value of full-fat raw or full-fat treated sunflower seed for poultry. A small amount of sunflower seed (1–2%) has been reported to have been included in commercial mixed scratch grains for poultry in the past because of the distinct and attractive appearance of the seeds (Morrison, 1956). The seeds contain 25–32% oil, about 16% protein and 12–28% fibre. The reason for this wide variation in oil is variety, soil and climatic conditions, and fibre content variation depends on variety and the extent to which the seed was cleaned prior to its analysis. Daghir et al. (1980b) tested the use of sunflower seed as full-fat raw or full-fat steamed or heated as a source of energy and protein for broilers. They concluded that, in practical-type broiler rations, sunflower seed fed in the full-fat raw form can constitute at least 10% of the ration without any adverse effect on performance. Cheva-Isarakul and Tagtaweewipat (1991) studied the utilization of sunflower seed fed at levels of 0–50% of the diet to broilers up to 7 weeks of age. These authors concluded that sunflower seed is a good source of crude protein and ME in broiler diets. Lysine supplementation improved diets containing high levels of sunflower seed. They recommended that, to avoid plugging due to high fat content, sunflower seed should be milled in two steps, the first through a small mesh-size sieve and the second through a sieve with a larger mesh. Also, to avoid rancidity, seeds should be ground just before feed mixing and the mixed diet kept no longer than 2 weeks. Rodriguez et al. (2005) reported that the inclusion of high oleic acid sunflower seed decreased digestibility of fat and fatty acids in broilers. They also observed reduced digestibility of certain amino acids and AME_n. Body weights were reduced at all levels ranging from 8 to 32% of the diet.

Few studies have dealt with the value of sunflower seed or meal for laying hens. The early literature was reviewed by Rose et al. (1972), who evaluated the replacement of soybean meal with sunflower seed meal in maize–soybean meal rations for laying hens. Sunflower seed meal replaced 50% of the soybean meal protein without adversely affecting hen performance. Lysine supplementation did not consistently improve utilization of the diets used. These workers reported that characteristic eggshell stains, which develop after egg laying when sunflower seed meal is used in mash rations, were markedly reduced when similar rations were fed in crumble form. They suggested that chlorogenic acid present in sunflower seeds is responsible for these stains. ME values of 2205 and 2139 kcal/kg were reported by these workers for the two sunflower seed meals used in their study. Uwayjan *et al.* (1983) evaluated unprocessed whole sunflower seed as a partial replacement for soybean meal and yellow maize in laying rations. Sunflower seed used at 10, 20, or 30% in the ration had no detrimental effect on performance. Neither lysine nor lysine plus methionine supplementation of sunflower seed-containing rations improved hen performance. Rations containing sunflower seed gave a significantly lower yolk colour score and a significant rise in yolk cholesterol content. Lee and Moss (1989), in two feeding trials with laying hens, observed that unhulled, confectionery-type sunflower seeds depressed egg production when fed at levels exceeding 20% of the diet. Feed efficiency, egg weight, albumen weight and eggshell thickness were not significantly different among treatments.

Safflower meal (*Carthamus tinctorius*)

Safflower is a crop grown mainly for its high-grade oil content (Pond *et al.*, 1995) but also provides meal for livestock and poultry and as bird seed. Safflower is grown in more than 60 countries, but over half of the world production is in India. For use in poultry rations, it needs to be decorticated to decrease the fibre content (Leeson and Summers, 1997). The protein quality of this meal is low because it is deficient in lysine. Dehulled safflower meal contains about 42–44% protein and 14% fibre. Hollaran (1961) reported that one-third of the protein in the diet could be supplied by safflower meal if the diet contained fish meal. Valadez *et al.* (1965) reported that safflower meals can replace 50% of the soybean meal in broiler diets if the diet is supplemented with lysine.

Sesame meal (*Sesamum indicum*)

Sesame is also known as benne, gingili and teel or til. The leading countries in the production of sesame are India, Iraq, Egypt and Pakistan. Sesame is known to be high in methionine, cystine and tryptophan but deficient in lysine; hence it cannot be fed as the major protein supplement in broiler rations (Daghir and Kevorkian, 1970). The very early work on the nutritive value of sesame meal and the extent to which it can replace soybean meal in broiler rations was reviewed by Daghir *et al.* (1967). These workers showed that sesame meal may replace 50% of the soybean meal in broiler rations supplemented with 2% fish meal without significant change in weight gains or feed efficiency. Chicks receiving a combination of soybean meal and sesame meal in the proportion of 10 : 20 supplemented with 0.32% lysine had significantly greater body weight than those receiving an all-soybean meal diet.

Cuca and Sunde (1967) reported that sesame meal binds dietary calcium owing to its high phytic acid content. Using maize–soyabean meal diets, they found normal bone ash values with calcium levels as low as 0.8%. With California sesame meal, 1.05% calcium was required for normal bone ash values and with Mexican sesame meal 1.5% calcium was needed. Lease *et al.* (1960) reported that sesame meal interferes with the biological availability of zinc. In later work, Lease (1966) showed that autoclaving sesame meal for 2 h caused an increase in chick growth, a decrease in leg abnormalities and increased bone ash as compared with chicks receiving similar diets containing unautoclaved sesame meal. Bell *et al.* (1990) reported that sesame seed meal may provide an acceptable alternative to soybean meal in broiler rations when the substitution level is 15% or less.

Linseed meal

Linseed meal is obtained from flax (*Linum usitatissimum*) and contains high levels of mucilage, which is almost completely indigestible by poultry. It also contains a small amount of a cyanogenetic glycoside called linamarin and an enzyme (linase) that is capable of hydrolysing this glycoside, producing hydrogen cyanide. The meal has a poor-quality protein that is low in both methionine and lysine.

Because of the above limitations, this meal is not considered a satisfactory protein supplement for poultry. Depressed growth has been reported when chicks were fed diets containing as little as 5% linseed meal. The adverse effects of this meal can be partially reduced by autoclaving and increasing the level of vitamin B_6 in the ration. McDonald *et al.* (1988) did not recommend using it at levels exceeding 3% of the diet.

In light of the importance of dietary n-3 fatty acids for human health, Ajuyah *et al.* (1993) investigated the effect of dietary full-fat linseed on the fatty acid composition of chicken meat. The white meat of birds fed this seed had elevated levels of n-3 fatty acids. Egg yolk n-3 fatty acid content can also be increased by feeding full-fat flax seeds to laying hens. Van Elswyk (1997) gave an extensive review on the feeding of flax to laying hens and its effects on the composition of eggs.

Mustard seed meal

Wild mustard is widely spread all over the world and grows heavily in wheat and barley fields. It is mainly the presence of sinigrin, a sulfur-containing glucoside, which has prevented the use of mustard seed meal as a protein source for poultry. The glucoside is also present in rapeseed, crambe seed and other seeds of the *Brassica* family. This glucoside, when digested by an endogenous enzyme (myrosinase), is hydrolysed to volatile allylisothiocyanate, an anti-nutritional goitrogenic factor (Fainan *et al.*, 1967). Sakhawat *et al.* (2003) were able to lower the allylisothiocyanate in mustard seed meal from 1.55% to traces and its phytic acid content from 3.54 to 0.40%. Attempts to detoxify sinigrin in mustard seed by soaking, boiling and/or roasting failed (Daghir and Mian, 1976). Detoxification of allylisothiocyanate in rapeseed has been achieved by chemical additives, particularly $FeSO_4.7H_2O$ (Bell *et al.*, 1971). Daghir and Charalambus (1978) tested this $FeSO_4$ treatment on mustard seed meal and found it to be an effective method of detoxification of the meal for use in broiler rations.

Single-cell protein

Microbial fermentation for the production of protein has been known for a long time. Single-cell organisms such as yeast and bacteria grow very rapidly and can double their cell mass in 3–4 h. A range of nutrient substrates from plant sources can be used, as well as unconventional materials such as methanol, ethanol, alkanes, aldehydes and organic acids. The protein content of bacteria is higher than that of yeasts and contains higher levels of the sulfur-containing amino acids but a lower level of lysine. Single-cell proteins contain high levels of nucleic acids, ranging from 5 to 12% in yeasts and 8 to 16% in bacteria. Yeast single-cell protein (YSCP) has been studied for use as a protein supplement in poultry rations since the early part of the 20th century. Clark (1931) was the first to report problems with feeding yeast and claimed that it was unsuitable for growing chicks. The early literature on feeding yeast to chickens has been reviewed by Saoud and Daghir (1980). These workers reported that molasses-grown yeast results in growth depression and lower feed efficiencies as compared with a ration not containing single-cell protein. The detrimental effects of single-cell protein were mainly due to changes in blood nitrogenous constituents. Daghir and Abdul-Baki (1977) found that yeast protein produced from molasses can be used at 5 and 10% levels of well-balanced broiler rations with adequate supplementation of lysine and methionine. The addition of 15% yeast protein to the ration depresses growth and feed efficiency in broilers, even with methionine and lysine supplementation. Daghir and Sell (1982) studied the amino acid limitations of YSCP when fed as the only source of protein in semi-purified diets for broilers. They showed that methionine and arginine were the first and second most limiting amino acids in the YSCP tested. Their data suggested that palatability of the diet containing YSCP was also an important factor in reducing feed intake of those diets. Based on these studies as well as others, single-cell protein is not recommended for use at levels exceeding 5% for broilers and 10% for layers.

Novel Feedstuffs

Cassava root meal

Cassava root meal, also known as tapioca, mandioca, manioca, yucca and manioc meal, is a readily digestible carbohydrate food that is made from the roots of the cassava plant. It is a perennial root crop native to the humid tropics.

The cortical layers of roots contain linamarin, a cyanogenic glycoside. These compounds yield hydrogen cyanide upon treatment with acid or appropriate hydrolytic enzymes. Because of this compound, many of the early studies showed growth depression in chicks fed increasing amounts of cassava. Sweet cassava is low in hydrocyanic acid and can be fed after drying. Bitter cassava, however, needs to be grated and pressed to remove the juice, which contains most of the glycoside. Therefore, effective processing of cassava reduces its cyanide content, thereby reducing its toxic effects (Reddy, 1991).

Studies on cassava root meal have shown a wide variation in response of broilers to this feedstuff. Recommended levels of the meal for broilers have ranged from 10% (Vogt, 1966) to 50% (Enriquez and Ross, 1967; Olson et al., 1969). This variability in results may be due to differences in experimental conditions, as well as in cassava-processing method. Cassava, if properly detoxified by sun-drying whole-root chips on a concrete floor, can be satisfactorily used in broiler diets (Gomez et al., 1983). Gomez et al. (1987), in Colombia, studied the effects of increasing the ME of diets containing 0, 20 and 30% cassava root meal by adding either vegetable oil or animal tallow. Diets containing 20% cassava root meal with either vegetable oil or animal tallow produced the greatest body weight of broilers. Eruvbetine and Afolami (1992) were able to feed as much as 40% cassava in place of maize to broilers in Nigeria without any significant effect on body weight, feed conversion or mortality. Fuentes et al. (1992) evaluated sun-dried cassava meal in Brazil as a replacement for maize in broiler diets. They concluded that cassava meal can replace up to 75% of the maize without a significant effect on weight gain or feed conversion.

According to Scott et al. (1982), the dried cassava root meal contains about 1.8% crude protein, 1.3% fat, 85% nitrogen-free extract and 1.8% fibre. The mineral content includes 0.35% phosphorus, 0.3% Ca and 16 p.p.m. manganese.

Dates and date by-products

Dates are produced in large quantities in many hot regions of the world. The total world production is estimated to be about 5582 thousand tonnes (FAOSTAT, 2005). The leading countries in the production of dates are Egypt, Saudi Arabia, Iran, Iraq, UAE, Algeria and Pakistan. These seven countries produce more than 90% of the world's supply. Dates are used for the production of molasses, alcohol, vinegar, yeast and several other products. Dates and date by-products are also used in animal feeds. Date pits, which make up about 15% of the total weight of dates, are rich in energy and fat but very high in fibre (18–24%) and very low in protein. Date by-products resulting from molasses production from dates are high in sugars and very low in protein. This product is usually very high in moisture and needs to be dried before use in ruminant feed. Although the use of date by-products has been mainly in ruminant feeds, a number of studies have been conducted on poultry. Kamel et al. (1981) tested different levels of date pits (5–15%) and found that one can go up to 10% without a reduction in weight or feed conversion. They reported the chemical composition of date pits from six different date varieties to be as follows: 6.40% protein, 5.06% fat, 33.78% fibre, 1.53% ash and 44.28% NFE. Jumah (1973) found that even 10% can reduce growth rate in broilers. Abdul-Ghani et al. (2003) fed a date molasses by-product at 4 or 8% of the diet and found that it had no effect on weight gain, feed conversion or dressing percentage. Khalil (2005) concluded that it is possible to use small quantities of certain date by-products in poultry feeds and thus contribute to reduced feed costs in areas where dates are plentiful.

Afzal et al. (2006) studied the effect of different levels of whole dates fed to broilers in the grower and finisher period (20–49 days). They did not detect any significant differences in feed intake, live weight, or feed conversion with levels ranging from 0 to 30%.

Palm kernel meal

Palm kernel meal (PKM) or cake is the residue of oil extraction of palm kernel, which is extensively cultivated in tropical countries. This meal has a fairly good-quality protein and a good balance of calcium and phosphorus. It has not been widely used for poultry because of its unpalatability and high fibre content (15%). The unpalatability is due mainly to the presence of β-mannan and to its grittiness. A number of studies conducted in Nigeria (Onwudike, 1986a,b,c) have shown that palm kernel meal can be a valuable source of protein for poultry. This view is not shared by all researchers (Ezieshi and Olomu 2004; Sundu et al., 2005a). Onwudike (1988) reported that the meal could be fed to laying hens at high levels if the diet contained fish meal. The same worker (Onwudike, 1992) was able to replace fish meal in the high PKM diet by blood meal, along with the addition of DL-methionine, without any significant drop in egg production. Perez et al. (2000) fed PKM to laying hens at levels ranging from 0 to 50%. They concluded that this meal could be up to 40% in the diet. Panigrahi (1991) observed behaviour changes in broiler chicks fed on diets containing high levels of palm kernel meal. This altered behaviour was associated with lower feed intake and reduced weight gains. When these diets containing high levels of palm kernel meal were supplemented with high levels of maize oil, behaviour was normal and weight gains were only 3.5% less than those of controls. Although Onwudike (1992) suggested that in the presence of an animal protein supplement palm kernel meal can be used at high levels in poultry rations, because of the palatability problem it probably should not be used at levels exceeding 20% of the ration. More recent studies indicate that because PKM is free of anti-nutritional factors, it can be included at levels up to 40% in the diet, provided both energy and amino acids are adequate and well balanced. There is a need for more information on this meal since its use in poultry rations in all warm regions of Asia, Africa and South America can lead to substantial savings on feed costs.

Mungbean (*Vigna radiata*)

Mungbean is also known as mungdal, moong dal, mash bean, munggo, green gram, golden gram and green soy. It is an important source of protein in human nutrition in several countries in the Asian continent. It is the most important grain legume in Thailand and the Philippines and ranks second in Sri Lanka and third in India, Burma, Bangladesh and Indonesia. Bhardwaj et al. (1999) reported that the annual mung production in the USA was at 50,000 ha, 90% mainly in Oklahoma, California and Texas. The mungbean contains about 24% crude protein, 1.5% crude fat, 6% crude fibre, 5% ash and 63.5% NFE (Ravindran and Blair, 1992). Thailand alone produces about 350 thousand tonnes of mungbean per year.

Breadfruit meal

Breadfruit (*Artocarpus altilis*) grows wild in many African countries. Achinewhu (1982) reported that it is prevalent in the southern parts of Nigeria and is a tropical fruit-bearing tree which stores mainly carbohydrates in its fruit. The pulp of the fruit used to be consumed by people, but this consumption has decreased in recent years and more of the fruit is available for other uses. Ravindran and Sivakanesan (1995) reported that breadfruit could be processed into meal that may be suitable for poultry. Worrell and Carrington (1997) believe that this fruit is as good nutritionally as tubers and root crops and rich in vitamins and minerals. Oladunjoye et al. (2004) reported that breadfruit contains 12.98% CP, 4.22% CF, 3.49% ash and 3870 kcal/kg gross energy. These workers also reported the presence of traces of anti-nutritional factors like oxalates, tannins and phytates. These factors may interfere with digestibility, absorption and utilization of nutrients (Bullard et al., 1980). Adekunle et al. (2006) investigated the response of broilers to diets containing raw or cooked breadfruit meal at levels ranging from 0 to 30%. The diets were isocaloric and isonitrogenous. They observed that cooking increased daily weight gains and there was a significant interaction between cooking and inclusion level for weight gain, feed conversion and protein efficiency ratio. They concluded that cooked breadfruit meal can adequately replace dietary maize at levels of about 10% of the diet.

Ipil-ipil leaf meal (*Leucaena leucocephala*)

Ipil-ipil is a leguminous tree that grows abundantly in the Philippines, Hawaii, Thailand and other tropical countries. The leaves of this tree, which are called *koa haole* in Hawaii, are used in animal feeds. This meal contains approximately 24% protein, 3.25% fat, 14% fibre and 530 mg β-carotene activity per kg (Scott et al., 1982). The major problem of this meal is the presence of a toxic alkaloid, mimosine. This alkaloid constitutes about 97% of the acid fraction extracted from the plant (Alejandrino et al., 1976). In studies with chickens (Librijo and Hathcock, 1974), feed intake declined 33% and egg production declined 49% when this meal was included at 30% of the diet. Berry and D'Mello (1981) showed that egg production and live weight gain of chickens on diets containing 20% of this meal were significantly reduced. Iji and Okonkwo (1991) reported that regional differences exist in the composition of ipil-ipil meals. These differences should be better known before levels higher than 3% can be used in poultry diets. Scott et al. (1982) reported that levels of ipil-ipil meal above 5% cause reduced growth of broilers and reduced egg production in laying hens. This is probably due to poor amino acid digestibility as well as the toxic alkaloid mimosine, since Picard et al. (1987) presented very low amino acid digestibility values for this meal. These workers concluded that ipil-ipil meals are not suitable for poultry protein nutrition and should be regarded only as a pigment source to be used at low levels of the diet.

Salseed (*Shorea robusta*)

Salseed is produced in large quantities in many tropical and subtropical regions, primarily for its oil. After removal of the oil, about 85% of the seed is available as meal. Panda (1970) reviewed the early work on salseed and salseed meal and pointed out the economic importance of using this meal in poultry feeds as a source of energy.

Zombade et al. (1979) reported that salseed meal contains 9.8% crude protein, 2.2% ether extract, 45.0% available carbohydrates and 11.7% tannins. These workers observed that more than 5.0% salseed meal in the diet of White Leghorn male chickens resulted in poor growth and food conversion. Verma and Panda (1972) also observed that 5.0% salseed caused growth depression in chicks. Most of this growth depression is due to the high tannin content of the meal since about 0.5% tannin in the diet has been found to affect chicks adversely (Vohra et al., 1966). Mahmoud et al. (2006) studied the effect of chemical treatments of salseed meal on nutrient digestibility and digestive enzymes in both hens and broilers. Both nutrient digestibility and pancreatic enzymes are depressed when salseed meal is used in the diet and bicarbonate was the most effective treatment to improve nutrient digestibility.

Dried poultry waste

Dried poultry waste (DPW) has been successfully used for feeding ruminants for many years past. Its use in poultry diets, however, has not been widely accepted. Interest in the use of DPW has developed because the energy-yielding components of high-energy poultry rations were digested and metabolized only to the extent of 70–80% (Young and Nesheim, 1972). Most of the early work on the use of DPW was conducted on laying hens at Michigan State University (Flegal and Zindel, 1969). Most of these studies indicated that including DPW in layer rations at relatively low levels had no adverse effects on performance as long as the low energy content of the product is considered in formulating diets. Nesheim (1972) and Young (1972) conducted several studies on DPW collected from cage layer operations and found that the apparent utilization of this material is not more than 30%. Several workers were able to show that laying hens can utilize some of the essential amino acids found in DPW (Blair and Lee, 1973; Rinehart et al., 1973). Daghir and Amirullah (1978) did not observe any differences in egg production or egg weight between birds receiving 10% DPW and those receiving a standard maize–soybean diet. Furthermore, they did not detect any differences in overall acceptability of boiled eggs produced by hens fed DPW at a 10% level. Vogt (1973), on the other hand, reported that incorporation of 10% in a laying ration significantly decreased egg production and feed conversion without affecting egg weight. In areas where feedstuffs are scarce and expensive, this product may contribute to reducing feed costs if it is properly handled and dried.

Buffalo gourd meal (*Cucurbita foetidissima*)

The buffalo gourd is a desert plant that has been recognized since the middle of the last century as a potential source of oil and protein. The oil portion has been found to be edible and rich in the essential fatty acids. The protein is similar to other oilseed proteins and is especially rich in arginine, aspartic acid and glutamic acid but low in lysine, threonine and methionine (Daghir and Zaatari, 1983).

The main problem limiting the use of buffalo gourd meal in practical poultry rations is the presence of naturally occurring toxins in the seed. Daghir et al. (1980a) reported that buffalo gourd meal, prepared by hexane extraction and fed at levels sufficient to contribute 8% protein to the diet, is toxic to growing chicks. It produces a neuromuscular condition that is characterized by an abnormal position of the neck and inability of the bird to keep its head up. Further work showed that the toxin is mainly located in the embryo of the seed since feeding the hulls or the oil produced zero incidence of this abnormality (Daghir and Sell, 1980). Several bitter principles have been isolated from the family *Cucurbitaceae* and have been given the name cucurbitacins. Some of these have been chemically identified as triterpenoid glycosides. Daghir and Flaifel (1984) evaluated several detoxification procedures on hexane-extracted buffalo gourd meal and tried to verify the adequacy of these methods by feeding the differently treated meals to growing chicks. The treatments, which consisted of water soaking, alcohol extraction, chloroform–methanol extraction or chloroform–methanol–alcohol extraction, had no positive effect on the meal. The detoxification of desert plant seeds like the buffalo gourd should have a tremendous economic impact on the poultry industry since all these plants are potential crops for the arid lands of the hot regions.

Guar meal (*Cyamopsis tetragonaloba*)

Guar is a drought-resistant legume that is widely cultivated in India and Pakistan. The endosperm of the guar seed is a rich source of a galactomannan polysaccharide called guar gum. Guar meal is obtained after mechanical separation of the endosperm from the hulls and germ of the ground seeds. The meal contains 4% fat, 45% crude protein, 6% fibre and 4.5% ash. It is rich in lysine and methionine. Vohra and Kratzer (1964a) showed that as little as 7.5% guar meal in the ration caused a significant depression in chick growth. Furthermore, they showed that a major part of this depression is attributed to the guar gum in the meal. Their results indicated that guar meal can be slightly improved by autoclaving. The same workers (Vohra and Kratzer, 1964b) found growth of chickens to be inhibited about 25–30% by the inclusion of guar gum at levels that contributed 2 or 4% pectin to the diet. This depression was overcome by treatment with enzymes capable of hydrolysing this gum, namely pectinase and cellulase or a preparation from the sprouted guar beans. Patel *et al.* (1980) studied the effects of γ-irradiation, penicillin and pectic enzyme on chick growth depression and faecal stickiness caused by guar gum and found that the faecal condition of birds fed guar gum was significantly improved by a combination of γ-irradiation and pectic enzyme supplement. The deleterious effects of guar meal on broilers are attributed mainly to residual gum fractions. Lee *et al.* (2003) found that the feeding of up to 7.5% of the germ fraction did not lower growth or feed conversion while the hull fraction reduced growth and increased intestinal viscosity at all inclusion levels. Lee *et al.* (2005) found that β-mannanase improved feed per gain ratio of diets containing 5% of each fraction of guar meal. Their results indicate that the upper feeding level of guar meal and germ and hull fraction of guar meal is 2.5%. Addition of β-mannanase increases the upper feeding level for the germ fraction to 5%.

Bambara groundnut meal (*Voandzeia subterranea*)

Bambara groundnut (BGN) is an indigenous African leguminous crop that has been described as the most resistant pulse (National Academy of Science, 1979). Its production is restricted to the African continent and is estimated to be about 330 thousand tonnes. BGN seeds contain a trypsin inhibitor (Poulter, 1981), which would limit its use at high levels in poultry diets. Onwudike and Eguakum (1992) evaluated meals from raw seeds and seeds boiled for 30 and 60 min. Raw BGN had trypsin inhibitory activity of 20.8 units/mg, and boiling for 30 min completely eliminated all that activity. Raw BGN meal contained 7.5% ether extract, 2.03% fibre and 20.6% protein. The ME value of the raw BGN was 2.65 kcal/g, and the ME value of the heat-treated meal was 3.88 kcal/g. Phosphorus was the element with the highest concentration in BGN (0.11%) and zinc was the highest microelement (36 p.p.m.).

Jojoba meal (*Simmondsia chinensis*)

This meal is a by-product of the oil produced from the jojoba seed. It contains 25–30% protein (Verbiscar and Banigan, 1978). Feeding it to chickens resulted in impaired body weight, and reduced feed intake and feed efficiency (Ngou

Ngoupayou *et al.*, 1982). This effect may be due to several compounds present in the meal, such as glycosides, polyphenols, phytic acids and trypsin inhibitors (Wiseman and Price, 1987). Arnouts *et al.* (1993) studied the possibility of using jojoba meal to inhibit feed intake of broiler breeder pullets and thus limit body-weight gain, as recommended by breeder companies. They found that the growth retardation required was obtained with a 4% level of jojoba meal in the diet and that reduced growth was the result of decreased feed intake linked with the simmondsin content of the meal, as well as other anti-nutritional compounds affecting digestibility.

Velvet beans (*Mucuna* spp.)

The beans contain 20–25% protein and 3–5% fat (Del Carmen *et al.*, 1999). The same workers reported that raw *Mucuna pruriens* has a true ME value of 2370 kcal/kg while Ukachukwu *et al.* (1999) gave values of 1050, 3000, 3190 and 3220 kcal/kg for raw, toasted, cooked and boiled *Mucuna cochinchinensis*. Iyayi and Taiwo (2003) fed laying hens 6% of either raw, wet autoclaved or roasted *Mucuna pruriens* during the early part of the production cycle. They observed that only the raw beans had a negative effect on performance. These workers believe that laying hens are not as sensitive to this source of protein as broilers. Several species of *Mucuna* are available as feed. *Mucuna pruriens* is most often used. Other species available are *Mucuna utilis, Mucuna monosperma, Mucuna gigantean, Mucuna urines, Mucuna solanei* and *Mucuna cochinchinensis*. Non-structural carbohydrates make up half of *Mucuna* while the neutral detergent fibre is about 10%. Amino acid profiles are similar to other legumes such as soybeans. The bean contains several anti-nutritional toxic substances that are harmful to poultry. These include antitrypsin factors, tannins, cyanide, anticoagulants, antipyretic and anti-inflammatory factors. Several attempts have been made to process the beans and remove the toxic factors by heating, water extraction or other methods but with limited success. Olaboro *et al.* (1991) reported that heat treatment of velvet beans by autoclaving improved growth in young chicks equal to that of control chicks fed roasted soybeans as the protein source. Emiola *et al.* (2007) tried different processing methods on *Mucuna* (*Mucuna pruriens* var. *utilis*) and found that heated meal gave better results than dehulled or toasted meals. They concluded that treated *Mucuna* can be used to replace 50% protein supplied by soybean meal in broiler starter and finisher diets without any adverse effects on performance of intestinal organs. Carew and Gernat (2006) have reviewed the literature on this bean and concluded that more research is needed to discover the proper processing techniques for making them useful in commercial poultry diets.

Conclusions

This chapter summarizes research work on the composition and nutritional value of feedstuffs that are available in the hot regions of the world.

Although a great deal of work has been done on the composition of agricultural and industrial by-products and several feeding tests have been conducted, more work is needed, particularly on methods to render these more palatable and free of anti-nutritional factors. Table 7.1 presents recommended levels of

Table 7.1. Recommended levels of inclusion of selected cereals and protein supplements in poultry feeds as % of diet.

Feedstuff	Broiler starter	Broiler finisher	Chick starter	Chick grower	Chick developer	Layer and breeder
Barley	0–5	0–5	0–10	0–10	0–20	0–60
Coconut meal	0–2	0–3	0–2	0–5	0–5	0–3
Cottonseed meal	0–5	0–15	0–10	0–15	0–15	0–10
Linseed meal	0–2	0–3	0–2	0–3	0–3	0–3
Millet	0–10	0–10	0–15	0–30	0–50	0–40
Mustard seed meal	0–3	0–5	0–3	0–5	0–5	0–3
Groundnut meal	0–5	0–10	0–5	0–10	0–15	0–10
Rice bran	0–5	0–5	0–10	0–15	0–20	0–15
Rice paddy	0–5	0–5	0–5	0–10	0–15	0–15
Rice polishings	0–5	0–5	0–10	0–15	0–20	0–15
Safflower meal	0–5	0–10	0–5	0–10	0–15	0–10
Sesame meal	0–5	0–10	0–5	0–10	0–15	0–10
Single-cell protein	0–3	0–5	0–3	0–5	0–10	0–10
Sorghum	0–20	0–40	0–20	0–30	0–40	0–50
Sunflower meal	0–5	0–10	0–5	0–10	0–15	0–10
Sunflower seed	0–10	0–15	0–10	0–15	0–15	0–20
Triticale	0–20	0–30	0–20	0–30	0–30	0–60

Table 7.2. Recommended levels of inclusion of selected novel feedstuffs in poultry feeds as % of diet.

Feedstuff	Broiler starter	Broiler finisher	Chick starter	Chick grower	Chick developer	Layer and breeder
Bambara groundnut meal	0–2	0–3	0–2	0–3	0–5	0–3
Breadfruit meal	0–5	0–10	0–5	0–10	0–10	0–5
Buffalo gourd meal	0–2	0–3	0–2	0–3	0–5	0–3
Cassava root meal	0–10	0–20	0–10	0–15	0–20	0–20
Dried poultry waste	0–2	0–3	0–2	0–3	0–5	0–5
Guar meal	0–2	0–3	0–2	0–3	0–5	0–3
Ipil–ipil leaf meal	0–2	0–3	0–2	0–3	0–5	0–3
Jojoba meal	0–2	0–3	0–2	0–3	0–3	0–3
Mung bean	0–2	0–3	0–2	0–3	0–5	0–5
Palm kernel meal	0–2	0–3	0–2	0–3	0–5	0–5
Salseed	0–3	0–5	0–3	0–5	0–5	0–3
Velvet beans	0–2	0–3	0–2	0–3	0–5	0–3

inclusion of cereals and protein supplements, while Table 7.2 presents recommended levels of inclusion of novel feedstuffs covered in this chapter. It is observed that, in general, recommended levels of inclusion for the novel feedstuffs are low because of problems with either the palatability of these feedstuffs or the presence of anti-nutritional factors. As pointed out by Farrell (2005), many countries in the hot regions of the world may not be able to continue importing maize and soybean as their industry grows and therefore may have to depend more on the locally produced ingredients.

Therefore, research on methods of improving the palatability and reducing anti-nutritional factors in those feedstuffs is badly needed. This type of research can lead to more extensive use of these novel feedstuffs, which would enhance further expansion of the poultry industry in the hot regions of the world without competing with humans for scarce and costly cereal grains. It should be pointed out that until more information is obtained on these novel feedstuffs and further improvements are made on their nutritional value the levels of their inclusion in poultry feed will continue to be relatively low. Furthermore, even with these low levels of inclusion, greater care should be taken in the balancing and formulation of poultry rations containing these ingredients than in those containing commonly used feedstuffs.

References

Abate, A.N. and Gomez, M. (1983) Substitution of finger millet (*Eleusine coracana*) and bulrush millet (*Pennisetum typhoides*) for maize in broiler feeds. *Animal Feed Science and Technology* 10, 291–299.

Abdul-Ghani, M.F, Sadek, H.L. and Khalil, R.T. (2003) Effect of date by-products in broiler rations on performance and meat quality. *Al-Anbar Journal of Agricultural Sciences* 10, 209–215.

Acen, A. and Varum, K. (1987) Beta-glucaner i bygg. *NLVF's fagutvalg for Korn fordling seminar. Norges Landbrukshoegskole*, Norway, pp. 16–17.

Achinewhu, S.C. (1982) Nutritive qualities of plant food. *Nutrition Reports International* 25, 643–647.

Adekunle, K.A., Farimo, A.O., Abiola, S.S. and Akegbejo-Samsons, Y. (2006) Potential of breadfruit meal as alternative energy source to maize in diet of broiler chickens. *Journal of Poultry Science* 43, 241–249.

Afzal, N., Nafemipou, H. and Riasi, A. (2006) The effect of different levels of surplus date in grower and finisher diets on broiler performance. *World's Poultry Science Journal. XII European Poultry Conference Proceedings*, pp. 372–373.

Ajuyah, A.O., Hardin, R.T. and Sim, J.S. (1993) Effect of dietary full fat flax seed with and without antioxidant on the fatty acid composition of major lipid classes of chicken meats. *Poultry Science* 72, 125–136.

Alejandrino, A.L., Concepcion, F. and Belone, B. (1976) A modified method of isolating and determining mimosine from ipil-ipil leaf meal. *Philippine Agriculture* (Suppl.) 9, 10.

Allen, D. (1990) Feedstuffs ingredient analysis table. *Feedstuffs* 62.

Ali, M.A. and Leeson, S. (1995) The nutritive value of some indigenous Asian poultry feed ingredients. *Animal Feed Science and Technology* 55, 227–237.

Anderson, J.O. and Warnick, R.E. (1965) Amino acid deficiencies in peanut meal and in corn and peanut meal rations. *Poultry Science* 44, 1066–1072.

Anderson, J.O., Wagstaff, R.K. and Dobson, D.C. (1961) Studies on the value of hull-less barley in chick diets and means of increasing this value. *Poultry Science* 40, 1571–1584.

Armanious, M.W., Britton, W.M. and Fuller, H.L. (1973) Effect of methionine and choline on tannic acid and toxicity in the laying hen. *Poultry Science* 52, 2160–2168.

Armstrong, W.D., Featherston, W.R. and Rogler, J.C. (1973) Influence of methionine and other dietary additions on the performance of chicks fed bird resistant sorghum grain diets. *Poultry Science* 52, 1592–1599.

Arnouts, S., Bugse, J., Cokelaere, M.U. and Decuypere, E. (1993) Jojoba meal in the diet of broiler breeder pullets. *Poultry Science* 72, 1714–1721.

Attia, Y.A., Qota, E.M.A., Aggoor, F.A.M and Kies, A.K. (2003) Value for rice bran, its maximal utilisation and its upgrading by phytase and other enzymes and diet-formulation based on available amino acids in the diet for broilers. *Archiv für Geflügelkunde* 67, 157–166.

Bailey, A.V., Harris, J.A., Skau, E.L. and Kerr, T. (1966) Cyclopropenoid fatty acid content and fatty acid composition of crude oils from twenty-five varieties of cottonseed. *Journal of the American Oil Chemists Society* 43, 107–110.

Barbour, G.W., Farran, M.T., Usayran, N.N., Darwish, A.H., Machlab, H.H., Hruby, M., Uwayjan, M.G. and Ashkarian, V.M. (2006) Nutritional evaluation of the local barley varieties and the impact of dietary enzymes on their apparent M.E. *Journal of Poultry Science* 43, 228–234.

Bedford, M.R. (1996) The effect of enzymes on digestion. *Journal of Applied Poultry Research* 5, 370–378.

Bell, D.E., Ibrahim, A.A., Denton, G.W., Long, G.G. and Bradley, G.L. (1990) An evaluation of sesame seed meal as a possible substitute for soyabean oil meal for feeding broilers. *Poultry Science* 69, 157 (Abstract).

Bell, J.M., Young, G.G. and Downey, R.K. (1971) A nutritional comparison of various rapeseed and mustard seed solvent-extracted meals of different glucocinolate composition. *Canadian Journal of Animal Science* 51, 259–269.

Berry, S. and D'Mello, J.P.F. (1981) A comparison of *Leucaena leucocephala* and grass meals as sources of pigments in diets for laying hens. *Tropical Animal Production* 6, 167–173.

Bhardwaj, H.L., Rangappa, M. and Hamama, A.A. (1999) Chickpea, Faba beans, lupin, mungbean and pigeonpea: potential new crops for the mid-Atlantic regions of the United States. In: Janick, J. (ed.) *Perspective on New Crops*. ASHS Press, Alexandria, Virginia, pp. 202–220.

Blair, R. and Lee, D.J.W. (1973) Supplementation of low protein layer diets. *British Poultry Science* 14, 9–16.

Bonino, M.F., Azcona, J.O. and Sceglio, O. (1980) The effect of ammoniation, methionine and gluten meal supplementation on the performance of laying hens fed high tannin sorghum grain diets. *Proceedings of the 6th European Poultry Conference* 3, 481–484.

Bragg, D.B. and Sharby, T.F. (1970) Nutritive value of triticale for broiler chick diets. *Poultry Science* 49, 1022–1027.

Briggs, K.G. (2001) A study in the growth potential of triticale in Western Canada. *Report for Alberta Agriculture, Food and Rural Development*, Alberta, Canada, p. 122.

Bullard, R.W., Garrisson, M.V., Kilburn, S.R. and York, J.O (1980) Laboratory comparisons of polyphenols and their repellent characteristics in bird-resistant sorghum grains. *Journal of Agriculture and Food Chemistry* 28, 1006–1011.

Cabel, M.C. and Waldroup, P.W. (1989) Ethoxyquin and ethylenediamine tetra acetic acid for the prevention of rancidity in rice bran stored at elevated temperature and humidity for various lengths of time. *Poultry Science* 68, 438–442.

Carew, L.B. and Gernat, A.G. (2006) Use of velvet beans (*Mucuna* spp.) as a feed ingredient for poultry, a review. *World's Poultry Science Journal* 62, 131–141.

Castanon, J.I., Ortiz, R.V. and Perez-Lansac, J. (1990) Effect of high inclusion levels of triticale in diets for laying hens containing 30% field beans. *Animal Feed Science and Technology* 31, 349–353.

Chang, S.I. and Fuller, H.L. (1964) Effect of tannin content of grain sorghums on their feeding value for growing chicks. *Poultry Science* 43, 30–36.

Chawla, J.S., Nagra, S.S. and Pannu, M.S. (1987) Different cereals for laying hens. *Indian Journal of Poultry Science* 22, 95–100.

Cheva-Isarakul, B. and Tagtaweewipat, S. (1991) Effect of different levels of sunflower seed in broiler rations. *Poultry Science* 70, 2284–2294.

Clark, M. (1931) Practical food evaluation. *Food Technology (London)* 1, 98–99.

Classen, H.L., Graham, H., Inborr, J. and Bedford, M.R. (1991) Growing interest in feed enzymes to lead to new products. *Feedstuff* 63(4), 22–25.

Coon, C.N., Shepler, R., McFarland, D. and Nordhein, J. (1979) The nutritional evaluation of barley selection and cultivars from Washington State. *Poultry Science* 58, 913–918.

Coon, C.N., Obi, I. and Hamre, M.L. (1988) Use of barley in laying hen diets. *Poultry Science* 67, 1306–1313.

Costa, E.F., Miller, B.R., Pasti, G.M., Bakalli, R.I. and Ewing, H.P. (2001) Studies on feeding peanut meal as a protein source for broiler chicken. *Poultry Science* 80, 306–313.

Cuca, M.G. and Avila, E.G. (1973) Preliminary studies on triticale in diets for laying hens. *Poultry Science* 52, 1973–1974.

Cuca, M. and Sunde, M.L. (1967) The availability of calcium from Mexican and Californian sesame meals. *Poultry Science* 46, 994–1002.

Daghir, N.J. and Abdul-Baki, T.K. (1977) Yeast protein in broiler rations. *Poultry Science* 56, 1836–1841.

Daghir, N.J. and Amirullah, I. (1978) Dehydrated poultry waste and urea as feed supplements in layer rations. *Iran Journal of Agricultural Research* 6, 91–97.

Daghir, N.J. and Charalambus, K. (1978) Metabolizable energy of $FeSO_4$-treated and untreated mustard seed meal. *Poultry Science* 57, 1081–1083.

Daghir, N.J. and Flaifel, F.A. (1984) Buffalo gourd protein for growing chickens. *Proceedings of the 17th World's Poultry Congress*, pp. 295–297.

Daghir, N.J. and Kevorkian, K.A. (1970) Limiting amino acids in sesame meal chick diets. *Proceedings of the 14th World's Poultry Congress*, Madrid, pp. 711–715.

Daghir, N.J. and Mian, W.A. (1976) Mustard seed meal as a protein source for chickens. *Poultry Science* 55, 1699–1703.

Daghir, N.J. and Nathanael, A.S. (1974) An assessment of the nutritional value of triticale for poultry. *Proceedings of the 15th World Poultry Congress*, pp. 612–614.

Daghir, N.J. and Rottensten, K. (1966) The influence of variety and enzyme supplementation on the nutritional value of barley for chicks. *British Poultry Science* 7, 159–163.

Daghir, N.J. and Sell, J.L. (1980) Buffalo gourd seed and seed components for growing chickens. *Nutrition Reports International* 22, 445–452.

Daghir, N.J. and Sell, J.L. (1982) Amino acid limitations of yeast single cell protein for growing chickens. *Poultry Science* 61, 337–344.

Daghir, N.J. and Zaatari, I.M. (1983) Detoxification and protein quality of buffalo gourd meal for growing chickens. *Nutrition Reports International* 27, 339–346.

Daghir, N.J., Hajj, R. and Akrabawi, S.S. (1966) Studies on peanut meal for broilers. *Proceedings of the 13th World's Poultry Congress*, Kiev, pp. 238–246.

Daghir, N.J., Ullah, M.F. and Rottensten, K. (1967) Lysine supplementation of sesame meal broiler rations. *Tropical Agriculture*, (Trinidad) 44, 235–242.

Daghir, N.J., Ayyash, B. and Pellet, P.L. (1969) Evaluation of groundnut meal protein for poultry. *Journal of the Science of Food and Agriculture* 20, 349–354.

Daghir, N.J., Mahmoud, H.K. and El-Zein, A. (1980a) Buffalo gourd meal: nutritive value and detoxification. *Nutrition Reports International* 21, 837–847.

Daghir, N.J., Raz, M.A. and Uwayjan, M. (1980b) Studies on the utilization of full fat sunflower seed in broiler rations. *Poultry Science* 59, 2272–2278.

Dale, N.M., Wyatt, R.D. and Fuller, H.L. (1980) Additive toxicity of aflatoxins and dietary tannins in broiler chicks. *Poultry Science* 59, 2417–2420.

Del Carmen, J., Gernat, A.G., Myhrman, R. and Carew, L.B. (1999) Evaluation of raw and heated velvet beans (*Mucuna pruriens*) as feed ingredients for broilers. *Poultry Science* 78, 866–872.

Douglas, J.H., Sullivan, T.W., Bond, P.L. and Steuwe, F.J. (1990a) Nutrient composition and metabolizable energy values of selected grain sorghum varieties and yellow corn. *Poultry Science* 69, 1147–1155.

Douglas, J.H., Sullivan, T.W., Bond, P.L., Struwe, F.J., Baier, J.G. and Robesone, L.G. (1990b) Influence of grinding, rolling, and pelleting on the nutritional value of grain sorghums and yellow corn for broilers. *Poultry Science* 69, 2150–2156.

Driver, J.P., Atencio, A., Edwards, H.M. and Pesti, G.M. (2006) Improvements in nitrogen-corrected apparent metabolizable energy of peanut meal in response to phytase supplementation. *Poultry Science* 85, 96–99.

El-Alaily, H., Soliman, H., Anwar, A., El-Zeiny, M. and Ibrahim, S. (1985) Sorghum grains as an energy source for chicks. *Egyptian Poultry Science* 5.

El-Boushy, A.R. and Raternick, R. (1989) Replacement of soyabean meal by cottonseed meal and peanut meal or both in low energy diets for broilers. *Poultry Science* 68, 799–804.

Elkin, R.G., Featherston, W.R. and Rogler, J.C. (1978) Investigation of leg abnormalities in chicks consuming high tannin sorghum grain diets. *Poultry Science* 57, 757–762.

Emiola, A.I., Anthony, O.D. and Robert, G.M. (2007) Influence of processing of mucuna (*Mucuna pruriens* var *utilis*) and kidney bean (*Phaseolus vulgaris*) on the performance and nutrient utilization of broiler chickens. *Journal of Poultry Science* 44, 168–174.

Enriquez, F.Q. and Ross, E. (1967) The value of cassava root meal for chicks. *Poultry Science* 46, 622–626.

Eruvbetine, D. and Afolami, C.A. (1992) Economic evaluation of cassava (*Manihot esculenta*) as a feed ingredient for broilers. *Proceedings of the 19th World's Poultry Congress*, Vol. 3, pp. 532–535.

Esteve-Garcia, E., Brufan, J., Fe'rez-Vendrell, M.A. and Duves, K. (1997) Bioefficacy of enzyme preparations containing β-glucanase and xylanase activities in broiler diets based on barley or wheat in combination with flavomycin. *Poultry Science* 76, 1728–1737.

Ezieshi, E.V. and Olomu, J.M. (2004) Comparative performance of broiler chickens fed varying levels of palm kernel meal and maize offal. *Pakistan Journal of Nutrition* 3(4), 254–257.

Fainan, C., Ryan, R.J. and Eickel, H.T. (1967) Endocrinology 81, 279.

FAOSTAT (2005) http://faostat.fao.org/site/340/default.aspx. viewed on 30/03/2007.

Farrell, D.J. (1994) Utilization of rice bran in diets of domestic fowl and ducklings. *World's Poultry Science Journal* 50, 115–131.

Farrell, D.J. (2005) Matching poultry production with available feed resources: issues and constraints. *World's Poultry Science Journal* 61, 298–307.

Farrell, D.J., and Martin, E.A. (1998) Strategies to improve the nutritive value of rice bran in poultry diets III. The addition of inorganic phosphorus and a phytase to duck diets. *British Poultry Science* 39, 601–611.

Feltwell, R. and Fox, S. (1978) *Practical Poultry Feeding*. Faber and Faber, London.

Fernandez, R. and McGinnis, J. (1974) Nutritive value of triticale for young chicks and effect of different amino acid supplements on growth. *Poultry Science* 53, 47–53.

Flegal, C.A. and Zindel, H.C. (1969) The utilization of dehydrated poultry waste by laying hens. *Poultry Science* 48, 1807–1810.

Flores, M.P., Castanon, J.I.R. and McNab, J.M. (1992) Use of enzymes to improve the nutritive value of triticale in poultry diets. *Proceedings of the 19th World's Poultry Congress*, Vol. 2, pp. 253–254.

Fuentes, M.F.F., Coelho, M.G.R., Souza, F.M., Lopes, I.R.V. and Pereira, L.I. (1992) Sun-dried cassava meal in tropical broiler diets. *Proceedings of the 19th World's Poultry Congress*, Vol. 3, p. 551.

Fuente, J.M, Perez de Ayala, P. and Villamide, M.J. (1995) Effect of dietary enzymes on the M.E. of diets with increasing levels of barley fed to broilers at different ages. *Animal Feed Science and Technology* 56, 45–53.

Garcia, R.G., Mendes, A.A., Sartori, J.R., Paz, I.C., Oliveira, R., Takashi, S.E., Pelicia, K., Quinteiro, R.R. and Pinheiro, D.F. (2004) Digestibility of rations containing sorghum, with or without tannin, for broilers submitted to three room temperatures. *Proceedings of the 22nd World Poultry Congress*, Istanbul, Turkey, p. 549.

Gippert, T., Halmagyi-Valler, T. and Gati, L. (1988) Utilization of differently treated, extracted sunflower coarse meal in the nutrition of broiler chickens. *Proceedings of the 18th World's Poultry Congress*, pp. 927–928.

Gomez, G., Valdirieso, M., Santos, J. and Hoyos, C. (1983) Evaluation of cassava root meal prepared from low- or high-cyanide containing cultivars in pig and broiler diets. *Nutrition Reports International* 28, 693–704.

Gomez, G., Tellez, G. and Caicedo, J. (1987) Effects of the addition of vegetable oil or animal tallow to broiler diets containing cassava root meal. *Poultry Science* 66, 725–731.

Hasselman, K., Elwinger, K., Nilsson, M. and Thomke, S. (1981) The effect of beta-glucanase supplementation, stage of ripeness, and storage treatment of barley in diets fed to broiler chickens. *Poultry Science* 60, 2664–2671.

Herstad, O. (1987) Cereals with higher fiber content (barley, oats, millet). *Proceedings of the Sixth European Symposium on Poultry Nutrition*, pp. A15–A23.

Heywang, B.W., Bird, H.R. and Altschul, A.M. (1955) Relationship between discolorations in eggs and dietary free gossypol supplied by different cotton-seed products. *Poultry Science* 34, 81–90.

Hollaran, H.R. (1961) High protein safflower meal for chickens. *Feedstuffs* 33(45), 70–71.

Hulse, J.H., Laing, E.M. and Pearson, O.E. (1980) *Sorghum and Millets: Their Composition and Nutritive Value*. Academic Press, London.

Hussein, A.S. and Kratzer, F.H. (1982) Effect of rancidity on the feeding value of rice bran for chickens. *Poultry Science* 61, 2450–2455.

Ibrahim, S., Fisher, C., El-Alaily, H., Soliman, H. and Anwar, A. (1988) Improvement of the nutritional quality of Egyptian and Sudanese sorghum grains by the addition of phosphates. *British Poultry Science* 29, 721–728.

ICARDA (2004) *Annual Report*. International Centre for Agricultural Research in Dry Areas, Aleppo, Syria.

Iji, P.A. and Okonkwo, A.C. (1991) Leucaena and neem leaf meals: potential protein sources for layers? *Feed International*, December, 29–32.

Iyayi, E.A. and Taiwo, V.O. (2003) The effect of diets incorporating mucuna (*Mucuna pruriens*) seed meal on the performance of laying hens and broilers. *Tropical and Sub-tropical Agroecosystems* 1, 239–246.

Jerock, H. (1987) Nutritional value of wheat, rye, and triticale in broiler chickens and laying hens. *Proceedings of the Sixth European Symposium on Poultry Nutrition*, pp. A4–A14.

Jeroch, H. and Danicke, S. (1995) Barley in poultry feeding: a review. *World's Poultry Science Journal* 51, 271–291.

Jones, L.A. (1981) Special cottonseed products report. *Feedstuff* 53(52), 19–21.

Karmajeewa, H. and Than, S.H. (1984) Choice feeding of the replacement pullet on whole grains and subsequent performance on laying diets. *British Poultry Science* 25, 99–109.

Jumah, B.S. (1973) Some nutritional aspects of feeding ground date pits for broilers. *Mesopotamia Journal of Agriculture* 8, 139–145.

Kamel, B.S, Diab, M.F., Ilian, M.A. and Salman, A.J. (1981) Nutritional values of whole date pits in broiler rations. *Poultry Science* 60, 1005–1011.

Khalil, R.T. (2005) Use of dates byproducts in broiler nutrition. *Poultry Middle East and North Africa* 184, 38–40.

Klain, G.J., Hill, D.C., Branion, H.D. and Gray, J.A. (1956) The value of rapeseed oil meal and sunflower seed oil meal in chicks starter rations. *Poultry Science* 32, 542–547.

Lease, J.G. (1966) The effect of autoclaving sesame meal on its phytic acid content and on the availability of its zinc to the chick. *Poultry Science* 45, 237–241.

Lease, J.G., Barnett, B.D., Lease, E.J. and Turk, D.E. (1960) The biological unavailability to the chick of zinc in a sesame meal ration. *Journal of Nutrition* 72, 66–71.

Lee, J.T., Bailey, C.A. and Cartwright, A.L. (2003) Guar meal germ and hull fractions differently affect performance of broiler chickens. *Poultry Science* 82, 1589–1595.

Lee, J.T., Connor-Appleton, S., Bailey, C.A. and Cartwright, A.L. (2005) Effect of guar meal by-products with and without β-mannanase on broiler performance. *Poultry Science* 84, 1261–1267.

Lee, K. and Moss, C.W. (1989) Performance of laying chickens fed diets containing confectionery-type sunflower seeds. *Poultry Science* 68 (Suppl. 1), 84 (Abstract).

Leeson, S. and Summers, J.D. (1987) Response of white Leghorns to diets containing ground or whole triticale. *Canadian Journal of Animal Science* 67, 583–585.

Leeson, S. and Summers, J.D. (1997) *Commercial Poultry Nutrition*, 2nd edn. University Books, Guelph, Canada.

Levi, R.S., Reilich, H.G., O'Neill, H.J., Cucullu, A.F. and Skau, E.L. (1967) Quantitative determination of cyclopropenoid fatty acids in cottonseed meal. *Journal of the American Oil Chemists Society* 44, 249–252.

Librijo, N.T. and Hathcock, J.N. (1974) Metabolism of mimosine and other compounds from *Leucaena lecocephala* by the chicken. *Nutrition Reports International* 9, 217–222.

Lordelo, M.M., Danis, A.J., Calhoun, M.C., Dowd, M.K. and Dale, N.M. (2005) Relative toxicity of gossypol enantiomers in broilers. *Poultry Science* 84, 1376–1382.

Lordelo, M.M., Calhoun, M.C., Dale, N.M., Dowd, M.K. and Davis, A.J. (2007) Relative toxicity of gossypol enantiomers in laying and broiler breeder hens. *Poultry Science* 86, 582–590.

Luckbert, J., Maitre, I. and Castaing, J. (1988) Using triticale in laying hen diets. *Proceedings of the 18th World's Poultry Congress*, pp. 797–798.

Luis, E.S., Sullivan, T.W. and Nelson, L.A. (1982) Nutrient composition and feeding value of proso millet, sorghum grain and corn in broiler diets. *Poultry Science* 61, 311–320.

Mahmoud, S., Khan, M.A., Sarwar, M. and Nisa, M. (2006) Chemical treatments to reduce antinutritional factors in salseed meal: effect on nutrient digestibility in colostomized hens and intact broilers. *Poultry Science* 85, 2207–2215.

McDonald, P., Edwards, R.A. and Greenbagh, J.F.D. (1988). *Animal Nutrition*, 4th edn. Wiley, New York, pp. 463–464.

McDonald, P., Edwards, R.A., Greenhalgh, J.F. and Morgan, C.A. (1995) *Animal Nutrition*, 5th edn, Wiley, New York, p. 607.

McNab, J.M. (1987) The energy value of roots and mill by-products. *Proceedings of the Sixth European Symposium on Poultry Nutrition*, pp. A26-A34.

McNaughton, J.L. and Deaton, J.W. (1981) Sunflowers: poultry applications. *Feed Management* 32(6), 27.

Mezoui, C.M. and Bird, F.H. (1984) Peanut as a protein source for growing broilers. *Revue Science et Technique, serie. Sciences de la Sante* 1(7), 45–51 (Abstract).

Mohammadain, G.M., Babiker, S.A. and Mohammad, T.A. (1986) Effect of feeding millet, maize and sorghum grains on performance, carcass yield and chemical composition of broiler meat. *Tropical Agriculture* (Trinidad) 63, 173–176.

Mohan Ravi Kumar, A., Reddy, V.R., Reddy, P.V.V.S. and Reddy, P.S. (1991) Utilization of pearl millet for egg production. *British Poultry Science* 32, 463–469.

Morrison, A.B., Clandinin, D.R. and Robblee, A.R. (1953) The effects of processing variables on the nutritive value of sunflower seed oil meal. *Poultry Science* 32, 492–496.

Morrison, F.B. (1956) *Feeds and Feeding*, 21st edn. Morrison, Ithaca, New York.

Mushtaq, T., Sarwar, M., Ahmad, G., Nisa, M.U. and Jamil, A. (2006) The influence of exogenous multienzyme preparation and graded levels of digestible lysine in sunflower meal-based diets on the performance of young broiler chicks two weeks posthatching. *Poultry Science* 85, 2180–2185.

Nagalaskshmi, D., Rama Rao, S.V., Panda, A.K. and Sastry, V.R.B. (2007) Cottonseed meal in poultry diets: a review. *Journal of Poultry Science* 44, 119–134.

National Academy of Science (1979) *Tropical Legumes*. Resources for the Future, Washington, DC.

National Research Council (1994) *Nutrient Requirements of Poultry*, 9th ed. National Academy Press, Washington, DC.

Nelson, T.S., Stephenson, E.L., Burgos, A., Floyd, J. and York, J.O. (1975) Effect of tannin content and dry matter digestion on energy utilization and average amino acid availability of hybrid sorghum grains. *Poultry Science* 54, 1620–1623.

Nesheim, M.C. (1972) Evaluation of dehydrated poultry manure as a potential poultry feed ingredient. *Animal Waste Conference Proceedings*, Cornell University, Ithaca, New York.

Newman, R.K. and Newman, C.W. (1988) Nutritive value of a new hull-less barley cultivar in broiler chick diets. *Poultry Science* 67, 1573–1579.

Ngou Ngoupayou, J.D., Maiorind, P.M. and Reid, B.L. (1982) Jojoba meal in poultry diets. *Poultry Science* 61, 1692–1696.

Nir, I., Melcion, J.P. and Picard, M. (1990) Effect of particle size of sorghum grains on feed intake and performance of young broilers. *Poultry Science* 69, 2177–2184.

Nyachoti, C.M., Atkinson, J.L. and Leeson, S. (1997) Sorghum tannins: a review. *World's Poultry Science Journal* 53, 5–21.

Olaboro,G., Okot, M.W., Mugerwa, J.S. and Latshaw, J.D. (1991) Growth depressing factors in velvet beans fed to broiler chicks. *East African Agriculture and Forestry Journal* 57, 103–110.

Oladunjoye, I.O., Ologhobo, A.D., Farinle, G.D., Emiola, I.A., Omao, A.O., Adedeji T.A. and. Salako, R.A. (2004) Chemical composition and energy value of differently processed breadfruit meal (*Artocarpus altilis*). *Proceedings of the 29th Annual Conference of the Nigerian Society for Animal Production*, pp. 144–146.

Olson, D.W., Sunde, M.L. and Bird, H.R. (1969) Amino acid supplementation of mandioca meal in chick diets. *Poultry Science* 48, 1949–1953.

Onwudike, O.C. (1986a) Palm kernel meal as a feed for poultry: 1. Composition and availability of its amino acids to chicks. *Animal Feed Science and Technology* 16, 179–186.

Onwudike, O.C. (1986b) Palm kernel meal as a feed for poultry. 2. Diets containing palm kernel meal for starter and grower pullets. *Animal Feed Science and Technology* 16, 187–194.

Onwudike, O.C. (1986c) Palm kernel meal as feed for poultry. 3. Replacement of groundnut cake by palm kernel meal in broiler diets. *Animal Feed Science and Technology* 16, 195–202.

Onwudike, O.C. (1988) Palm kernel meal as a feed for poultry: 4. Use of palm kernel meal by laying hens. *Animal Feed Science and Technology* 20, 279–286.

Onwudike, O.C. (1992) Effect of blood meal on the use of palm kernel meal by laying birds. *Proceedings of the 19th World's Poultry Congress*, Vol. 3, pp. 514–519.

Onwudike, O.C. and Eguakum, A. (1992) Effect of heat treatment on the composition, trypsin inhibitory activity, ME level and mineral bioavailability of Bambara groundnut meal with poultry. *Proceedings of the 19th World's Poultry Congress*, Vol. 3, pp. 508–511.

Panda, B. (1970) Processing and utilization of agro-industrial by-products as livestock and poultry feed. *Indian Poultry Gazette* 54, 39–71.

Panigrahi, S. (1991) Behaviour changes in broiler chicks fed on diets containing palm kernel meal. *Applied Animal Behaviour Science* 34, 277–281.

Panigrahi, S. (1992) Energy-deficit induced behaviour changes in broiler chicks fed on copra meal based diets. *Proceedings of the 19th World's Poultry Congress*, Vol. 3, pp. 503–507.

Panigrahi, S. and Hammonds, T.W. (1990) Egg discoloration effects of including screw-press cottonseed meal in laying hen diets and their prevention. *British Poultry Science* 31, 107–120.

Panigrahi, S. and Morris, T.R. (1991) Effects of dietary cottonseed meal and iron treated cottonseed meal in different laying hen genotypes. *British Poultry Science* 32, 167–184.

Panigrahi, S. and Plumb, V.E. (1996) Effects on dietary phosphorus of treating cotton seed meal with crystalline ferrous sulfate for the production of brown yolk discoloration. *British Poultry Science* 37, 403–411.

Panigrahi, S., Machin, D.H., Parr, W.H. and Bainton, J. (1987) Responses of broiler chicks to dietary copra cake of high lipid content. *British Poultry Science* 28, 589–600.

Panigrahi, S., Plumb, V.E. and Machin, D.H. (1989) Effects of dietary cottonseed meal, with and without iron treatment on laying hens. *British Poultry Science* 30, 641–651.

Patel, M.B., Jami, M.S. and McGinnis, J. (1980) Effect of gamma irradiation, penicillin and/or pectic enzyme on chick growth depression and fecal stickiness caused by rye, citrus pectin and guar gum. *Poultry Science* 59, 2105–2110.

Perez, J.F., Gernat, A.G. and Murillo, J.G. (2000) The effect of different levels of palm kernel meal in layer diets. *Poultry Science* 79, 77–79.

Perez-Maldonado, R.A., Barram, K.M. and Singh, D.N. (2003) Estimating amino acid availability from digestibility coefficient: application to poultry diets. *Asian Pacific Journal of Clinical Nutrition* 12 (Suppl,), S41.

Pesti, G.M., Bakalli, R.I., Driver, J.P., Sterling, K.G., Hall, L.E. and Bell, E.M. (2003) Comparison of peanut meal and soybean meal as protein supplements for laying hens. *Poultry Science* 82, 1274–1280.

Phelps, R.A. (1966) Cottonseed meal for poultry: from research to practical application. *World's Poultry Science Journal* 22, 86–112.

Phelps, R.A., Shenstone, F.S., Kemmerer, A.R. and Evans, R.J. (1965) A review of cyclopropenoid compounds: biological effects of some derivatives. *Poultry Science* 44, 358–394.

Picard, M., Angulo, I., Antoine, H., Bouchot, C. and Sauveur, B. (1987) Some feeding strategies for poultry in hot and humid environments. *Proceedings of the 10th Annual Conference of the Malaysian Society of Animal Production*, pp. 110–116.

Pond, W.G., Church, D.C. and Pond, K.R. (1995) *Basic Animal Nutrition and Feeding*, 4th edn, Wiley, New York.

Poulter, N.H. (1981) Properties of some protein fractions from bambara ground-nut. *Journal of the Science of Food and Agriculture* 32, 44–50.

Proudfoot, F.G. and Hulan, H.W. (1988) Nutritive value of triticale as a feed ingredient for broiler chickens. *Poultry Science* 67, 1743–1749.

Rad, F.H. and Keshavarz, K. (1976) Evaluation of the nutritional value of sunflower meal and the possibility of substitution of sunflower meal for soyabean meal in poultry diets. *Poultry Science* 55, 1757–1760.

Randall, J.M., Sayre, R.N., Schultz, W.G., Fong, R.Y., Mossman, A.P., Triberlhorne, R.E. and Saunders, R.M. (1985) Rice bran stabilization by extrusion cooking for extraction of edible oil. *Journal of Food Science* 50, 361–364.

Rao, D.R., Johnson, W.M. and Sunki, G.R. (1976) Replacement of maize by triticale in broiler diets. *British Poultry Science* 17, 269–274.

Ravindran, V. and Blair, R. (1992) Food resources for poultry production in Asia and the pacific region. II Plant protein sources. *World's Poultry Science Journal* 48, 205–231.

Ravindran, V. and Sivakanesan, R. (1995) Breadfruit (*Artocarpus altilis*) meal. Nutrient composition and feeding value for broilers. *Journal of the Science of Food and Agriculture* 69, 379–383.

Razool, Z. and Athar, M. (2007) Pakistan feed milling industry: a current overview on the poultry feed sector, http://www.engormix.com Pakistan-feed-milling-industry-e-articles-212-Bal.htm.

Reddy, C.V. (1991) Tapioca instead of maize. *Poultry International*, October, 42–44.

Reddy, D.R. and Reddy, C.V. (1970) Influence of source of grain on the performance of laying stock. *Indian Veterinary Journal* Al, 157–163.

Reid, B.L., Galaviz-Moreno, S. and Maiorino, P.M. (1987) Evaluation of isopropanol-extracted cottonseed meal for laying hens. *Poultry Science* 66, 82–89.

Rinehart, K.E., Snetsinger, D.C., Rogland, W.W. and Zimmerman, R.A. (1973) Feeding value of dehydrated poultry waste. *Poultry Science* 52, 2079–2083.

Rodriguez, M.L., Ortiz, L.T., Alzueta, C., Rebole, A. and Trevino, J. (2005) Nutritive value of high-oleic acid sunflower seed for broiler chickens. *Poultry Science* 84, 395–402.

Rojas, S.W. and Scott, M.L. (1969) Factors affecting the nutritive value of cottonseed meal as a protein source in chick diets. *Poultry Science* 48, 819–835.

Rose, R.J., Coit, R.N. and Sell, J.L. (1972) Sunflower seed meal as a replacement for soyabean meal protein in laying hen rations. *Poultry Science* 51, 960–967.

Rosen, G.D. (1958) Groundnuts (peanuts) and groundnut meal. In: Altshul, A.M. (ed.) *Processed Plant Protein Foodstuffs*. Academic Press, New York, pp. 419–448.

Rotter, B.A., Friesen, O.D., Guenter, W. and Marawardt, R.R. (1990) Influence of enzyme supplementation on the bio available energy of barley. *Poultry Science* 69, 1174–1181.

Ruiz, N., Marion, J.E., Miles, R.D. and Barnett, R.B. (1987) Nutritive value of new cultivars of triticale and wheat for broiler chicken diets. *Poultry Science* 66, 90–97.

Sakhawat, A., Kausor, T., Shah, W.H. and Niazi, A.H. (2003) Studies on detoxification of feeds and feed ingredients. *Pakistan Journal of Scientific and Industrial Research* 46, 300–303.

Saoud, N.B. and Daghir, N.J. (1980) Blood constituents of yeast fed chicks. *Poultry Science* 59, 1807–1811.

Sayre, R.N., Earl, L., Kratzer, F.H. and Saunders, R.M. (1987) Nutritional qualities of stabilized and raw rice bran for chicks. *Poultry Science* 66, 493–499.

Scott, M.L., Nesheim, M.C. and Young, R.J. (1982). *Nutrition of the Chicken*. M.L. Scott, Ithaca, New York.

Scott, T.A., Silvers, F.G., Classen, M.L., Swift, M.L. and Bedford, M.R. (1999) Prediction of the performance of broiler chicks from apparent M.E. and protein digestibility values obtained using a broiler chick bioassay. *Canadian Journal of Agricultural Sciences* 79, 59–64.

Sell, D.R., Rogler, J.C. and Featherston, W.R. (1983) The effects of sorghum tannin and protein level on the performance of laying hens maintained in two temperature environments. *Poultry Science* 62, 2420–2428.

Sell, J.L., Hodgson, G.C. and Shebeski, L.H. (1962) Triticale as a potential component of chick rations. *Canadian Journal of Animal Science* 42, 158–166.

Senkoylu, N. and Dale, N. (1999) Sunflower meal in poultry diets: a review. *World's Poultry Science Journal* 55, 153–174.

Sharma, B.D., Sadagopan, V.R. and Reddy, V.R. (1979) Utilization of different cereals in broiler diets. *British Poultry Science* 20, 371–378.

Sundu, B., Kumar, A. and Dingle, J. (2005) Response of birds fed increasing levels of palm kernel meal supplemented with enzymes. *Australian Poultry Science Symposium* 17, 227–228.

Torki, M. and Falahati, M. (2006) Effects of different levels of rice bran supplemented by enzyme on performance of broiler chicks. *World's Poultry Science Journal XII European Poultry Conference Proceedings*, pp. 351–352.

Ukachukwu, S.N., Obioha, F.C. and Madubuike, R.C. (1999) Determination of the true metabolizable energy (TME) of raw and heat-treated *Mucuna cochinchinensis* using adult broilers. *Tropical Journal of Animal Sciences* 3, 25–31.

Uwayjan, M.G., Azar, E.J. and Daghir, N.J. (1983) Sunflower seed in laying hen rations. *Poultry Science* 62, 1247–1253.

Valadez, S., Fetherston, W.R. and Pichett, R.A. (1965) Utilization of safflower meal by the chick and its effect upon plasma lysine and methionine concentrations. *Poultry Science* 44, 909–915.

Valdivie, M.L., Sardinas, O. and Garcia, J.A. (1982) The utilization of 20% sunflower seed meal in broiler diets. *Cuban Journal of Agricultural Sciences* 16, 167–171.

Van Elswyk, M.E. (1997) Nutritional and physiological effects of flax seed in diets for laying fowl. *World's Poultry Science Journal* 53, 253–264.

Verbiscar, A.J. and Banigan, T.F. (1978) Composition of jojoba seeds and foliage. *Journal of Agricultural Food Chemistry* 26, 1456–1459.

Verma, S.V.S. and Panda, B. (1972) Studies on the metabolizable energy values of salseed (*Shorea robusta*) and salseed cake by chemical and biological evaluation in chicks. *Indian Journal of Poultry Science* 7, 5–12.

Verma, S.V.S., Tyagi, P.K. and Singh, B.P. (1992) Nutritional value of broken rice (rice kani) for chicks. *Proceedings of the 19th World's Poultry Congress*, pp. 524–527.

Villamide, M.J. and San Juan, L.D. (1998) Effect of chemical composition of sunflower seed meal on its true metabolizable energy and amino acid digestibility. *Poultry Science* 77, 1884–1892.

Vogt, H. (1966) The use of tapioca meal in poultry rations. *World's Poultry Science Journal* 22, 113–125.

Vogt, H. (1973) The utilization of production and processing waste from egg production. *World's Poultry Science Journal* 29, 157–158.

Vohra, P. and Kratzer, F.H. (1964a) The use of guar meal in chick rations. *Poultry Science* 43, 502–503.

Vohra, P. and Kratzer, F.H. (1964b) Growth inhibitory effect of certain polysaccharides for chickens. *Poultry Science* 43, 1164–1170.

Vohra, P., Kratzer, F.H. and Joslyn, M.A. (1966) The growth depressing and toxic effects of tannins to chicks. *Poultry Science* 45, 135–142.

Vohra, P., Hafez, Y., Earl, L. and Kratzer, F.H. (1975) The effect of ammonia treatment of cottonseed meal on its gossypol-induced discoloration of egg yolks. *Poultry Science* 54, 441–447.

Waldroup, P.W. (1981) Cottonseed meal in poultry diets. *Feedstuffs* 53(52), 21–24.

Wilson, B.J. and McNab, J.M. (1975) The nutritive value of triticale and rye in broiler diets containing field beans (*Vicia faba* L.). *British Poultry Science* 16, 17–22.

Wiseman, M.O. and Price, R.L. (1987) Characterization of protein concentrates of jojoba meal. *Cereal Chemistry* 64, 91–93.

Worrell, D.B. and Carrington, C.M. (1997) Breadfruit. In: Mitra, S.K. (ed.) *Post-Harvest Physiology and Storage of Tropical and Sub-tropical Fruit*. CAB International, pp. 347–348.

Young, G.J. and Nesheim, M.C. (1972) Dehydrated poultry waste as feed ingredient. *Proceedings of the Cornell Nutrition Conference*, p. 46.

Young, R.J. (1972) Evaluation of poultry waste as a feed ingredient and recycling waste as a method of waste disposal. *Proceedings of the Texas Nutrition Conference*, p. 1.

Zatari, I. and Sell, J.L. (1990) Sunflower meal as a component of fat-supplemented diets for broiler chickens. *Poultry Science* 69, 1503–1507.

Zhang, Y. and Parsons, C.M. (1996) Effects of overprocessing on the nutritional quality of peanut meal. *Poultry Science* 75, 514–518.

Zombade, S.S., Lodhi, G.N. and Ichhponani, J.S. (1979) The nutritional value of salseed (*Shorea robusta*) meal for growing chicks. *British Poultry Science* 20, 433–438.

8 Mycotoxins in Poultry Feeds

N.J. DAGHIR

Faculty of Agricultural and Food Sciences, American University of Beirut, Lebanon

Introduction	198
Aflatoxins	199
Occurrence	200
Aflatoxicosis in broilers	200
Aflatoxicosis in layers and breeders	202
Aflatoxin residues in eggs and poultry meat	202
Influence of aflatoxin on resistance and immunity	203
Aflatoxin and vitamin nutrition	205
Aflatoxin and zinc in feeds and feedstuffs	206
Cyclopiazonic acid	206
Sterigmatocystin	207
Citrinin	207
Fumonisins	208
Ochratoxins	209
Oosporein	210
T-2 toxin	211
Vomitoxin (deoxynivalenol – DON)	212
Zearalenone	213
Detection of mycotoxins	213
Control of mycotoxins	214
Conclusions	217
References	218

Introduction

The term mycotoxin is used to refer to all toxins derived from fungi. The names of most mycotoxins are based on the names of the fungi that produce them. In some cases, the name is the actual chemical name, or it may be based on a toxic manifestation of the toxin. Information on the existence and spread of these toxins in the hot regions of the world has been accumulating very rapidly, and a comprehensive review of the literature since the first edition of this book was published in 1995 is beyond the scope of this chapter. Selected references will, however, be used to cover new developments in this field. A recent review of the effect of mycotoxins on poultry was presented by Devegowda and Murthy (2005), who also gave some practical solutions to minimize the effects of mycotoxins. Methods of harvesting, storage, processing and handling of feedstuffs and complete feeds in hot regions of the world are conducive for fungal growth and therefore for the existence of these toxins in them. Mycotoxin contamination of feedstuffs was first recognized in the early 1960s in the UK, with the discovery of aflatoxin in imported groundnut meals as the cause of 'turkey X-disease'. The author detected aflatoxin poisoning in Lebanon as far back as 1965 in birds receiving contaminated groundnut meals (Daghir *et al.*, 1966). In a small survey conducted on groundnut meals used by feed manufacturers during that year, samples were found to vary from 0.5 to 3.0 p.p.m. of aflatoxin. Since that date, aflatoxins have been reported to occur in poultry feeds in several countries of the hot regions.

The early literature on the degree of contamination of feeds in various countries has been reviewed by Jelinek *et al.* (1989). The incidence of aflatoxins in poultry feeds in countries of the hot regions in which it was studied, such as Egypt (Shaaban *et al.*, 1988), India (Raina and Singh, 1991), Indonesia (Purwoko *et al.*, 1991), Pakistan (Rizvi *et al.*, 1990), Saudi Arabia (Shaaban *et al.*, 1991) and Turkey (Babila and Acktay, 1991), varied from 18.9 to 94.4%. The level of aflatoxin reported by these countries exceeds the Food and Drug Administration and European Community permissible level of 20 p.p.b. An FAO/WHO survey showed that about 100 countries regulate aflatoxins in foods and feeds (FAO, 1995). The legal limits vary from one country to another. They all seem to vary from 10 to 20 p.p.b. of total aflatoxin. A survey was conducted in the USA (Anonymous, 1988) on seven large poultry companies in the south-east (Georgia, North Carolina and Virginia). Approximately 1000 samples were gathered from feed troughs at regular intervals and checked for levels of aflatoxin, vomitoxin, zearalenone and T-2 toxin. Table 8.1 shows the results of this survey. Although these companies were all located in the south-east, they all stated that they were buying Midwestern maize for their rations. It is interesting that the vomitoxins were more of a problem on these farms than the aflatoxins. Blaney and Williams (1991) reported that, in Australia, the mycotoxins that produce substantial losses are the aflatoxins, zearalenone and deoxynivalenol, occurring mainly in grains. Beg *et al.* (2006) conducted a survey of mycotoxin contamination in

Table 8.1. Tests for mycotoxin contamination of feed on poultry farms in the south-eastern USA (from Anonymous, 1988).

Toxins	Levels considered positive (p.p.b.)	% Positive
Aflatoxin	> 5	35
Vomitoxin	> 250	64
Zearalenone	100	11
T-2 toxin	> 50	5

yellow maize, soybean meal, wheat bran and finished poultry feeds in Kuwait. Average aflatoxin concentration in maize was 0.27 p.p.b., soybean 0.20 p.p.b., wheat bran 0.15 p.p.b., broiler starter 0.48 p.p.b., finisher 0.39 p.p.b. and layer 0.21 p.p.b. Ochratoxin A ranged from 4.6 to 9.6 p.p.b. and fumonisins from 46 to 67 p.p.b. in various feeds. In a recent survey conducted in Lebanon, it was found that 14% of commercial corn shipments used on poultry farms in Lebanon contain between 6 and 30 p.p.b. of aflatoxin B, with no presence of aflatoxin B_2, G_1 or G_2 (Barbour et al., 2008).

In 1995, when the first edition of this book was published, there were only six mycotoxins important to the poultry industry. Today there are nine, namely aflatoxins, ochratoxins, citrinin, cyclopizanoic acid, sterigmatocystin, T-2 toxins, fumonisins, maniliformins and zearalenone; and therefore this chapter will cover these toxins only. The biological effects of mycotoxins, other than aflatoxin, on poultry have been reviewed by Wyatt (1979). The occurrence of these toxins varies with the season. For example, in the southern parts of India, Chandrasekaran (1996) found that the most prevalent in the summer are aflatoxins, ochratoxins and T-2 toxins, while in the winter zearalenone, DON and T-2 toxin are more prevalent. Wang et al. (2003) conducted an extensive survey in China on complete feeds and individual ingredients and found that 88, 84, 77 and 60% of maize samples contained T-2 toxins, aflatoxins, fumonisins and OTA, respectively. All maize samples contained zearalenone and DON. Reports also exist on the presence of aflatoxins, OTA, T-2 toxin, DON, zearalenone and others in Latin American countries such as Brazil, Peru, Mexico, Colombia, Venezuela and Argentina. Mycotoxins are no longer restricted to the hot and humid regions of the world and are widely distributed because of the international trade (Devegowda et al., 1998).

Aflatoxins

Aflatoxin is the common name for a group of structurally related compounds that include B_1, B_2, G_1, G_2 produced by fungi of the *flavus parasiticus* group of the genus *Aspergillus*. These toxins have received a great deal of attention since their discovery in the UK in 1960. They are by far the most widespread in poultry feeds. Besides their occurrence in groundnut meals, they have been found in cottonseed meals, soybean meals, maize, barley, wheat and oats, as well as other feedstuffs (Goldblatt, 1969). Aflatoxins are common in warm and humid regions of the world like those existing in Africa, Asia, Latin America and the southern USA.

Occurrence

Aflatoxicosis has been reported to occur in several avian species besides the chicken. Ducks, geese, pheasants, quail and turkey have all been shown to be affected by aflatoxin. It is believed that all domestic birds are susceptible, but some are more sensitive than others. Chickens are least and ducks most susceptible. There are apparently even breed and strain differences in susceptibility to aflatoxin. The New Hampshire breed is susceptible to diets containing as little as 0.5 p.p.m., which normally does not seriously affect other chicken breeds (Gumbman et al., 1970).

Aflatoxicosis in broilers

Most studies on the effects of aflatoxin in broiler chickens have been designed as subacute studies, in which aflatoxin was fed at moderate to high levels (0.625–10 p.p.m.) from 0 to 3 weeks of age (Smith and Hamilton, 1970; Tung et al., 1975; Huff, 1980). In these studies, about 2.5 p.p.m. aflatoxin was usually required to decrease body weight significantly. Even in turkey poults, which are considered more sensitive than broilers, 2 p.p.m. are required to decrease body weight significantly (see Fig. 8.1). Besides reduction in body weight and feed conversion, birds show a variety of symptoms, such as enlarged livers, spleen and pancreas, repressed bursa, and pale combs,

Fig. 8.1. Effect of aflatoxin on body weight in turkey poults. Control is on far left.

shank and bone marrow. Aflatoxin B and fumonisin B, fed to broilers from 21 to 42 days of age at levels of 50–2450 µg and 10 mg respectively, do not change histological or serological parameters, but they cause lesions in liver and kidney of these birds (Del Bianchi *et al.*, 2005). Aflatoxin has also been shown by Osborne *et al.* (1975) to inhibit fat digestion in broilers by decreasing the enzymes and bile acids required for fat digestion. A high-fat diet made aflatoxicosis less severe in broilers, and so did high-protein diets. Raju and Devegowda (2000) found that aflatoxins decrease the activities of several enzymes important for digestion. They also reduce total serum proteins, cholesterol and blood urea nitrogen. Table 8.2 shows that 300 p.p.b. of aflatoxin significantly reduced all three of these important serum components.

Several studies have shown that aflatoxin is a hepatotoxin in broilers. The hepatotoxicity is reflected in elevated liver lipid levels (Tung *et al.*, 1973), as well as disruption of hepatic protein synthesis (Tung *et al.*, 1975). This hepatotoxicity also induces severe coagulopathies (Doerr *et al.*, 1974; Doerr and Hamilton, 1981) and anaemia (Tung *et al.*, 1975). Aflatoxin has also been reported to increase the susceptibility of young broiler chickens to bruising (Tung *et al.*, 1971; Girish and Devegowda, 2004). Huff *et al.* (1983) observed elevated prothrombin times and increased incidence and severity of thigh and breast bruises in broilers fed 2.5 p.p.m. aflatoxin up to 6 weeks of age. These authors suggested that aflatoxin is implicated in increased susceptibility of broilers to bruising during live-haul and processing. It is now well established that this relationship of aflatoxin to bruising in broilers is very prevalent, and normal events, such as catching of birds prior to processing, physical insults in the broiler house and removal from the coops at the processing plant, may cause rupture of fragile capillaries. This is referred to as haemorrhages of the musculature and skin and often leads to condemnations or downgrading of these birds. Wyatt (1988a) concluded that this bruising effect will occur with aflatoxin concentrations of less than 100 p.p.b. in the feed. Hamilton (1990) confirmed this and showed that the level of aflatoxin in the feed that can cause bruising is much lower than that causing growth inhibition (Table 8.3).

In another study, by Doerr *et al.* (1983), it was shown that the abnormalities normally encountered in broilers fed moderate to high levels of aflatoxin can be produced with much lower levels of the toxin (0.075–0.675 p.p.m.) if fed from day-old to market. These workers showed reduced growth, poor pigmentation and fatty livers resulting from this chronic low-level

Table 8.2. Effect of feeding an aflatoxin-contaminated diet on serum biochemical parameters in 5-week-old broilers (from Devegowda and Murthy, 2005).

Aflatoxin 300 p.p.b.	Total proteins (g %)	Total cholesterol (mg %)	Blood urea nitrogen (mg %)
−	2.73^a	114.8^a	2.22^a
+	1.45^b	57.3^b	1.01^b

^{a,b} Means in a column with different superscripts are significantly different ($P < 0.01$).

Table 8.3. Aflatoxin and bruising in chickens (from Hamilton, 1990).

Aflatoxin (ng/g)	Body weight (g)	Minimal bruising energy (joules)	Capillary fragility (mm Hg)
0	444[a]	0.43[a]	350[a]
625	451[a]	0.37[b]	260[b]
1250	438[a]	0.31[c]	185[c]
2500	380[b]	0.23[d]	–

[a–d] Values in a column with different superscripts differ significantly ($P < 0.05$).

aflatoxicosis. Their data demonstrate that fatty livers may result in broiler chickens consuming a very low dose of aflatoxin and thus may explain the occasional idiopathic occurrence of fatty livers in broilers at processing. Their work also demonstrates the difficulty in establishing a 'safe' level of aflatoxin-contaminated feed, one which can be fed to broiler chickens with no adverse effects on performance, because toxicity in their experiments varied from one experiment to the other, depending on the stress factors present in each case.

Aflatoxicosis in layers and breeders

The effects of aflatoxin on laying birds have been studied by Garlich *et al.* (1973). Aflatoxin decreased egg production about 2–4 weeks after administration of the toxin. The toxin decreased egg weight but had no significant effect on shell thickness or on percentage of shell (Hamilton and Garlich, 1971). These workers also suggested that dietary aflatoxin can cause a fatty liver syndrome in laying hens. This was confirmed later by Pettersson (1991), who reported that more than 2 p.p.m. aflatoxin in the feed decreases egg production and egg weight and increases the incidence of fatty livers in laying hens.

The effects of aflatoxin in the diet were also investigated with broiler breeder hens. Hens exhibited typical symptoms of aflatoxicosis, including enlarged livers and spleens. Although fertility was not affected, both egg production and hatchability of fertile eggs decreased (Dalvi, 1986; Afzali and Devegowda, 1999). Qureshi *et al.* (1998) reported significant decreases in fertility and hatchability 4 days after feeding the contaminated feed and observed an increase in the concentration of aflatoxin and its metabolites in the serum and in eggs from hens fed high levels of aflatoxin.

Aflatoxin residues in eggs and poultry meat

Aflatoxin B_1 is the mycotoxin that has been most studied for possible residues in tissues. Practical analytical methods have been developed for aflatoxin B_1 in tissues and for only one of the aflatoxin B_1 transformation products, aflatoxin M_1. Residues of aflatoxin B_1 have been found in eggs and in tissues from hens and broilers fed aflatoxin-contaminated rations (Jacobson and Wiseman, 1974).

Some workers have been concerned about the potential of finding aflatoxicol (Ro), which is the most toxic of the known metabolites, in eggs or meat. Trucksess et al. (1983) demonstrated that aflatoxin B_1 and its metabolite aflatoxicol can be detected in eggs and edible tissues from hens given feed contaminated with aflatoxin B_1 at a level of 8 p.p.m. The levels of aflatoxin residues increased steadily for 4–5 days to a plateau and decreased after toxin withdrawal at about the same rate that it increased. At 7 days after withdrawal, only trace amounts could be found in eggs.

Kan et al. (1989) did not detect any residues of aflatoxin B_1 in breast muscle of broilers and laying hens or in eggs or liver of broilers after feeding these birds diets with maize contaminated with 150 or 750 p.p.b. aflatoxin B_1 for 6 and 3 weeks, respectively. These authors concluded that aflatoxin B_1 levels in feed of about 100 p.p.b., which is much higher than the current EC tolerance of 20 p.p.b., do not impair performance of broilers and laying hens.

Influence of aflatoxin on resistance and immunity

There is now enough experimental evidence that aflatoxins can diminish innate resistance and immunogenesis in birds. It is also fairly well established that aflatoxin affects the production of certain non-specific humoral substances, the activity of thymus-derived lymphocytes and the formation of antibodies. The early literature on this subject has been reviewed by Richard et al. (1978) and more recently by Devegowda and Murthy (2005). Susceptibility to several infectious diseases, including salmonellosis, candidiasis and coccidiosis, has been shown to be enhanced by aflatoxin consumption.

Studies on the effects of aflatoxin on susceptibility to infection in the chicken are summarized in Table 8.4. It is seen that, in all diseases tested, aflatoxins enhanced susceptibility to the parasite except in the case of *Salmonella gallinarum* and *Candida albicans* (Pier, 1986).

Elzanaty et al. (1989) studied the effect of aflatoxins on Newcastle disease (ND) vaccination in broiler chicks. Different concentrations of aflatoxin (0, 8, 16, 32 and 48 µg/bird) were given daily for 2 weeks. The birds were vaccinated against ND at 4 and 18 days of age with Hitchner B_1 and Lasota strains, respectively, and at 35 days with Komarov strain. Adverse effects on humoral immunity to ND virus were observed, as measured by a haemagglutination inhibition test at 10, 17 and 24 days post-treatment, and no effect on protection, as indicated by a challenge test at the end of the experiment. Swamy and Devegowda (1998) reported decreased antibody titres against Infectious Bursal Disease and Newcastle disease in broiler chicks receiving 100–400 mg of aflatoxin B_1 per kg diet.

Under experimental conditions, the effects of aflatoxin on immune responsiveness in the chicken have been studied by several workers (Pier, 1986). Data

Table 8.4. Enhancement of pathogenicity of certain infectious agents in chicken by aflatoxin.

Agent	Dose	Enhanced susceptibility*	Reference
Salmonella gallinarum	5 p.p.m.	–	Pier, 1986
Candida albicans	0.625–10 p.p.m.	=	Pier, 1986
Eimeria tenella	0.2 p.p.m. B_1	+	Edds, 1973
Eimeria tenella	2.5 p.p.m. B_1	+	Wyatt et al., 1975
Salmonella spp. S. worthington, S. derby and S. thompson	10 p.p.m.	+	Boonchuvit and Hamilton, 1975
S. typhimurium	0.625–10 p.p.m.	+	Boonchuvit and Hamilton, 1975
Marek's disease virus	0.2 p.p.m. B_1	+	Edds, 1973
Infectious bursal disease	0.625–10 p.p.m.	+	Chang and Hamilton, 1982

* Increased incidence of infection or severity of disease.

show that several cell products participating in non-specific defence mechanisms in the chicken are affected. The effects of aflatoxin on γ-globulin levels and antibody titres are less constant than the effects on non-specific humoral substances. Consumption of moderate levels of aflatoxin does not decrease the level of immunoglobulin. However, a lag or decrease in antibody titre to antigens tested has been reported (Thaxton et al., 1974; Boonchuvit and Hamilton, 1975). Decreased levels of certain immunoglobulins (IgG and IgA) have also been reported (Tung et al., 1975; Giambrone et al., 1978). These decreases in IgG and IgA have been observed only when relatively high doses of aflatoxin were administered (2.5–10 p.p.m.). Mohiuddin (1992) reported that the feeding of aflatoxin to poultry resulted in a decrease in antibody and cell-mediated immune responses, resulting in severe disease outbreak, even after vaccination. The decrease in humoral immunity was established by determining the antibody titres in the serum. The decrease in cell-mediated immune response was established by decreases in phagocytic activity and in T lymphocytes, as demonstrated by the α-naphthyl esterase (ANAE) reaction and adenosine deaminase esterase (ADA) reaction in separated lymphocytes. This author suggested that several vaccination failures in the chicken might be the effect of aflatoxin on the immune system. Aflatoxin contributes to immune suppression in many disease outbreaks, to vaccination failures and to poor antibody titres (Raju and Devegowda, 2002).

The effects of aflatoxin on the different lymphatic tissues in poultry seem to depend on the dose given. Moderate levels of intake (0.5 p.p.m. B_1) caused thymic involution, but the bursa was not affected (Pier, 1986). Higher doses (2.5–10 p.p.m.), however, caused both thymic and bursal hypoplasia (Thaxton et al., 1974).

Aflatoxin has been shown to affect other aspects of the immune system in poultry. Complement activity has been shown to be suppressed by this toxin in a study by Campbell et al. (1983). These workers evaluated the combined effects of aflatoxin and ochratoxin on immunity in young broilers. They observed anaemia, hypoproteinaemia, lymphocytopenia, heterophilia and decreased bursa weight when both toxins were present. They concluded, however, that the most pronounced effect of aflatoxin on the immune system was depression of complement activity.

Vasenko and Karteeva (1987) studied the influence of feeds containing fungi on the immune system in hens housed in laying cages. Hens showed symptoms of mycotoxicosis. Neutropenia and lymphopenia were significant in blood samples. Serum lysozyme activity decreased by 1.2 µg/ml compared with controls. The complement activity decreased twofold. The phagocytosis index decreased by 0.54–1.2. Total serum immunoglobulins decreased by 3.03 mg/ml and IgM by 1.11 mg/ml compared with controls.

Several field observations have been made relating aflatoxicosis with increased incidence and severity of several poultry diseases. Samvanshi et al. (1992) reported on enhanced susceptibility to aflatoxicosis in Infectious Bursal Disease on broiler farms in Uttar Pradesh, India. Rao et al. (1985) reported on a severe outbreak of aspergillosis in Khaki Campbell ducklings receiving maize containing 0.28 p.p.m. aflatoxin. They suggested that the presence of aflatoxin in the feed may have predisposed the ducklings to *Aspergillus* infection.

Hegazy et al. (1991) studied the interaction of naturally occurring aflatoxins in poultry feed and immunization against fowl cholera. Out of 1175 poultry feed samples examined, 30.7% proved positive for aflatoxin. Outbreaks of fowl cholera were diagnosed on two farms where aflatoxin was detected in the rations used. The impact of aflatoxins in the feed on the efficacy of immunization against fowl cholera was monitored by a haemagglutination test and found in chickens of the involved farms, which were compared with groups of chickens and ducks fed on aflatoxin-free rations and vaccinated with the same polyvalent fowl cholera bacterin. The antibody titre of the chickens fed aflatoxin-free diets was 4–15 times higher than in those of the involved farms.

In summary, aflatoxicosis reduces the ability of the chicken to synthesize proteins, and thus the ability to synthesize antibodies is also reduced. This will result in very low antibody titres if aflatoxin has been consumed prior to, during or after antigen administration or exposure. Aflatoxin ingestion has been shown to cause atrophy of the bursa and the thymus. Primary lymphatic tissue, such as the bursa and thymus, is required for development of immunity. Thus, because of the effect on these tissues, chickens will show deficiencies in both humoral and cell-mediated immunity. Aflatoxicosis has been demonstrated to cause an impairment in the development of immunity to Infectious Bursal Disease (gumboro), Newcastle disease, Marek's disease, salmonellosis and coccidiosis. Furthermore, aflatoxicosis can result in vaccine failures even though the quality of the vaccine may be excellent and the vaccination techniques adequate.

Aflatoxin and vitamin nutrition

Aflatoxins interact with the fat-soluble vitamins. They have been shown to depress hepatic storage of vitamin A and to increase the dietary requirement for vitamin D_3 by 6.6 IU/kg of diet for each 1 p.p.m. of aflatoxin B_1 in the ration (Bird, 1978). We have previously mentioned that aflatoxin-fed chicks show increased prothrombin times, capillary fragility and bruise susceptibility, which indicates a relationship to vitamin K. Bababunmi and Bassir (1982) reported a delay in blood clotting of chickens and ducks induced by aflatoxin treatment.

These studies all indicate that a marginal deficiency of these vitamins may be aggravated in the presence of aflatoxins. One exception seems to be thiamine, whose deficiency had a protective effect on aflatoxicosis (Wyatt et al., 1972).

Vitamin requirements of the chicken are therefore increased during aflatoxicosis, as well as certain amino acid requirements and protein in general. Mycotoxins are detoxified in the liver utilizing the glutathione system. Glutathione contains cystine, a derivative of methionine. Fortification of poultry rations with synthetic methionine, over and above National Research Council (NRC, 1994) requirements, has been shown to alleviate the growth depression usually seen during aflatoxicosis. In general, the supplementation of poultry feeds with high levels of methionine, selenium and vitamins during aflatoxicosis has been found to be beneficial.

Aflatoxin and zinc in feeds and feedstuffs

Ginting and Barleau (1987) studied 48 samples of poultry feeds and 53 maize samples in Bogor, India. Data showed a significant correlation between aflatoxin and zinc content in both maize and poultry feed. The zinc content of the feed samples was about ten times the reported nutrient requirements for chickens. Jones and Hamilton (1986) studied factors influencing fungal activity in feeds. As feed moved from the feed mill to the feeder pans, there was an increase in fungal activity, and this increase was associated with an increase in zinc concentration. Jones et al. (1984) analysed feed samples from five commercial chicken operations and found that aflatoxin content correlated with zinc content. They suggested that stricter control of zinc levels during manufacture could reduce aflatoxin contamination of chicken feed.

Cyclopiazonic Acid

Cyclopiazonic acid is a toxic indole-tetrameric monobasic acid that was first isolated from cultures of *Penicillium cyclopium* (Aolzapfel, 1968). It occurs naturally in *Aspergillus flavus*-contaminated maize. CPA and AF represent a potential problem in poultry feeds. Cullen *et al.* (1988) were among the first to report on the toxic effect of CPA in broilers. Widiastuti *et al.* (1988a,b) detected CPA in 81% of samples analysed at a poultry feed mill in Indonesia. Since CPA and AF often occur together, they present a potential threat to poultry through contamination of feeds.

Smith *et al.* (1992) demonstrated that both AF and CPA alone or in combination limit broiler performance and their effects are additive. CPA is associated with haemorrhagic lesions seen on thigh muscles and some believe that it is the cause of haemorrhagic syndrome. Chelation of calcium, magnesium and iron may be the main cause of CPA toxicity and the reason for cracks or thin shells in hens fed CPA.

Sterigmatocystin

Sterigmatocystin is a precursor of aflatoxin B_1 in the biological pathway. It is produced by *Aspergillus versicolor*, *Aspergillus nidulas* and other *Aspergillus* species. Sterigmatocystin is a carcinogen, with the liver as the target organ. It is found in heavily moulded grains, rice bran, sunflower meal, soybean meal, groundnut meal, etc. (Devegowda and Murthy, 2005).

Begum and Samajpati (2000) studied the occurrence of mycotoxins in rye, mustard, soybeans and groundnuts in India and found that 195 toxic fungi existed in eight different oilseed samples. Fourteen of these were able to synthesize aflatoxin B_1 and sterigmatocystin.

Citrinin

Citrinin is a mycotoxin produced by various species of the genera *Penicillium* and *Aspergillus*. It is a nephrotoxin whose toxicity is also characterized by reduced growth, decreased feed consumption and increased water consumption (Wyatt, 1988a). Increased water intake is followed by acute diarrhoea, and within 6 h after removal of citrinin from the diet, water consumption is normal (Ames *et al.*, 1976). Citrinin is similar to ochratoxin, and some of the fungi that produce ochratoxin also produce citrinin. These two mycotoxins therefore may be present in the feed at the same time. It is interesting to note that the increase in water consumption caused by citrinin is counteracted by the presence of ochratoxin (Devegowda and Murthy, 2005).

Citrinin at dietary levels of 250 p.p.m. was shown by Ames *et al.* (1976) to cause significant increases in the size of the liver and kidney, by 11 and 22% respectively. There were also alterations in serum sodium levels. Necropsy of birds with citrinin toxicity revealed the presence of pale and swollen kidneys. Roberts and Mora (1978) confirmed the effect of citrinin on water consumption and the resulting diarrhoea and observed haemorrhagic jejunums and mottled livers with dietary levels of 130 and 160 p.p.m. These workers also reported that *Penicillium citrinum*-contaminated maize free of citrinin was also toxic to young chicks. This demonstrates that *P. citrinum* may be producing toxic principles other than citrinin.

Dietary citrinin at levels as high as 250 p.p.m. fed to laying hens for 3 weeks had no effect on body weight, feed consumption or egg production (Ames *et al.*, 1976). Laying hens, like broiler chicks, however, developed acute diarrhoea after initiation of citrinin feeding.

Smith *et al.* (1983) studied the effects of temperature, moisture and propionic acid on mould growth on whole maize and subsequent citrinin production. They found that temperature was the predominant factor in determining the degree of mould on maize. At each moisture level studied (10, 15 and 20% added H_2O), mould growth was greater on samples incubated at room temperature than on samples incubated at a lower temperature. The effectiveness of propionic acid depended on the moisture content of the maize. Higher levels of propionic acid were needed to prevent mould growth and toxin production as the moisture content of the maize increased.

Fumonisins

Fusarium contamination of poultry feeds came to the forefront in the USA in the late 1980s when it was affecting several poultry operations and causing severe growth depression, decreased feed intake, high mortality, oral lesions and proventricular and gizzard erosion in young broilers (Wyatt, 1988a). The predominant species found in those contaminated feeds was *Fusarium moniliforme* (Wu *et al.*, 1991), and this species produces a group of toxins known as fumonisins.

Fumonisins are water-soluble toxins that include fumonisin A_1, A_2, B_1, B_2, B_3 and B_4 (Gelderblom *et al.*, 1992). Studies with broilers have shown that a high concentration of fumonisin B_1 present in *Fusarium moniliforme* cultures caused poor performance and increased organ weights, diarrhoea, multifocal hepatic necrosis, biliary hyperplasia and rickets (Brown *et al.*, 1992; Ledoux *et al.*, 1992). Weibking *et al.* (1993) showed that broiler chicks fed 450 mg fumonisin B_1 or over per kg diet had significantly lower feed intakes and body-weight gains, increased liver and kidney weights and increased mean cell haemoglobin concentration. These workers reported that levels of fumonisin B_1 as low as 75 mg per diet may have an effect on the physiology of chicks because of its inhibition of sphingolipid biosynthesis. Wu *et al.* (1991) found that some isolates of *Fusarium moniliforme*, when fed to broiler chicks, resulted in decreased antibody responses to sheep red blood cells. Feeding grains naturally contaminated with *Fusarium* mycotoxins, even at levels regarded as high, exerted only minor adverse effects on plasma chemistry and haematology of ducklings (Chowdhury *et al.*, 2005) and production parameters were unaffected in this avian species.

Yegani *et al.* (2006) found that feeding grains contaminated with *Fusarium* mycotoxins could affect performance and immunity in broiler breeder hens. Fumonisins are found all over the world in warm as well as temperate regions. They have also been found as co-contaminants with other toxins such as aflatoxin, DON, ochratoxins and zearalenone. Sydenham et al. (1992) tested 21 samples of *Fusarium moniliforme*-contaminated feeds associated with outbreaks of mycotoxicoses in the state of Parana in Brazil. Fumonisin 1 and 2 were detected in 20 and 18 of the 21 feed samples, respectively. Finally, tibial dyschondroplasia, recognized as a developmental abnormality in broilers subjected to dietary mineral imbalance, can also be caused by *Fusarium*-produced toxins (Haynes et al., 1985; Haynes and Walser, 1986). Krogh et al. (1989) reported a 56% incidence of tibial dyschondroplasia on broiler farms in Denmark using feed contaminated with *Fusarium* spp. and containing fusarochromanone.

Ochratoxins

Ochratoxins were isolated as a result of feeding mouldy maize cultures to ducklings, rats and mice. They were first identified by South African researchers in 1965 (Devegowda and Murthy, 2005). Several species of *Aspergillus* and *Penicillium* fungi produce ochratoxin. The World Health Organization (WHO, 2001) proposed a maximum limit of 5 mg/kg in cereals and cereal products. Devegowda et al. (1998) reported that OA is produced mainly by *Aspergillus* species in warm countries and by *Penicillium* species in temperate countries. Ochratoxin A (OA) is the major toxic principle from *Aspergillus ochraceus*. It has been detected in maize, barley, oats, groundnuts, white beans and mixed feeds. Dwivedi and Burns (1986) reviewed the occurrence of OA and its effect in poultry and reported that OA-induced nephropathy has been recorded in natural field outbreaks in poultry from many countries. Ochratoxins do not appear to be as widely spread in the hot regions of the world as aflatoxins. O'zpnar et al. (1988) examined 74 samples by thin-layer chromatography from 15 feed factories in the Marmara area in Turkey and found 32 samples containing aflatoxin but none contained ochratoxins. Ochratoxicosis has been reported in broilers, layers and turkeys. Field outbreaks covering seven different states in the USA were recorded, with levels of ochratoxin in feed ranging from 0.3 to 16.0 p.p.m. Doerr et al. (1974) reported that, on a comparative basis, ochratoxin is more toxic to chicks than either aflatoxin or T-2 toxins. The minimal dietary growth-inhibitory dose for young broiler chicks is 2 p.p.m. for ochratoxin, whereas 2.4 and 4 p.p.m. are required for growth inhibition by aflatoxin and T-2 toxin, respectively.

In broilers, gross pathological findings include severe dehydration and emaciation, proventricular haemorrhages and visceral gout, with white urate deposits throughout the body cavity and internal organs. The target organ is the kidney, since OA is primarily a nephrotoxin affecting kidney function through disruption of the proximal tubules (Wyatt, 1988a).

Kidney damage mediated through nephrotoxic agents can result in hypertension. Several studies have shown deleterious effects on broiler chicks from feeding graded levels of OA up to 8 p.p.m. (Huff and Hamilton, 1975; Huff and Doerr, 1981). Huff and Hamilton (1975) observed poor weight gains, decreased pigmentation and very poor efficiency of feed utilization. These workers found reduced levels of plasma carotenoids in chicks fed diets containing 4 or 8 mg/kg OA. This depression of pigmentation is of economic importance as underpigmented chickens have decreased market value in many countries. Another factor of economic importance in broiler production is bruising. Huff et al. (1983) compared the effects of OA with those of aflatoxin and found that both are implicated in increased susceptibility to bruising in broilers. Furthermore, they observed that the effects of OA are longer-lasting than those of aflatoxin. Bone strength in broilers has been studied in ochratoxicosis, and Huff et al. (1974) demonstrated that ochratoxin causes a decrease in bone strength and a rubbery condition of the bones related to increased tibial diameters and possibly poor mineralization of bone tissue.

Mature laying hens fed OA show lowered egg production, growth depression and high morbidity. Prior and Sisodia (1978) reported that levels as low as 0.5 p.p.m. of OA cause a decrease in egg production and feed consumption. Fertility and hatchability of eggs, however, were unaffected by OA.

Scholtyssek et al. (1987) fed levels of OA ranging from 0 to 5.2 p.p.m. to laying hens and observed a decrease in body weight, feed consumption and egg production. They also observed some changes in functional traits of the eggs produced by those hens.

Ochratoxin has also been shown to affect the immune system in poultry. Reduced lymphoid organ size (bursa, spleen, thymus) has been observed after feeding the toxin. Chang et al. (1979) found a reduced number of lymphocytes in chickens fed 0.5 p.p.m. ochratoxin. Campbell et al. (1983) found a reduced number of lymphocytes when they fed 2 p.p.m. ochratoxin.

Dwivedi and Burns (1984) found a reduced number of immunoglobulin-containing cells in the lymphoid organs and reduced concentrations of immunoglobulins in sera from chickens fed 2 p.p.m. ochratoxin. Campbell et al. (1983) could not find any effect on antibody titres in chickens fed the same level of ochratoxin.

Residues of OA were found in liver, kidney and muscle, but not in eggs, fat or skin. Within 24 h after ending ochratoxin feeding, no residues were found in muscle. Residues in liver and kidney persisted after 48 h post-ochratoxin feeding (Huff et al., 1974). Bohn (1993) suggested that ochratoxin content in feed should be kept as low as possible because it accumulates in poultry tissues and its carcinogenic effects cannot be excluded.

Oosporein

Oosporein is one of the mycotoxins that is considered nephrotoxic. In other words, affected birds will show marked swelling of the kidneys. It is also known to cause severe growth depression and high mortality. During oosporein toxicosis, urate accumulation in the joints and covering the visceral organs is very common. Proventriculitis with loss of proventricular rigidity has also been reported, along with slight haemorrhage at the junction of proventriculus and gizzard. Wyatt (1988a) reported on a case that was confirmed in Venezuela, as well as several cases of natural outbreaks in the eastern USA. Manning and Wyatt (1984) found that the salts of oosporein (particularly K salt) are more toxic to broilers than the organic acid, and the natural occurrence of the salt form of oosporein increases the toxicity of *Chaetonium*-contaminated maize.

Pegram *et al.* (1982) fed turkey poults diets containing oosporein at 0, 500, 1000 and 1500 µg/g from hatching until three weeks of age. Dietary oosporein caused severe visceral and articular gout. A low level of 500 µg/g did not produce gout or cause mortality. These results confirmed the classification of oosporein as a nephrotoxin and aetiological agent of gout in birds.

Ross *et al.* (1989) used chemical ionization mass spectrometry for the purposes of confirming thin-layer chromatography tests for the presence of oosporein in poultry feeds. Oosporein was isolated in pure form from a culture of *Chaetonium trilaterale*. This method was successfully applied to extracts of feed containing 5 p.p.m. oosporein. Rottinghaus *et al.* (1989) reported on a rapid screening procedure, involving extraction, column clean-up and detection by thin-layer chromatography, for oosporein in feed.

T-2 toxin

This is a trichothecene produced by *Fusarium tricinctum*. In initial studies dealing with trichothecenes in poultry, Wyatt *et al.* (1972) found that necrotic lesions were produced in young broilers fed a diet containing purified T-2 toxin. These studies showed that the degree of oral necrosis was dose-related and that oral necrosis was the primary effect of T-2 toxicosis. Oral inflammation occurred with dietary levels of T-2 toxin that did not reduce growth or feed efficiency (Wyatt *et al.*, 1973a). The presence of oral lesions in poultry is regarded as the primary means of field diagnosis of T-2 toxicosis and other trichothecene toxicities. The oral lesions appear first on the hard palate and along the margin of the tongue, and later there are large caseous discharges around the corner of the mouth and along the margin of the beak. They are raised lesions, yellowish-white in colour and caseous in texture (Wyatt *et al.*, 1972).

T-2 toxin also affects the nervous system of poultry (Wyatt *et al.*, 1973b). Abnormal positioning of wings, hysteroid seizures and impaired righting reflex have all been reported by these workers. Another symptom of T-2 toxicosis is abnormal feathering. At or above growth-inhibitory doses (4 p.p.m.), the feathering process is disturbed and the feathers present are short and protruding at odd angles. The mechanism responsible for this abnormal feathering is not known, but malnutrition may be a cause since feed intake in these birds is severely reduced.

Laying hens are also affected by T-2 toxin. Besides the oral response, there is reduced feed intake and body-weight loss. Egg production drops in

7–14 days after start of T-2 toxin feeding. Shell thickness and strength are reduced (Chi *et al.*, 1977). Speers *et al.* (1971) also demonstrated that, when laying hens are fed balanced rations with 2.5 and 5% maize invaded by *F. tricinctum* (8 and 16 p.p.m.), feed consumption, egg production and weight gain are reduced.

Chi *et al.* (1978) demonstrated the transmission of T-2 toxin to the egg, 24 h after dosing. Hatchability and egg production were reduced by diets containing 4, 6 and 8 p.p.m. T-2 toxin. Allen *et al.* (1982) studied the effects of *Fusarium* cultures, T-2 toxin and zearalenone on reproduction of turkey females. Their results show that, although fertility was not affected by *Fusarium* mycotoxins, hatchability was decreased by each of the three *Fusarium* cultures fed. Pure zearalenone or T-2 toxin, however, did not affect hatchability, which suggests that other mycotoxins in these cultures are potent embryotoxic agents. In the same study, they undertook an assessment of the effects of *Fusarium* cultures on the immune responses of female turkeys to a killed ND virus vaccine. There was no indication of suppression of the immune response from feeding *Fusarium* mycotoxins. Richard *et al.* (1978) reported that T-2 toxin causes a reduction in lymphoid organ size (bursa, thymus, spleen). The effect on the immune system has been noted by Boonchuvit *et al.* (1975) to be an increase in sensitivity to *Salmonella* infections.

Vomitoxin (deoxynivalenol – DON)

Vomitoxin is a potent mycotoxin to swine that causes vomiting in affected animals. It is a member of the trichothecenes, the same group of toxins to which the T-2 toxin belongs, and is thus produced by moulds belonging to the *Fusarium* genus.

Wyatt (1988a) summarized the most typical effects that are associated with tricothecene contamination of feedstuffs observed in the commercial poultry industry as follows:

1. Necrotic cream-coloured lesions in the oral cavity of affected birds.
2. Severe decreases in feed intake during the period of time the toxin is being consumed.
3. Increases in feed intake during recovery from the toxicosis.
4. Very small spleen size in affected birds.
5. Mild enteritis.

Harvey *et al.* (1991) investigated the effects of deoxynivalenol (DON)-contaminated wheat diets on haematological measurements, cell-mediated immune responses and humoral immune responses of Leghorn chickens. The authors concluded that subtle changes in haematological and immunological parameters could affect productivity or increase susceptibility to infection and thus caution should be exercised when utilizing DON-contaminated feedstuffs in poultry diets.

Bergs *et al.* (1993) reported that DON increases the incidence of developmental anomalies in chicks hatched from hens receiving the toxin in feed at levels ranging from 120 to 4900 μg/kg. The most frequent major malformations were cloacal atresia and cardiac anomalies, while minor malformations included unwithdrawn yolk sac and delayed ossification.

Behlow (1986) reported that DON was costing the poultry industry millions of dollars in the USA alone, due to depression of feed efficiency and growth rate, reproductive disorders and low product output.

Muirhead (1989) showed that, out of 1018 feed samples tested in the USA, 406 (40%) were positive for vomitoxin.

Zearalenone

Zearalenone is a mycotoxin that is produced by many species of *Fusarium*, but the most common and well-known producers are *Fusarium graminearum* and *Fusarium culmorum*. They produce the toxin in the field during the growing stage and during storage of cereals at high moisture. Zearalenone is known to possess a strong oestrogenic activity in swine. Besides cereals, other feedstuffs, such as tapioca, manioc and soybeans, can be contaminated (Gareis *et al.*, 1989).

Zearalenone has very low acute toxicity in chickens. Feeding high zearalenone levels (> 300 p.p.m.) in the feed to female broiler chicks results in increased comb, ovary and bursa weights (Speers *et al.*, 1971; Hilbrich, 1986). The combs in male chicks were decreased when high levels of the toxin were fed.

The effects of zearalenone on laying hens are minimal, even when it is fed at high concentration (Allen *et al.*, 1981). Egg production is rarely reduced, with the exception of a few field cases where a reduction in egg production has been reported when both zearalenone and DON have been found in the feed. Bock *et al.* (1986), however, reported on a case of high mortality in broiler breeders due to salpingitis, with a possible role for zearalenone. The histopathological examination showed chronic salpingitis and peritonitis. Examination of feed extracts by radio-receptor assay indicated a high degree of oestrogenic activity. Examination by thin-layer and high-pressure liquid chromatography indicated that zearalenone was present in concentrations of up to 5 μg/g of feed.

Detection of Mycotoxins

Although mycotoxins have been shown to produce numerous physiological and production changes in various species of poultry, none of these changes are specific enough to be used in a differential diagnosis. Furthermore, the continued effects of mycotoxins under field conditions in warm regions make the clinical diagnosis very difficult. For example, Huff *et al.* (1986) demonstrated an additive effect of the simultaneous administration of aflatoxin and vomitoxin to broiler chicks. Ultrastructural lesions in the kidneys of Leghorn chicks were exacerbated when both citrinin and ochratoxin were fed and the severity of lesions was modified by the duration of administration (Brown *et al.*, 1986). Therefore final diagnosis of toxicity should depend on measurement of the mould and the amount of toxin present in the feed.

In the case of aflatoxin, rapid screening methods of detection are now available. These include examination for the presence of the *Aspergillus flavus* mould in damaged groundnut kernels and examination of cottonseed and maize for the bright greenish-yellow fluorescence frequently observable when material contaminated with aflatoxin is exposed to ultraviolet illumination. There are now also some rapid minicolumn chemical tests that can detect specified levels of aflatoxin in various materials. Tutour *et al.* (1987) gave limits of detection for aflatoxin B_1, using the mini-column technique, of 5 μg/kg in wheat, maize and soyabean meal, 7 μg/kg in flax cake, 10 μg/kg in wheat bran and 15 μg/kg in mixed feeds.

Test kits for detection of a range of mycotoxins are available for thin-layer chromatography and can be used in diagnostic laboratories and for quality control of ingredients in hot countries (Gimeno, 1979).

Wyatt (1988b) studied the black light test for bright greenish-yellow fluorescence and concluded that there are major limitations to this test and that the ELISA technique is more accurate.

Immunoassay offers considerable promise in detecting, identifying and quantifying mycotoxins in feed ingredients. Commercial kits employing specific monoclonal antibodies are now available, which simplify the processes of extraction and detection.

Biological assay using susceptible species such as ducklings can be used in areas where analytical equipment is unavailable. These methods, however, are often expensive and time-consuming. Furthermore, to confirm a diagnosis, a sufficient quantity of a suspect feed needs to be available for a controlled-feeding trial.

In areas where a quick evaluation needs to be made, visual examination of maize can be used and the absence of abnormal kernels indicates higher-quality grain since the highest levels of mycotoxins occur in abnormal kernels.

It has been observed that mould development in stored ground grain significantly reduces the fat content of these grains (Bartov *et al.*, 1982). The fat content of diets containing these grains also decreased after a short period of storage. Therefore, the determination of dietary fat level, which is a relatively simple procedure, could be used as an additional estimate to evaluate fungal activity in poultry diets.

Control of Mycotoxins

The best control of mycotoxin formation is to prevent the development of fungi in feedstuffs and complete feeds. Good conditions during harvest, transportation and storage of the feedstuffs are important in preventing the growth of moulds. Use of mould inhibitors is sometimes warranted when good management and handling conditions are not easy to attain. Propionic acid has been extensively used for that purpose and the level needed depends on the storage conditions of the feedstuffs and their moisture content. Smith *et al.* (1983) reported that, as the moisture content of maize increased, higher levels of propionic acid were needed to prevent mould growth and toxin production.

Propionic acid is now commonly used as an antifungal agent for maize and does not reduce the value of this grain as an animal feed. Both ammonia and propionic acid significantly reduce mould growth and subsequent formation of aflatoxin and ochratoxin and they both should have practical application for preventing the formation of mycotoxins in stored maize. Ammoniation treatment has been reported by several workers to be effective in minimizing the effects of aflatoxins (Ramadevi *et al.*, 1990; Phillips *et al.*, 1991). Maryamma *et al.* (1991) tested five different substances for the inactivation of aflatoxin, both *in vivo* and *in vitro*, and found that kaolin and bleaching powder were most effective and did not cause any deleterious effects in the White Pekin ducks that they used in those tests. Jansen van Rensburg *et al.* (2006) tested the efficacy of a humic acid (oxihumate) as an aflatoxin binder in broilers. This acid was effective in diminishing the adverse effects of aflatoxin on body weight, liver damage, and stomach and heart enlargement, as well as some haematological and serum changes.

Different extraction procedures have been tried for the removal of aflatoxin. It has been shown that the most successful application is the removal of aflatoxin from oils during normal commercial processing. A variety of polar solvents have been found to be effective in extracting aflatoxins. Examples of these are 95% ethanol, 90% aqueous acetone, 80% isopropanol, hexane-ethanol, hexane-methanol, etc. Although these solvents are effective in removing all the aflatoxin from the meal, they add to the processing costs.

One approach to the detoxification of aflatoxin is the use of non-nutritive sorptive materials in the diet to reduce aflatoxin absorption from the gastrointestinal (GI) tract. Dalvi and Ademoyero (1984) and Dalvi and McGowan (1984) reported improvement in feed consumption and weight gain when activated charcoal was added to poultry diets containing aflatoxin. Many clays have been used for counteracting mycotoxin toxicity. Those that have been used commercially are the aluminosilicates, the bentonites and the zeolites. Hydrated sodium calcium aluminium silicate at a concentration of 0.5% of the diet significantly diminished many of the adverse effects caused by aflatoxin in chickens (Kubena *et al.*, 1990a,b) and turkeys (Kubena *et al.*, 1991). Kubena *et al.* (1993a), in a study on the efficacy of hydrated sodium calcium aluminium silicate to reduce aflatoxin, found that this compound gave total protection against the effects caused by aflatoxin. Scheideler (1993) compared four types of aluminosilicates for their ability to bind aflatoxin B_1 and prevent the effects of aflatoxicosis. Three of the four aluminosilicates tested by this worker alleviated the growth depression caused by aflatoxin B_1. Kubena *et al.* (1993b) confirmed the fact that hydrated sodium calcium aluminosilicates are protective against the effects of aflatoxin in young broilers and further emphasized that silicate-type solvents are not all equal in their ability to protect against aflatoxicosis. These authors concluded that sorbent compound use may be another tool for the preventive management of mycotoxin-contaminated feedstuffs in poultry. The mechanism of action appears to involve aflatoxin sequestration and chemisorption of hydrated sodium calcium aluminosilicates in the GI tract of poultry, resulting in a major reduction in bioavailability and toxicity (Phillips et al., 1989). Sodium bentonite (0.3%) reduced the incidence and severity of hepatic lesions associated with aflatoxicosis in broilers fed 2.5 mg/kg aflatoxin B_1 and 200 mg/kg fumonisin (Miazzo et al., 2005).

Another group of compounds that have been studied for their efficacy on aflatoxin toxicity are the zeolitic ores. Harvey et al. (1993) evaluated six of these compounds and found that zeolite mordenite ore reduced the toxicity of aflatoxin by 41%, as indicated by weight gains, liver weight and serum biochemical measurements. Effectiveness of zeolites in inhibiting absorption of aflatoxin offers promise for hot climates to alleviate the effects of mycotoxins that are really damaging in those regions. In certain areas of the world it may be necessary to use a combination of antifungal compounds in the feed as well as a physical adsorbent. Ramkrishna et al. (1992) concluded that a combination of organic acids with physical adsorbent is most effective in combating toxicosis in India. Saxena (1992) feels that one of the major causes of aflatoxicosis in India is the use of contaminated groundnut (peanut) meals and, by monitoring this ingredient or the use of alternative protein sources in poultry diets, the incidence can be drastically reduced.

In recent years there has been a great deal of work on the use of yeast-derived products to overcome the harmful effects of mycotoxin-contaminated feeds. Modified glucomannan has been found to adsorb several mycotoxins better than inorganic binders (Mahesh and Devegowda, 1996; Volkl and Karlovsky, 1998). Besides improving performance, Raju and Devegowda (2000) found that modified glucomannan improved serum biochemical and haematological parameters. The addition of modified glucomannan protected broilers from feeds contaminated with both T-2 toxin and DON (Smith et al., 2001). Modified glucomannan also rapidly adsorbs aflatoxin and T-2 toxin in the GI tract at different intervals after ingestion by broilers (Murthy and Devegowda, 2004). Aravind et al. (2003) studied the efficacy of esterified glucomannan counteracting the toxic effects of mycotoxins in naturally contaminated diets (aflatoxin, 168 p.p.b., ochratoxin 8.4 p.p.b., zearalenone 54 p.p.b. and T-2 toxin 32 p.p.b.). These workers found that the addition of dietary esterified glucomannan reduces the toxic effects of these mycotoxins.

Mannan-oligosaccharides were also found to be effective in adsorbing and degrading aflatoxin B_1, reducing gastrointestinal absorption of the toxin in laying hens and its levels in eggs and liver tissues (Zaghrini et al., 2005).

Several other methods of control have been under investigation. Heating, roasting and chemical inactivation have been studied, and some of these methods appear to be promising for the future. Nearly all of these have dealt with aflatoxin and very few studies are being conducted on other mycotoxins. Gonzalez et al. (1991) evaluated the possible reduction of *Aspergillus flavus* contamination of grains by ozone and the effect of ozone on the nutritional value of maize. Treatment with ozone (60 mg/l) for 4 min killed conidial suspensions of *Aspergillus flavus*. These workers did not detect any significant difference in live weight of chickens consuming ozonized maize compared with control birds, nor did they find any pathological changes in those birds. Another possibility for alleviation of unfavourable effects of mould-contaminated feeds was the use of antibiotics in broiler diets. Ivandija (1989) tested several antibiotics in feeds that were contaminated with aflatoxins and zearalenone. Of the antibiotics tested, oxytetracycline at 50 p.p.m. gave the most favourable results.

Plant geneticists are working on the development of commercially acceptable varieties of grains and oilseeds that would resist toxin-producing moulds or possibly inhibit the production of the toxin. Such a task may be more difficult than breeding for regular plant disease resistance.

Since in mycotoxicoses there is a decrease in energy utilization by birds, owing to an impairment of lipid absorption or metabolism and in part to a decrease in energy content of the mouldy feed, Bartov (1983) studied the effects of adding both fat and propionic acid to diets containing mouldy maize. He concluded that the nutritional value of diets containing mouldy grain can be completely restored if their fat content is increased in proportion to the amount lost in the mouldy grains and propionic acid is added. Recently, Hazzele *et al.* (1993) showed that some of the detrimental effects of OA in the diet can be counteracted by dietary supplementation of 300 mg ascorbic acid/kg diet.

Chickens rapidly recover from chronic poor performance when changed to diets free from moulds or mycotoxins. Growing chickens recover in about 1 week and production returns to normal within 3–4 weeks in adult birds.

Conclusions

This chapter has dealt with the problems associated with the contamination of feeds with various mycotoxins. The mycotoxins presented were aflatoxins, cyclopiazonic acid, sterigmatocystin, citrinin, fumonisins, ochratoxins, oosporein, T-2 toxin, vomitoxin and zearalenone. Since aflatoxins have been researched most and since they appear to be the most prevalent in hot climates, a detailed description of their occurrence, effects on broilers, layers and breeders, residues in eggs and poultry meat, influence on resistance and immunity, and relationship to certain vitamins and trace minerals has been presented. A general description of methods of detection and control of these mycotoxins has also been included.

The most significant conclusion of this chapter is that all those concerned in the production, processing and marketing of feedstuffs, as well as those concerned in the formulation, production, processing and marketing of complete poultry feeds, should make every effort so that the poultry producer will be provided with feeds that not only are complete from a nutrient requirement standpoint but also have no or minimal levels of mycotoxins in them. Furthermore, poultry producers in those regions should purchase ingredients from reputable suppliers against presentation of analytical certificates from reliable laboratories.

References

Afzali, N. and Devegowda, G. (1999) Ability of modified mannan-oligosaccharide to counteract aflatoxicosis in broiler breeder hens. *Poultry Science* 78 (Suppl. 1), 228 (Abstract).

Allen, N.K., Mirocha, C.J., Aakhus, A.S., Bitgood, J.J., Wedner, G. and Bates, F. (1981) Effect of dietary zearalenone on reproduction of chickens. *Poultry Science* 60, 1165–1174.

Allen, N.K., Jerne, R.L., Mirocha, C.J. and Lee, Y.W. (1982) The effect of a *Fusarium roseum* culture and diacetoxyscirpenol on reproduction of White Leghorn females. *Poultry Science* 61, 2172–2175.

Ames, D.D., Wyatt, R.D., Marks, H.L. and Washburer, K.W. (1976) Effect of citrinin, a mycotoxin produced by *Penicillium citrinum*, on laying hens and young broiler chicks. *Poultry Science* 55, 1294–1301.

Anonymous (1988) Mycotoxin testing: not 'whether' grain is contaminated but 'how much'. *Feed Management* 39, 11.

Aolzapfel, C.W. (1968) The isolation and structure of cyclopiazonic acid, a toxic metabolite of *Penicillium cyclopium* Westling. *Tetrahedrov* 24, 2101–2119.

Aravind, K.L., Patil, V.S., Devegowda, G., Umakantha, B. and Ganpule, S.P. (2003) Efficacy of esterified glucomannan to counteract mycotoxicosis in naturally contaminated feed on performance and serum biochemical and hematological parameters in broilers. *Poultry Science* 82, 571–576.

Bababunmi, E.A. and Bassir, O. (1982) A delay in blood clotting of chickens and ducks induced by aflatoxin treatment. *Poultry Science* 61, 166–168.

Babila, A. and Acktay, B. (1991) Determination of total fungal numbers and toxins in feeds and raw feed materials in outbreaks of mycotoxicosis in poultry. *Pendik Hayran Hastalklar Merkez* 22, 63–85.

Barbour, G.W., Farran, M.T. Usayran, N. and Daghir, N.J. (2008) Evaluation of the physical and chemical characteristics of corn and soybean meal in major poultry feed operations in Lebanon. *World's Poultry Science Journal* (in press).

Bartov, I. (1983) Effects of propionic acid and copper sulfate on the nutritional value of diets containing moldy corn for broilers. *Poultry Science* 62, 2195–2200.

Bartov, I., Paster, N. and Lisker, N. (1982) The nutritional value of moldy grains for broiler chicks. *Poultry Science* 61, 2247–2254.

Beg, M.U., Al-Mutairi, M., Beg, K.R., Al-Mazeedi, H.M., Ali, L.N. and Saeed, T. (2006) Mycotoxins in poultry feed in Kuwait. *Archives of Environmental Contamination and Toxicology* 50, 594–602.

Begum, F. and Samajpati, N. (2000) Mycotoxin production on rice, pulses and oil seeds. *Naturwissen schaften* 87, 275–277.

Behlow, R.F. (1986) Deoxynivalenol said to be costing millions. *Poultry Digest* 45, 512–513.

Bergs, B., Herstad, O. and Naystad, I. (1993) Effects of feeding deoxynivalenol contaminated oats on reproduction performance in White Leghorn hens. *British Poultry Science* 34, 147–159.

Bird, F.H. (1978) The effect of aflatoxin B_1 on the utilization of cholecalciferol by chicks. *Poultry Science* 57, 1293–1296.

Blaney, B.J. and Williams, K.C. (1991) Effective use in livestock feeds of mouldy and weather damaged grain containing mycotoxins – case histories and economic assessments pertaining to pig and poultry industries of Queensland. *Australian Journal of Agriculture Research* 42, 993–1012.

Bock, R.R., Shore, L.B., Samberg, Y. and Perl, S. (1986) Death in broiler breeders due to salpingitis: possible role of zearalenone. *Avian Pathology* 15, 495–502.

Bohn, J. (1993) On the significance of mycotoxins deoxynivalenol, zearalenone and ochratoxin A for livestock. *Archives of Animal Nutrition* A2, 95–111.

Boonchuvit, B. and Hamilton, P.B. (1975) Interaction of aflatoxin and para-typhoid infection in broiler chickens. *Poultry Science* 54, 1567–1573.

Boonchuvit, B., Hamilton, P.B. and Burmiester, H.R. (1975) Interaction of T-2 toxin with salmonella infections in chickens. *Poultry Science* 54, 1693–1696.

Brown, T.P., Manning, R.O., Fletcher, O.J. and Wyatt, R.D. (1986) The individual and combined effects of citrinin and ochratoxin A on renal ultrastructure in layer chicks. *Avian Diseases* 30, 191–196.

Brown, T.P., Rottinghaus, G.E. and Williams, M.E. (1992) Fumonisin mycotoxicosis in broilers: performance and pathology. *Avian Diseases* 36, 450–454.

Campbell, M.L., May, J.D., Huff, W.E. and Doerr, J.A. (1983) Evaluation of immunity of young broilers during simultaneous aflatoxicosis and ochratoxicosis. *Poultry Science* 62, 2138–2144.

Chandrasekaran, D. (1996) Survey on the presence of T-2 toxin and ochratoxin in feed/ingredients in Namakkal area. *Proceedings of the 20th World Poultry Congress*, New Delhi, India, Vol. 4, p. 268.

Chang, C.F. and Hamilton, P.B. (1982) Increased severity and new symptoms of IBD during aflatoxicosis in broiler chicks. *Poultry Science* 61, 1061–1068.

Chang, C.F., Huff, W.E. and Hamilton, P.B. (1979) A leucocytopenia induced in chickens by dietary ochratoxin A. *Poultry Science* 58, 555–558.

Chi, M.S., Mirocha, C.J., Kurtz, H.F., Weaver, G., Bates, F. and Shimoda, W. (1977) Effects of T-2 toxin on reproductive performance and health of laying hens. *Poultry Science* 56, 628–637.

Chi, M.S., Robinson, T.S., Mirocha, C.J. and Reddy, K.R. (1978) Acute toxicity of 12, 13-epoxytrichothecenes in one-day-old broiler chicks. *Applied Environmental Microbiology* 35, 636–640.

Chowdhury, S.R., Smith, T.K., Boermans, H.J., Sefton, A.E., Downey, R. and Woodward, B. (2005) Effects of feeding blends of grains naturally contaminated with *Fusarium* mycotoxins on performance, metabolism, hematology, and immunocompetence of ducklings. *Poultry Science* 84, 1179–1185.

Cullen, J.M., Wilson, M., Hogler, W.M., Ort, J.F. and Cole, R.J. (1988) Histologic lesions in broiler chicks given cyclopiazonic acid orally. *American Journal of Veterinary Research* 49, 728–731.

Daghir, N.J., Hajj, R. and Akrabawi, S.S. (1966) Studies on peanut meal for broilers. *Proceedings of the 13th World's Poultry Congress*, pp. 238–246.

Dalvi, R.R. (1986) An overview of aflatoxicosis of poultry: its characteristics, prevention and reduction. *Veterinary Research Communication* 10, 429–443.

Dalvi, R.R. and Ademoyero, A.A. (1984) Toxic effects of aflatoxin B1 in chickens given feed contaminated with *Aspergillus flavus* and reduction of the toxicity by activated charcoal and some chemical agents. *Avian Diseases* 28, 61–69.

Dalvi, R.R. and McGowan, C. (1984) Experimental induction of chronic aflatoxicosis in chickens by purified aflatoxin B1 and its reversal by activated charcoal, phenobarbitol and reduced glutathione. *Poultry Science* 63, 485–491.

Del Bianchi, M., Oliveira, C.A., Albuquerque, R., Guerra, J.L. and Correa, B. (2005) Effects of prolonged oral administration of aflatoxin B_1 and fumonisin B1 in broiler chickens. *Poultry Science* 84, 1835–1840.

Devegowda, G. and Murthy, T.N.K. (2005) Mycotoxins: their effects in poultry and some practical solutions. In: Diaz, D.E. (ed.) *The Mycotoxin Blue Book*. Nottingham University Press, pp. 25–56.

Devegowda, G., Raju, M.V.L.N., Afzali, N. and Swamm, H.V.L.N. (1998) Mycotoxin picture worldwide: novel solutions for their counteractions. *Feed Compounder* 18(6), 22–27.

Doerr, J.A. and Hamilton, P.B. (1981) Aflatoxicosis and intrinsic coagulating function in broiler chickens. *Poultry Science* 60, 1406–1411.

Doerr, J.A., Huff, W.E., Tung, A.T., Wyatt, R.D. and Hamilton, P.B. (1974) A survey of T-2 toxin, ochratoxin and aflatoxin for their effects on the coagulation of blood in young broilers. *Poultry Science* 53, 1728–1734.

Doerr, J.A., Huff, W.E., Wabeck, C.J., Choloupka, G.W., May, J.D. and Murkley, J.W. (1983) Effects of low level chronic aflatoxicosis in broiler chickens. *Poultry Science* 62, 1971–1977.

Dwivedi, P. and Burns, R.B. (1984) Effect of ochratoxin A on immunoglobulins in broiler chicks. *Research in Veterinary Science* 36, 117–121.

Dwivedi, P. and Burns, R.B. (1986) The natural occurrence of ochratoxin A and its effects in poultry. A review. I. Epidemiology and toxicity. *World's Poultry Science Journal* 42, 32–47.

Edds, G.T. (1973) Acute aflatoxicosis: a review. *Journal of the American Veterinary Medical Association* 162, 304–309.

Elzanaty, K., El-Maraghy, S.S.M., Abdel-Motelib, T. and Salem, B. (1989) Effect of aflatoxins on Newcastle disease vaccination in broilers. *Assiut Veterinary Medical Journal* 22, 184–189.

FAO (Food and Agriculture Organization) (1995) Worldwide regulations for mycotoxins, 1995 FAO Food and Nutrition Paper 64, FAO, Viale delle Terme di Caracalla, odos. Rome, Italy.

Gareis, M., Bauer, J., Enders, C. and Gedek, B. (1989) Contamination of cereals and feed with *Fusarium* mycotoxins in European countries. In: Chalkowski, J. (ed.) *Fusarium Mycotoxins, Taxonomy and Pathogenicity*. Elsevier, Amsterdam, pp. 441–472.

Garlich, J.D., Tung, H.T. and Hamilton, P.B. (1973) The effects of short term feeding of aflatoxin on egg production and some plasma constituents of the laying hen. *Poultry Science* 52, 2206–2211.

Gelderblom, W.C.A., Marasas, W.F.O., Vieggar, R., Thiel, P.G. and Cawooe, M.E. (1992) Fumonisins: isolation, chemical characterization and biological effects. *Mycopathologia* 117, 11–16.

Giambrone, J.J., Ewert, D.L., Wyatt, R.D. and Eidson, C.S. (1978) Effect of aflatoxin on the humoral and cell-mediated immune system of the chicken. *American Journal of Veterinary Research* 39, 305–308.

Gimeno, A. (1979) Thin layer chromatographic determination of aflatoxins, ochratoxins, sterigmatocystin, zearalenone, citrinian, T-2 toxin, diacetoxyscirpenol, penicillic acid, patulin and penitren. *Journal of the Association of Official Analytical Chemists* 62, 579–586.

Ginting, N. and Barleau, B.I. (1987) Correlation between aflatoxin and zinc in maize and chicken feed. *Penyakit Hewan* 19, 94–96.

Girish, C.K. and Devegowda, G. (2004) Evaluation of modified glucomannan (mycosorb) and hydrated sodium calcium aluminosilicate to ameliorate the individual and combined toxicity of aflatoxin and T-2 toxin in broiler chickens. *Australian Poultry Science Symposium* 16, 126–129. Sydney, Australia.

Goldblatt, L.A. (1969) *Aflatoxins*. Academic Press, New York.

Gonzalez, M., Malerio, J. and Muroz, C. (1991) Ozone action on *Aspergillus flavus* cultures and on feeds for broiler chickens. *Revista de Salud Animal* 13, 193–198.

Gumbman, M.R., Williams, G.N., Booth, A.N., Vohra, P., Ernst, R.A. and Bettard, M. (1970) Aflatoxin susceptibility in various breeds of poultry. *Proceedings of the Society of Experimental Biology and Medicine* 34, 683–686.

Hamilton, P.B. (1990) Problems with mycotoxins persist, but can be lived with. *Feedstuffs* 62, 22–23.

Hamilton, P.B. and Garlich, J.D. (1971) Aflatoxin as a possible cause of fatty liver syndrome in laying hens. *Poultry Science* 50, 800–804.

Harvey, R.B., Kubena, L.F., Huff, W.E., Elissalde, M.H. and Phillips, T.D. (1991) Hematologic and immunologic toxicity of deoxynivalenol-contaminated diets to growing chickens. *Bulletin of Environmental Contamination and Toxicology* 46, 410–416.

Harvey, R.B., Kubena, L.F., Elissalde, M.H. and Phillips, T.D. (1993) Eficacy of zeolitic ore compounds on the toxicity of aflatoxin to growing broiler chickens. *Avian Diseases* 37, 67–73.

Haynes, J.S. and Walser, M.M. (1986) Ultrastructure of *Fusarium*-induced tibial dyschondroplasia in chickens: a sequential study. *Veterinary Pathology* 23, 499–502.

Haynes, J.S., Walser, M.M. and Lowler, E.M. (1985) Morphogenesis of *Fusarium* spp.-induced tibial dyschondroplasia in chickens. *Veterinary Pathology* 22, 629–633.

Hazzele, F.M., Guenter, W., Marquart, R.R. and Frohlich, A.A. (1993) Beneficial effects of dietary ascorbic acid supplement on hens subjected to ochratoxin A toxicosis under normal and high temperature. *Canadian Journal of Animal Science* 73, 149–157.

Hegazy, S.M., Azzam, A. and Gabal, M.A. (1991) Interaction of naturally occurring aflatoxins in poultry feed and immunization against fowl cholera. *Poultry Science* 70, 2425–2428.

Hilbrich, P. (1986) Abnormal comb growth in female chicks due to gonadotropic substances in the feed. *Deutch Tierarztliche Wochenschrift* 93, 39–40.

Huff, W.E. (1980) Evaluation of tibial dyschondroplasia during aflatoxicosis and feed restriction in young broiler chickens. *Poultry Science* 59, 991–995.

Huff, W.E. and Doerr, J.A. (1981) Synergism between aflatoxin and ochratoxin A in broiler chickens. *Poultry Science* 60, 550–555.

Huff, W.E. and Hamilton, P.B. (1975) The interaction of ochratoxin A with some environmental extremes. *Poultry Science* 54, 1659–1662.

Huff, W.E., Wyatt, R.D. and Tucker, T.L. (1974) Ochratoxicosis in the broiler chicken. *Poultry Science* 53, 1585–1591.

Huff, W.E., Doerr, J.A., Wabeck, C.J., Choloupka, G.W., May, J.D. and Merkley, J.W. (1983) Individual and combined effects of aflatoxin and ochratoxin A on bruising in broiler chickens. *Poultry Science* 62, 1764–1771.

Huff, W.E., Kubena, L.F., Harvey, R.B., Hagler, W.M., Swanson, S.P., Phillips, T.D. and Greger, C.R. (1986) Individual and combined effects of aflatoxin and deoxynivalenol (DON – vomitoxin) in broiler chickens. *Poultry Science* 65, 1291–1295.

Ivandija, L. (1989) Possibilities for alleviation of unfavourable effects of mould contaminated feeds in broiler chicks. *Krmiva* 31, 39–45.

Jacobson, W.C. and Wiseman, H.G. (1974) The transmission of aflatoxin B_1 into eggs. *Poultry Science* 53, 1743–1745.

Jansen van Rensburg, C., Van Rensburg, C.E.J., Van Ryssen, J.B.J., Casey, N.H. and Rottinghaus, G.E. (2006) In vitro and in vivo assessment of humic acid as an aflatoxin binder in broiler chickens. *Poultry Science* 85, 1576–1583.

Jelinek, C.F., Pohland, A.E. and Wood, G.E. (1989) Worldwide occurrence of mycotoxins in foods and feeds: an update. *Journal of the Association of Analytical Chemists* 72, 223–230.

Jones, F.T. and Hamilton, P.B. (1986) Factors influencing fungal activity in low moisture poultry feeds. *Poultry Science* 65, 1522–1525.

Jones, F.T., Hogler, W.M. and Hamilton, P.B. (1984) Correlation of aflatoxin contamination with zinc content of chicken feed. *Applied and Environmental Microbiology* 47, 478–480.

Kan, C.A., Rump, R. and Kosutzky, J. (1989) Low level exposure of broilers and laying hens to aflatoxin Bl from naturally contaminated corn. *Archiv fur Geflugelkunde* 53, 204–206.

Krogh, P., Christensen, D.H., Hald, B., Harlou, B., Larsen, C., Pedersen, E. and Thane, U. (1989) Natural occurrence of the mycotoxin fusarochromanone, a metabolite of *Fusarium equiseti* in cereal feed associated with tibial dyschondroplasia. *Applied and Environmental Microbiology* 55, 3184–3188.

Kubena, L.F., Harvey, R.B., Huff, W.E., Corrier, D.E., Phillips, T.D. and Rottinghaus, G.E. (1990a) Efficacy of a hydrated sodium calcium aluminosilicate to reduce the toxicity of aflatoxin and T-2 toxin. *Poultry Science* 69, 1078–1086.

Kubena, L.F., Harvey, R.B., Phillips, T.D., Corrier, D.E. and Huff, W.E. (1990b) Diminution of aflatoxicosis in growing chickens by dietary addition of a hydrated sodium calcium aluminosilicate. *Poultry Science* 69, 727–735.

Kubena, L.E., Huff, W.E., Harvey, R.B., Yersin, A.G., Elissaide, M.H., Witzel, D.A., Giroir, L.E., Phillips, T.D. and Petersen, H.D. (1991) Effects of a hydrated sodium calcium aluminosilicate on growing turkey poults during aflatoxicosis. *Poultry Science* 70, 1823–1830.

Kubena, L.F., Harvey, R.B., Phillips, T.D. and Clement, B.A. (1993a) Effect of hydrated sodium calcium aluminosilicates on aflatoxicosis in broiler chicks. *Poultry Science* 72, 651–657.

Kubena, L.E., Harvey, R.B., Huff, W.E., Elissalde, M.H., Yersin, A.G., Phillips, T.D. and Rottinghaus, G.E. (1993b) Efficacy of a hydrated sodium calcium aluminosilicate to reduce the toxicity of aflatoxin and diacetoxy-cirpenol. *Poultry Science* 72, 51–59.

Ledoux, D.R., Brown, T.P., Weibking, T.S. and Rottinghaus, G.E. (1992) Fumonisin toxicity in broiler chicks. *Journal of Veterinary Diagnostic Investigation* 4, 330–333.

Mahesh, B.K. and Devegowda, G. (1996) Ability of aflatoxin binders to bind aflatoxin in contaminated poultry feeds: an invitro study. *Proceedings of the 20th World Poultry Congress*, New Delhi, India, Vol. 4, p. 296.

Manning, R.O. and Wyatt, R.D. (1984) Comparative toxicity of *Chaetonium* contaminated corn and various chemical forms of oosporein in broiler chicks. *Poultry Science* 63, 251–259.

Maryamma, K.I., Rajan, A., Gangadharan, B. and Manmohan, C.B. (1991) In-vivo and in-vitro studies on aflatoxin Bl neutralization. *Indian Journal of Animal Science* 61, 58–60.

Miazzo, R., Peralta, M.F., Salvano, M., Magnoli, C., Ferrero, S., Chiacchiera, S.M., Carvallo, E.C., Rosa, C.A. and Dalcero, A. (2005) Efficacy of sodium bentonite as a detoxifier of broiler feed contaminated with aflatoxin and fumonisin. *Poultry Science* 84, 1–8.

Mohiuddin, S.M. (1992) Effects of aflatoxin on immune response in viral diseases. *Proceedings of the 19th World's Poultry Congress*, Vol. 2 (Suppl.), pp. 50–53.

Muirhead, S. (1989) Studies show cost of mycotoxin to poultry firms. *Feedstuffs* 61, 10.

Murthy, T.N.K. and Devegowda, G. (2004) Efficacy of modified glucomannan (Mycosorb) to absorb aflatoxin B_1 in gut conditions of broiler chickens. *Proceedings of the 22nd World's Poultry Congress*, Istanbul, Turkey, p. 471.

National Research Council (NRC) (1994) Nutrient Requirements of Domestic Animals, No. 7. *Nutrient Requirements of Poultry*, 9th edn. NAS-NRC, Washington, DC.

Osborne, D.J., Wyatt, R.D. and Hamilton, P.B. (1975) Fat digestion during aflatoxicosis in broiler chickens. *Poultry Science* 54, 1802 (Abstract).

O'zpnar, H., O'zpnar, A. and Seael, H.S. (1988) Determination of aflatoxin and ochratoxin in poultry feed components and raw feedstuffs in the Marmara area. *Veteriner Facultesi Dergisi* (Istanbul) 14, 11–18.

Pegram, R.A, Wyatt, R.D. and Smith, T.L. (1982) Oosporein-toxicosis in the turkey poult. *Avian Disease* 26, 47–59.

Pettersson, H. (1991) Mycotoxins in poultry feed and detoxification. *Proceedings of the 8th European Symposium on Poultry Nutrition*, pp. 27–41.

Phillips, T.D., Clement, B.A., Kubena, L.F. and Harvey, R.B. (1989) Prevention of aflatoxicosis in animals and aflatoxin residues in food of animal origin with hydrated sodium calcium aluminosilicates. Healthy animals, safe foods, healthy man. *Proceedings of the World Association of Veterinary Food Hygienists Xth International Symposium*, pp. 103–108.

Phillips, T.D., Sarr, B.A., Clement, B.A., Kubena, L.F. and Harvey, R.B. (1991) Prevention of aflatoxicosis in farm animals via selective chemosorption of aflatoxin. In: Bray, G.A. and Ryan, D.H.J (eds) *Mycotoxins, Cancer and Health*. Louisiana State University, Baton Rouge, pp. 223–237.

Pier, A.C. (1986) Immunomodulation in aflatoxicosis. In: Richard, J.L. and Thurston, J.R. (eds) *Diagnosis of Mycotoxicosis*. Martinus Nijhoff, Dordrecht, the Netherlands, pp. 143–146.

Prior, M.G. and Sisodia, C.S. (1978) Ochratoxicosis in white Leghorn hens. *Poultry Science* 57, 619–623.

Purwoko, H.M., Hold, B. and Wolstrup, J. (1991) Aflatoxin content and number of fungi in poultry feedstuffs from Indonesia. *Letters in Applied Microbiology* 12, 212–215.

Qureshi, M.A., Brake, J., Hamilton, P.B., Hagler, W.M. and Nesheim, S. (1998) Dietary exposure of broiler breeders to aflatoxin results in immune dysfunction in progeny chicks. *Poultry Science* 77, 812–819.

Raina, J.S. and Singh, B. (1991) Prevalence and pathology of mycotoxicosis in poultry in Punjab. *Indian Journal of Animal Sciences* 61, 671–676.

Raju, M.V.L.N. and Devegowda, G. (2000) Influence of modified glucomannan on performance and organ morphology, serum biochemistry and haematology in broilers exposed to individual and combined mycotoxins (aflatoxin, ochratoxin and T-2 toxin). *British Poultry Science* 41, 640–650.

Raju, M.V.L.N. and Devegowda, G. (2002) Esterified glucomannan in broiler chicken diets contaminated with aflatoxin, ochratoxin and T-2 toxin: evaluation of its binding ability (in vitro) and efficacy as immunomodulator. *Australian Journal of Animal Science* 15, 1051–1056.

Ramadevi, V., Rao, P.R. and Moorthy, V.S. (1990) Pathological effects of ammoniated and sundried aflatoxin contaminated feed in broilers. *Journal of Veterinary and Animal Sciences* 21, 108–112.

Ramkrishna, G.T., Devegouda, G., Umesh, D. and Gagendragod, M.R. (1992) Evaluation of mold inhibitors in broiler diets and their influence on the performance of broilers. *Indian Journal of Poultry Science* 27, 91–94.

Rao, D.G., Naidu, N.R.G. and Rao, R.R. (1985) Observations on the concomitant incidence of aflatoxicosis and aspergillosis in khaki Campbell ducklings. *Indian Veterinary Journal* 62, 461–464.

Richard, J.L., Cysewski, S.J., Pier, A.C. and Booth, G.D. (1978) Comparison of effects of dietary T-2 toxin on growth, immunogenic organs, antibody formation, and pathologic changes in turkeys and chickens. *American Journal of Veterinary Research* 39, 1674–1679.

Rizvi, A., Shakoori, A.R. and Rizvi, S.M. (1990) Aflatoxin contamination of commercial poultry feeds in Punjab. *Pakistan Journal of Zoology* 22, 387–398.

Roberts, W.T. and Mora, E.C. (1978) Toxicity of *Penicillium citrinum* AUA 532 contaminated corn and citrinin in broiler chicks. *Poultry Science* 57, 1221–1226.

Ross, P.F., Osheim, D.L. and Rottinghaus, G.E. (1989) Mass-spectral confirmation of oosporein in poultry rations. *Journal of Veterinary Diagnostic Investigation* 1, 271–272.

Rottinghaus, G.E., Sklebor, H.T., Seuter, L.H. and Brown, T.P. (1989) A rapid screening procedure for the detection of the mycotoxin oosporein. *Journal of Veterinary Diagnostic Investigation* 1, 174.

Samvanshi, R., Mohanty, G.C., Verma, K.C. and Kataria, J.M. (1992) Spontaneous occurrence of aflatoxicosis, infectious bursal disease and their inter-action in chicken. *Indian Veterinary Medical Journal* 16, 11–17.

Saxena, H.C. (1992) Two major problems of poultry farms in India: colibacillosis and mycotoxicosis. *Misset World Poultry* 8, 47.

Scheideler, S.E. (1993) Effects of various types of aluminosilicates and aflatoxin Bl on aflatoxin toxicity, chick performance, and mineral status. *Poultry Science* 72, 282–288.

Scholtyssek, S., Niemiec, J. and Baver, J. (1987) Ochratoxin A in layer's feed. 1 Report: influence on laying performance and egg quality. *Archiv fur Geflugelkunde* 51, 234–240.

Shaaban, A.A., Ibrahim, T.A., Mokhtar, Z. and Shehata, A. (1988) Aflatoxin residues in rations of chicken laying flocks in Assiut province. *Assiut Veterinary Medical Journal* 19, 195–199.

Shaaban, F.E., Abou-Hadeed, A.H., El-Shazly, M.O. and Gammel, A.A. (1991) Detection and estimation of aflatoxin in stored grains and the pathological effect of its feeding in rats and chickens. *Egyptian Journal of Comparative Pathology and Clinical Pathology* 4, 339–357.

Smith, E.E., Kubena, L.F., Braithwaite, R.B., Harvey, R.B., Phillips, T.D. and Reine, A.U. (1992) Toxicological evaluation of aflatoxin and CPA in broilers. *Poultry Science* 71, 1136–1144.

Smith, J.W. and Hamilton, P.B. (1970) Aflatoxicosis in the broiler chicken. *Poultry Science* 49, 207–215.

Smith, P.A., Nelson, T.S., Kirby, L.K., Johnson, Z.B. and Beasly, J.N. (1983) Influence of temperature, moisture and propionic acid on mold growth and toxin production in corn. *Poultry Science* 62, 419–423.

Smith, T.K., MacDonald, E.J. and Haladi, S. (2001) Mycosorb: growth and neurological response of pigs given fusariotoxin contaminated diets. Poster presented at Alltech's 17th Annual Symposium, Lexington, Kentucky.

Speers, G.M., Meronuck, R.A., Ames, D.M. and Mirocha, C.J. (1971) Effect of feeding *Fusarium roseum* f. sp. *graminearum* contaminated corn and the mycotoxin F-2 on the growing chick and laying hen. *Poultry Science* 50, 627–633.

Swamy, H.V.L.N. and Devegowda, G. (1998) Ability of Mycosorb to counteract aflatoxicosis in commercial broilers. *Indian Journal of Poultry Science* 33, 273–278.

Sydenham, E.W., Marasas, W.F.O., Shephard, G.S., Thiel, P.G. and Hirooka, E.Y. (1992) Fumonisin concentrations in Brazilian feeds associated with field outbreaks of confirmed and suggested animal mycotoxicoses. *Journal of Agricultural and Food Chemistry* 40, 994–997.

Thaxton, J.P., Tung, H.T. and Hamilton, P.B. (1974) Immunosuppression in chickens by aflatoxin. *Poultry Science* 53, 721–725.

Trucksess, M.W., Staloff, L. and Young, K. (1983) Aflatoxicol and aflatoxin B1 and M1 in eggs and tissues of laying hens consuming aflatoxin contaminated feed. *Poultry Science* 62, 2176–2182.

Tung, H.T., Smith, J.W. and Hamilton, P.B. (1971) Aflatoxicosis and bruising in the chicken. *Poultry Science* 50, 795–800.

Tung, H.T., Donaldson, W.E. and Hamilton, P.B. (1973) Decreased plasma carotenoid during aflatoxicosis. *Poultry Science* 52, 80–83.

Tung, H.T., Cook, F.W., Wyatt, R.D. and Hamilton, P.B. (1975) The anaemia caused by aflatoxin. *Poultry Science* 54, 1962–1969.

Tutour, B., El-Yazgi, R. and Tantawi-Elaraki, A. (1987) Rapid detection by minicolumn of aflatoxin B1 in poultry feeds. *Actes de l'institut Agronomique de Veterinaire Hassan II* 7, 91–100.

Vasenko, S.V. and Karteeva, E.A. (1987) The influence of feeds containing fungi on the immune system in poultry. *Stornik Nonchnyhk Trudov – Moskovskaya Veterinarnaya Akademiya* 126, 26–29.

Volkl, A. and Karlovsky, P. (1998) Biological detoxification of fungal toxins and its use in plant breeding, feed and food production. *Natural Toxins* 7, 1–23.

Wang, R., Miao, C., Chong, R., Sawatwiroj, N., Dawoson, K., Fu, S. and Fung, D. (2003) Feed and feed ingredient mycotoxin contamination in China: a survey. In: *Proceedings of Alltech's 17th Annual Asia Pacific Lecture Tour.*

Weibking, T.S., Ledoux, D.R., Bermudez, A.J., Turk, J.R., Rottinghaus, G.E., Wang, E. and Merrill, A.H. (1993) Effects of feeding *Fusarium moniliforme* culture material, containing known levels of fumonisin B1, on the young broiler chick. *Poultry Science* 72, 456–466.

WHO (World Health Organization) (2001) Safety evaluation of certain mycotoxins in food. WHO, Geneva, Switzerland.

Widiastuti, R., Maryam, R., Blaney, B.J. and Stoltz, D.R. (1988) Cyclopiazonic acid in combination with aflatoxin, zearalenone and ochratoxin A in Indonesian corn. *Mycopathologia* 104, 153–156.

Widiastuti, R., Maryam, R., Blaney, B.J., Staltz, S. and Stoltz, D.R. (1988) Corn as a source of mycotoxins in Indonesian poultry feeds and the effectiveness of visual examination methods for detecting contamination. *Mycopathologia* 102, 45–49.

Wu, W., Cook, M.E. and Smalley, E.B. (1991) Decreased immune response and increased incidence of tibial dyschondroplasia caused by fusaria grown on sterile corn. *Poultry Science* 70, 293–301.

Wyatt, R.D. (1979) Biological effects of mycotoxins on poultry. In: NAS. *Interaction of Mycotoxins in Animal Production.* National Academy of Sciences, Washington, DC, pp. 87–95.

Wyatt, R.D. (1988a) Mycotoxins in feedstuffs, a hazard to flock health. *Vineland Update*, No. 23.

Wyatt, R.D. (1988b) How reliable is the black light test in the detection of aflatoxin contamination? *Feedstuffs* 60, 12–14.

Wyatt, R.D., Weeks, B.A., Hamilton, P.B. and Burmeister, H.R. (1972) Severe oral lesions in chickens caused by ingestion of dietary fusariotoxin T-2. *Applied Microbiology* 24, 251–257.

Wyatt, R.D., Hamilton, P.B. and Burmeister, H.R. (1973a) The effects of T-2 toxin in broiler chickens. *Poultry Science* 52, 1853–1859.

Wyatt, R.D., Harris, J.R., Hamilton, P.B. and Burmeister, H.R. (1973b) Possible outbreaks of fusariotoxicosis in avians. *Avian Diseases* 16, 1123–1130.

Wyatt, R.D., Hamilton, P.B. and Burmeister, H.R. (1975) Altered feathering of chicks caused by T-2 toxin. *Poultry Science* 54, 1042–1045.

Yegani, M., Smith, T.K., Leeson, S. and Boermans, H.J. (2006) Effects of feeding grains naturally contaminated with *Fusarium* mycotoxins on performance and metabolism of broiler breeders. *Poultry Science* 85, 1541–1549.

Zaghrini, A., Martelli, G., Roncada, P., Simioli, M. and Rizzi, L. (2005) Mannanoligosaccharides and aflatoxin B_1 in feed for laying hens: effects on egg quality, aflatoxin B_1 and M_1 residue in eggs and aflatoxin B_1 level in liver. *Poultry Science* 84, 825–832.

9 Broiler Feeding and Management in Hot Climates

N.J. Daghir

Faculty of Agricultural and Food Sciences, American University of Beirut, Lebanon

Introduction	227
Nutritional manipulations during heat stress	228
Energy and protein	229
Minerals and vitamins	231
Seasonal effects on broiler performance	235
Temperature and body composition	236
Broiler management in hot climates	238
Feeding programmes	238
Feed withdrawal	240
Dual feeding programme	240
Early feed restriction	240
Drug administration	241
Vaccination	242
Beak trimming	243
Broiler-house management	243
Water consumption	247
Acclimatization to heat stress	249
Lighting programmes	251
Conclusions	252
References	253

Introduction

As shown in Chapter 1, the broiler industry has undergone very rapid developments during the past two decades and it is expected that this development will continue and be more pronounced in the warm regions of the world than in the temperate regions. This is because most temperate regions have already reached self-sufficiency, while most countries in the warm regions have not. Furthermore, the implications of good health practices in relation to poultry meat consumption are spreading more and more in the developing warm regions of the world and therefore it is expected that this will greatly enhance the growth of the broiler industry in those areas.

This chapter deals mainly with feeding and management aspects of broilers that are unique to hot climates or the summer months in the temperate regions. Even in that respect, it is in no way exhaustive of everything that has been researched on the subject. Recent developments on the feeding of broilers in hot regions have been covered, as well as the effects of temperature and season on broiler performance, and the effects of temperature on body composition and carcass quality. Certain management practices, such as feed withdrawal, drug administration, vaccination, beak trimming, litter management and others, are discussed. The importance of water quality and quantity as well as lighting programmes to reduce detrimental effects of heat stress are reviewed. Finally, a section on the possible benefits of gradual acclimatization of broilers to heat stress is included.

Nutritional Manipulations During Heat Stress

Environmental temperature is considered to be the most important variable affecting feed intake and thus body-weight gain of broilers. Figure 9.1, adapted from data presented by North and Bell (1990), shows the effects of house temperature on growth and feed consumption of straight-run broilers. The increase in house temperature from 32 to 38°C causes a drop in feed intake of 21.3 g per bird per day and a reduction in body weight at 8 weeks of age of 290 g. Several workers have tried to quantify the extent to which reduced feed intake limits broiler performance at high environmental temperatures. Fuller and Dale (1979) studied two diurnal temperature cycles, a hot one consisting of 24–35°C and a cool one of 13–24°C. Birds were fed the experimental diets *ad libitum* in both environments, and an additional group of birds in the cool environment were limited to the amount of feed consumed by the birds in the hot environment. Their results showed that growth was depressed by 25% in birds maintained in the hot environment. When birds maintained in the cool environment were fed the same amount of feed as that consumed by birds in the hot environment, their performance was reduced by only 16% compared with those fed on an *ad libitum* basis.

The results of this study indicate that reduced feed intake is not the only factor causing reduced broiler performance in hot climates. Fattori *et al.* (1990) demonstrated this in a comprehensive study in which they simulated broiler grower conditions by incorporating environmental variation into the research design. Thus, they were able to partition growers into homogeneous production environments, which allowed a more accurate evaluation of different feed treatments. This resulted in feeding recommendations appropriate

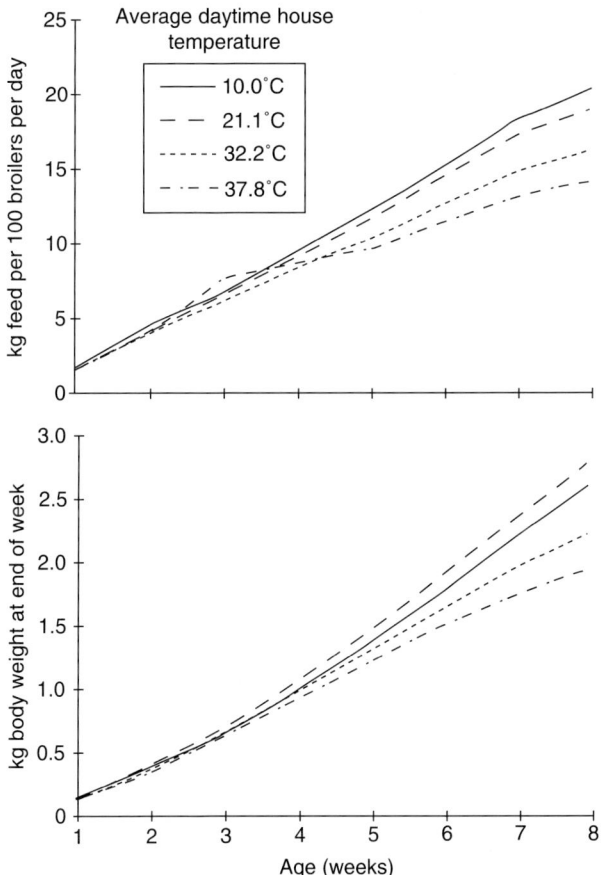

Fig. 9.1. Effect of house temperature on growth (lower graph) and feed consumption (upper graph) of straight-run broilers. (Adapted from North and Bell, 1990.)

to specific groups of broiler growers. Environmental temperature was the most important variable found by these workers to affect feed intake and thus weight gain.

Energy and protein

Farrell and Swain (1977a,b) conducted a series of studies on several aspects of energy metabolism in broilers. They varied environmental temperature from 2 to 35°C and determined the influence of environmental temperature on fasting heat production, metabolizable energy intake and efficiency of metabolizable energy utilization. They found that heat production in both fasting and fed birds decreased as environmental temperature increased from 2 to 30°C and plateaued at about 30–35°C. Although the efficiency of metabolizable energy (ME) retention was not affected by temperature, the amount of ME retained was maximum at 16–22°C. It declined at both higher and lower temperatures.

Hoffman et al. (1991) conducted several studies on the energy metabolism of growing broilers kept in groups in relation to environmental temperature. Energy metabolism was measured by indirect calorimetry by these workers. It was found that energy utilization was dependent on dietary protein content and decreased from 75 to 69% with an increase in protein content from 20 to 40%. Energy requirement (ME) for maintenance was independent of dietary protein and increased from 433 kJ/kg daily at 35°C to 693 kJ/kg daily at 15°C. With equal live weight, feed intake increased with decreasing temperature (Hoffman, 1991). This worker evaluated temperatures of 15–35°C and found feed intake to be similar between 20 and 25°C up to 57 days of age.

It has been known for some time that the reduction in energy intake of broilers is responsible for the reduced growth rate associated with heat stress (Pope, 1960). Several researchers have shown that increasing the energy content of the diet can partially overcome this growth depression. It is common practice now in formulating broiler feeds for hot regions to boost the energy level of these diets by the addition of fat. This practice not only increases the energy intake but also reduces the specific dynamic effect of the diet, which helps birds to cope better with heat stress (Fuller and Rendon, 1977; Fuller, 1978). Dale and Fuller (1979) demonstrated a beneficial effect of isocaloric substitution of fat for carbohydrate in broilers reared from 4 to 7 weeks at constant temperatures of either 20 or 30°C. These results were confirmed by the same workers (Dale and Fuller, 1980) in broilers subjected to cyclic temperature fluctuation (22 to 33°C). Miraei-Ashtiani et al. (2004) showed that high-fat diets (6%) have helped in reducing the detrimental effect of heat stress in broilers raised at 30–38°C. McCormick et al. (1980) noted that survival time of chicks fed a carbohydrate-free diet was greater than those fed a diet containing glucose.

Besides energy, consideration must be given to the amino acid balance of the diet during heat stress. If the energy content of the diet is increased, then it must be kept in mind that there must be a proportional increase in all other nutrients. Minimizing excesses of amino acids usually improves feed intake. Waldroup et al. (1976) improved the performance of heat-stressed broilers by minimizing amino acid excesses in their diets. During hot periods, diets containing lower protein levels and supplemented with the limiting amino acids, methionine and lysine, gave better results than high-protein diets. Ait-Tahar and Picard (1987), in a study on protein requirements of broilers at high temperature, showed that raising the protein level of the diet to 26% does not improve performance at 33°C versus 20°C. Furthermore, if the protein level of the diet is raised at high temperature without raising the level of the essential amino acid lysine, there is a significant reduction in body-weight gain and in feed conversion.

These workers concluded that increasing the essential amino acid levels of low-protein diets in hot climates does not compensate for the reduced growth. This does not seem to be in agreement with Waldroup (1982), who recommended such a practice to counter the effect of high temperature.

Yaghi and Daghir (1985) showed that, if National Research Council (NRC) recommended levels of methionine and lysine are met, the protein levels of starter, grower and finisher rations can be reduced to 20, 17 and 15%, respectively, without a significant reduction in performance. Lowering the protein level does not seem to be always helpful in hot environments. Filho et al. (2004) found that at 32°C low-protein diets (18–16.6–15%) for broilers impairs performance during the finisher period. These protein levels used were obviously very low for our modern broiler strain. Manfueda (2004) in a commercial test in Morocco reduced the protein levels by only 2% and increased the levels of lysine, methionine, threonine and tryptophan and were able to get as good a performance as with the standard diet. Baghel and Pradhan (1989a) studied energy and protein requirements during the hot and humid season in India at maximum temperatures of 42–45°C and minimum temperatures of 14–19°C with relative humidities of 63–68%. They reported that the best feed conversion was obtained when starter diets had 3200 kcal/kg and 24.9% protein, grower diets 3000 kcal/kg and 24.1% protein and finisher diets 2800 kcal/kg and 21.1% protein. The same workers in another study (Baghel and Pradhan, 1989b) found optimum daily gain and feed conversion to be at 2800 kcal/kg and 25, 24 and 21% protein with lysine levels of 1.2, 1.0 and 0.85% and methionine and cystine at 0.93, 0.72 and 0.6%. Koh et al. (1989) studied the effects of environmental temperature on protein and energy requirements of broilers in Taiwan. Feed intake and weight gain in the hot season (30°C) were 15–20% lower than in the cool season (16–18.6°C). Higher dietary energy improved feed efficiency and lowered energy and protein requirements per unit weight gain. Abdominal fat content was higher in broilers reared in the cool season and those fed on the high-protein diet had decreased abdominal fat.

Research at the Pig and Poultry Research and Training Institute in Singapore demonstrated that the most favourable growth and profit were obtained from broilers fed rations with 3100 kcal/kg and 22% protein in the starter and 3000 kcal/kg and 19.5% protein in the finisher (Anonymous, 1982). Lysine and methionine levels recommended were 1.34 and 0.44% in the starter and 1.08 and 0.38% in the finisher, respectively.

Minerals and vitamins

Several acid–base imbalances occur in heat-stressed broilers. The occurrence of alkalosis in heat-stressed birds has been known for a long time. Bottje and Harrison (1985) corrected the alkalotic state in birds by carbonating the drinking water and observed improved weight gain and survival compared with those birds given tap water. Teeter et al. (1985) improved weight gain by adding ammonium chloride to the diet of broilers. They were also able to accomplish the same thing by supplementing the drinking water with ammonium chloride (Teeter and Smith, 1986). These authors also observed beneficial effects on mortality by ammonium chloride supplementation. Smith and Teeter (1987) studied potassium retention in heat-stressed broilers and observed that potassium output was significantly increased in birds at 35°C versus those at 24°C. Drinking-water supplementation with potassium chloride significantly increased survival from 15 to 73%. Gorman (1992) evaluated the performance of finishing broilers (3–6 weeks of age) at high temperature (32°C), using micromineral supplementation at either normal or half-normal levels, with or without sodium bicarbonate supplementation (17 g sodium bicarbonate/kg). The results of this study indicated that commercial finisher rations may contain microminerals in excess of the requirement of broilers grown at high temperature. Furthermore, dietary supplementation with sodium bicarbonate at high temperature stimulated water and feed intakes and improved weight gain. Hayat *et al.* (1999) studied the effects of sodium bicarbonate and potassium bicarbonate supplements for broilers at levels of 2 and 8 g/l in the drinking water. The 2 g level at 31°C increased growth after 35 days of age and increased food intake after 42 days of age compared with those not supplemented. Sodium bicarbonate gave better food conversion and slightly better growth than supplements of potassium bicarbonate. High levels of both of these supplements were toxic to broilers, the potassium being more toxic than the sodium bicarbonate.

Smith (1994) investigated the possible interactions between electrolytes and photoperiod in broilers grown at elevated temperatures. They found that male birds that received sodium chloride gained 10.5% more weight than those receiving no water additive. The use of electrolytes had no effects on carcass characteristics.

McCormick *et al.* (1980) reported that much of the heat-induced mortality could be alleviated if the proper Ca:P ratio is maintained in the diet. Garlich and McCormick (1981) showed a direct relationship with plasma calcium. Survival time in fasted chicks was greater when their previous diet contained 0.3% calcium and 0.55% phosphorus rather than 1% calcium and 0.50% phosphorus. This may be important to consider when feed withdrawal is practised to reduce mortality during high temperature spells. More recently, Orban and Roland (1990) reported on the response of four broiler strains to dietary phosphorus when brooded at two temperatures (29.4 versus 35°C). There was a temperature–phosphorus interaction on bone strength, bone ash and mortality. Cier *et al.* (1992b) evaluated the effects of different levels of available phosphorus on broiler performance in a hot region in Israel. Available phosphorus levels above 0.45% in the starter and 0.40% in the grower did not improve growth rate or feed utilization. Levels of 0.55% in the starter and 0.51% in the grower caused very wet litter.

Ait-Boulahsen *et al.* (1992), in studying the relationship between blood ionized calcium and body temperature of broiler chickens during acute heat stress, observed that exposure to high temperature elicited in these birds a significant elevation in plasma sodium to calcium ion ratio. The magnitude of this elevation appeared to determine the degree of hyperthermia. A solution of potassium biphosphate produced the highest plasma sodium to calcium ion ratio. This was followed in decreasing order by ammonium chloride solution, sodium chloride solution, unsupplemented water (control) and, finally, potassium chloride solution. There was a highly significant positive correlation between plasma sodium to calcium ion ratio and body temperature. These authors concluded that changes in body temperature response to heat stress parallel those of the plasma sodium to calcium ion ratio. In other words, a high sodium to calcium ion ratio impairs heat tolerance and vice versa.

Based on the above, mineral therapy and manipulation appear to be effective means of reducing detrimental effects of heat stress in broilers.

Since heat stress always depresses appetite and therefore reduces nutrient intake, the use of a vitamin and electrolyte pack in the drinking water for 3–5 days, or until a heatwave passes, has been commonly practised. Morrison et al. (1988) reported the results of a survey conducted on 132 broiler flocks that either used or did not use a vitamin and electrolyte water additive during heat stress. They found that additives in water resulted in significantly lower mortality of broilers during heat stress. Although the feeding of extra amounts of vitamins during heat stress has been suggested for a long time (Faber, 1964), investigations into the effects of heat stress on specific vitamin requirements have not been very conclusive (see Chapter 6). Vitamin C supplementation during heat stress has been shown to have beneficial effects. Blood ascorbic acid decreases with an increase in environmental temperature (Thornton, 1961). Ascorbic acid limits the increase of body temperature during heat stress (Lyle and Moreng, 1968). Njokn (1984) reported that broilers reared under tropical conditions benefited from dietary supplementation of vitamin C during heat stress. Nakamura et al. (1992) exposed 24–26-day-old chickens to temperatures of 20 and 40°C and gave them ascorbic acid supplements of 150 or 1500 mg/kg for 7 days. Ascorbic acid supplements at 40°C reduced the severity of heat stress as measured on body-weight gain, feed intake and feed efficiency. Rajmane and Ranade (1992) tested the effect of a mixture of 1% ammonium chloride, 1000 p.p.m. ascorbic acid and 0.25 mg/100 g α-methyl-p-tyrosine on performance of broilers during the summer season in India. Temperatures ranged from 38 to 42°C and relative humidity was about 75%. Their results showed a significant decrease in mortality rate and improved performance. They also observed reduced plasma corticosteroids and increased plasma potassium and ascorbic acid. They proposed that ascorbic acid supplements might have altered plasma corticosterone levels and helped to maintain concentration of potassium. α-Methyl-p-tyrosine probably acts as a catecholamine-synthesis blocker and helps to maintain higher values of ascorbic acid. Cier et al. (1992a) investigated the use of ascorbic acid at 150, 300 and 600 p.p.m. in the feed for broilers under summer conditions in Israel. They concluded that ascorbic acid supplements significantly improved growth of both males and females, but the improvement was greater in males. No significant improvement was obtained in feed conversion. Several nutritionists, therefore, recommend the administration of 1 g ascorbic acid/l in the drinking water all through the heatwave period. A vitamin pack of A, D, E and B complex supplementation of drinking water was found by Ferket and Qureshi (1992) to be beneficial for both performance and immune function of heat-stressed broilers.

Since many countries in hot regions import their vitamin and trace mineral mixes, and since there are often delays in acquisition and transport of ingredients, the problem of vitamin stability is of primary concern. Various data show that appreciable loss of vitamin activity in feed can occur during storage. Temperature, moisture and oxidation by polyunsaturated fatty acids, peroxides and trace minerals are the most critical factors that affect vitamin stability in poultry feeds. Coelho (1991) presented data on vitamin stability in both complete feeds and vitamin–trace mineral premixes. Table 9.1 presents average industry vitamin stability in vitamin–trace mineral premixes as well as in complete feeds. It is observed that the use of a premix without choline considerably improves its vitamin stability. This is why it has been industry practice for many years to add the choline separately to the feed. The data in Table 9.1 show that vitamin stability in feeds is somewhat similar to vitamin stability in trace mineral premixes. In either case, there is substantial loss with time and, considering the conditions that feed ingredients and complete feeds are subjected to in hot climates, the percentage loss of vitamin activity would be much greater than that shown in Table 9.1. Therefore, vitamin activity in feeds can be preserved by the incorporation of antioxidants, selecting gelatin-encapsulated vitamins, appropriate storage conditions, adding choline separate from the vitamin and trace mineral premix, delaying the

Table 9.1. Average industry vitamin stability in vitamin–trace mineral premixes and in complete feeds (adapted from Coelho, 1991).

	Loss per month (%)		
	Premixes		
Vitamin	With choline	Without choline	Complete feeds
A 650 (beadlet)	8.0	1.0	9.5
D3 325 (beadlet)	6.0	0.6	7.5
E acetate	2.4	0.2	2.0
E alcohol	57.0	35.0	40.0
MSBC[a]	38.0	2.2	17.0
MPB[b]	34.0	1.6	15.0
Thiamine HCl	17.0	0.5	11.0
Thiamine mono	9.6	0.4	5.0
Riboflavin	8.2	0.3	3.0
Pyridoxine	8.8	0.4	4.0
B_{12}	2.2	0.2	1.4
Ca pantothenate	8.4	0.3	2.4
Folic acid	12.2	0.4	5.0
Biotin	8.6	0.3	4.4
Niacin	8.4	0.3	4.6
Ascorbic acid	40.0	3.6	30.0
Choline	2.0	–	1.0

[a]MSBC = menadione sodium bisulfite complex.
[b]MPB = menadione dimethylpyrimidol bisulfate.

addition of fats until just before the use of the feed and, finally, using feeds as soon as possible after mixing.

Finally, it is important to emphasize that nutritional manipulations can reduce the detrimental effects of high environmental temperature on broilers but cannot fully correct them.

Seasonal Effects on Broiler Performance

Nir (1992) reported that in a relatively mild hot climate, like that of certain areas of the Middle East where the days are hot and the nights are cool, a strong relationship was found between the growing season and broiler body weight. Two years' data from an experimental farm in the Jerusalem area showed a similar curvilinear pattern. The best performance was obtained during the winter and the worst during the summer. Those of spring and autumn were intermediate. Nir (1992) suggested that there are factors other than environmental temperature involved in growth regulation throughout the year. Within the ambient temperature range recorded in the poultry house, a 1°C increase in mean maximal ambient temperature from 4 to 7 weeks of age was accompanied by a decrease of 23 g in body weight. Corresponding values for the average and minimal ambient temperatures were 32 g and 43 g, respectively. Nir's observations agree with other workers (Hurwitz *et al.*, 1980; Cahaner and Leenstra, 1992), who reported that maximal growth of broilers and turkeys is achieved at ambient temperatures much below the thermoneutral zone. The response of broilers at high temperatures differs with different relative humidities. High temperature accompanied by high humidity is more detrimental to broiler growth than high temperature with low humidity. Nir (1992) reported that the response of broilers to excessive heat differed between Jerusalem and Bangkok. While a 1°C increase in maximal ambient temperature was accompanied by a 23 g decrease in body weight in the Jerusalem area, in Bangkok the corresponding decrease was 77 g. The higher relative humidity in the latter area must have partially contributed to this difference. It is well established that constant high temperature is more deleterious to poultry than cyclic or alternating temperatures. Osman *et al.* (1989) studied the influence of a constant high temperature (30–32°C) compared with an alternating temperature of 30 32°C by day and 25°C by night on broiler growth up to 12 weeks of age. High temperature decreased growth significantly starting at 4 weeks of age and the effect increased with age. Constant high temperature decreased growth more than an alternating temperature. Deaton *et al.* (1984) conducted five trials with commercial broilers to determine whether lowering the temperature during the cool portion of the day in the summer affected their performance. Treatments used were 24-h linear temperature cycles ranging from 35°C to 26.7°C or 21.1°C. Results obtained showed that lowering the low portion of the temperature cycle from 26.7°C to 21.1°C significantly increased broiler body weight at 48 days of age.

Feed conversion in broilers is subject to marked fluctuations because of seasonal as well as ambient temperature changes. Poultry producers in the state of Florida found that 0.09 kg more feed was required to produce a unit of gain in broilers in the period from June to August than from November to April (McDowell, 1972). The influence of high temperature on efficiency of feed utilization has been studied by several workers, and these studies all indicate that high temperatures bring about a reduction in efficiency in the utilization of feed energy for productive purposes. Animals eat less and they return less per unit of intake. McDowell (1972) summarized the reason for reduced feed efficiency at high temperature by saying 'in warm climates, generally, chemical costs for a unit of product are higher than in cooler climates because a portion is siphoned off for the processes required to dissipate body heat'.

Temperature and Body Composition

Kleiber and Dougherty (1934) were among the earliest workers to report on the effects of environmental temperature on body composition in the chicken. They observed that maximum fat synthesis occurred at an environmental temperature of 32°C, while no effect was noted on protein synthesis. This has been confirmed by Chwalibog and Eggum (1989). Olson et al. (1972) reported that carcass dry matter, fat and energy were increased with increasing temperature or dietary energy level, but protein decreased. The ME required per kcal of carcass was increased with lower temperature, reflecting the increased requirement for body temperature maintenance. These workers observed that a 1°C drop in temperature reduced ME efficiency by about 1%.

El-Husseiny and Creger (1980) studied the effects of raising broilers at 32°C versus 22°C on their carcass energy gain. Their results indicated that energy per g of dry tissue increased with decreasing ambient temperature and the energy-gained value for birds reared at 22°C was higher than at 32°C. This was confirmed by Sonaiya (1989), who showed further that broilers reared at 21°C retained more energy in the carcass as fat than did broilers reared at 30°C. Sonaiya (1988) studied the effects of temperature, dietary ME, age and sex on fatty acid composition of broiler abdominal fat. Chickens reared at high temperature (30°C) had a significantly lower proportion of polyunsaturated fatty acids in their abdominal fat between 34 and 54 days than birds reared at low temperature (21°C). The depot fat contents of oleic, linoleic and linolenic acids were all reduced by high temperature at 54 days. For chickens slaughtered at 54 days, saturated fatty acid content was much higher in females than in males at high temperature, while at low temperature polyunsaturated fatty acids were much lower in males than in females. Sonaiya (1988) recommended the early finishing of broilers from the viewpoint of fatty acid composition because the polyunsaturated fatty acid to saturated fatty acid ratio declines significantly with age, regardless of temperature.

Environmental temperature has been shown to influence carcass amino acid content of broilers. Tawfik et al. (1992) reported that broilers kept at 18°C from 4 to 12 weeks of age had higher glycine and proline concentrations in breast muscle than those kept at 32°C. Bertechini et al. (1991) did not observe any changes in carcass characteristics of broilers raised at 17.1, 22.2 or 27.9°C and given a maize–soybean meal base diet. Sonaiya *et al.* (1990) reported that broiler meat flavour was improved by age (34 versus 54 days) and high environmental temperature (21°C versus cycling 21 to 30°C).

Today's broilers are selected and managed with the aim of increasing meat yield and decreasing fat deposition. Several studies have shown that environmental temperature has an effect on carcass composition and meat yield. At high temperature, meat yield, particularly yield of breast meat, is reduced (Howlider and Rose, 1989; Tawfik *et al.*, 1989; Yalcin *et al.*, 1997, 2001). Leenstra and Cahaner (1992) found temperature to have a negative effect on breast meat yield. Males were more affected by high temperature than females.

Geraert *et al.* (1992) studied the effect of high ambient temperature and dietary protein concentration on growth performance, body composition and energy metabolism of genetically lean (IX) and fat (FL) male chickens between 3 and 9 weeks of age. Heat exposure reduced feed intake in both lines (−28%). Lean broilers gained significantly more weight between 5 and 7 weeks of age when exposed to 32°C than fat chickens (+20%). At 9 weeks of age, body dry matter, and mineral and lipid contents were not affected by high temperature, while body protein content was significantly decreased.

Lu *et al.* (2007) investigated the effect of chronic heat stress on fat deposition and meat quality in male chickens from a commercial, fast-growing strain (Arbor Acres – AA) and a local, slow-growing species (Beijing You Chicken – BJY). Abdominal fat deposition of BJY chickens was enhanced by heat exposure while fat deposition by AA broilers was decreased in heat-exposed and pair-fed chickens. They concluded that the impact of heat stress was breed dependent and that the BJY chickens showed higher resistance to high ambient temperature, which could be related to their increased feed efficiency and deposition of abdominal fat under heat exposure. Yalcin *et al.* (1999) in a study on heterozygous naked-neck birds found that breast protein content was depressed in the summer. The main effect of high temperature is a reduction in ME intake to decrease metabolic heat production, leading to a lower energy retention. Gross protein efficiency was depressed in hot conditions in the IX compared with the FL. Smith (1993) conducted a study in which carcass parts from broilers reared under different growing temperature regimens were examined for crude protein, fat, calcium, phosphorus, potassium and sodium. Protein content of thighs and drumsticks from birds grown at elevated temperatures were higher than those from birds grown in a constant temperature environment. The fat content of the parts was higher from birds in the cooler environment. The sodium content of the breast was higher at elevated temperatures. The breast portion of carcasses from heat-stressed birds had the greatest amount of potassium. Akit *et al.* (2006) studied the effect of temperature on meat quality during rearing and crating of broilers and found that high temperature had an adverse effect on meat quality. The crating temperature effect, however, was not as bad as the rearing temperature effect.

Broiler Management in Hot Climates

Feeding programmes

Recommended nutrient levels of broiler starter, grower and finisher rations are shown in Table 9.2. They have been modified to better suit conditions in hot climates. Protein levels recommended are 1–2% lower than what is normally used in the temperate regions because of what has been presented earlier in this chapter as well as in Chapter 6. Energy levels have also been adjusted to protein levels, but kept higher than those currently used in many hot regions. The potassium level has been increased to 0.6%, versus the 0.4% normally recommended in the temperate regions. Levels of certain amino acids have been raised higher than those normally used at that protein level.

Table 9.2 specifies that a starter is fed for the first 3 weeks of life to straight-run chicks and a grower from 3 to 6 weeks of age. The finisher is used from 6 weeks to market. Under certain conditions in hot climates, it may be beneficial to extend the feeding of the grower to 7 weeks, if growth rate is slow. Decuypere et al. (1992) studied different protein levels for broilers in Zaire on isocaloric diets and found that, in the hot and humid conditions of that

Table 9.2. Recommended broiler ration specifications.

	Starter	Grower	Finisher
Protein (%)	22	20	18
Metabolizable energy (kcal/kg)	3000	3050	3100
Calcium (%)	1.0	0.90	0.80
Available phosphorus (%)	0.45	0.42	0.40
Sodium (%)	0.18	0.18	0.18
Chloride (%)	0.18	0.18	0.18
Potassium (%)	0.60	0.60	0.60
Lysine (%)	1.20	1.15	0.95
Methionine (%)	0.50	0.45	0.40
Methionine + cystine (%)	0.85	0.80	0.70
Tryptophan (%)	0.23	0.20	0.18
Arginine (%)	1.45	1.35	1.15
Histidine (%)	0.35	0.35	0.28
Phenylalanine (%)	0.75	0.75	0.65
Phenylalanine + tyrosine (%)	1.35	1.35	1.15
Threonine (%)	0.80	0.75	0.70
Leucine (%)	1.35	1.35	1.15
Valine (%)	0.85	0.80	0.70
Glycine + serine (%)	1.50	1.25	1.00
Males	0–21 days	3–6 weeks	6 weeks to market
Females	0–14 days	2–5 weeks	5 weeks to market
Straight run	0–21 days	3–6 weeks	6 weeks to market

country, slight increases in dietary protein late in the broiler cycle are beneficial for growth and feed efficiency. Another advantage of increases in dietary protein observed by these workers is that abdominal fat content is reduced with these higher protein levels.

Providing adequate levels of each nutrient and a good balance of these nutrients is a prerequisite to successful feeding. Using feed with good-quality ingredients is a second prerequisite, and using only fresh feed is the third prerequisite. It is very important to order fresh feed, preferably once a week, in a hot, humid climate, where nutrient deterioration is more rapid and fats tend to go rancid quickly.

Broilers are usually fed either crumbles or pellets. The broiler starter is provided as crumbles, while grower and finisher rations are provided as pellets. Crumbles and pellets tend to reduce feed wastage and improve feed efficiency. In hot climates, broilers prefer to eat feed with larger particle size. Yo *et al.* (1997) found that when maize was fed as whole grain birds consumed more of the protein concentrate (43.7% CP) in a self-selected diet and gave better feed conversion.

Broilers are normally fed *ad libitum*. However, in recent years, intermittent feeding programmes have been used on modern, well-managed broiler farms. If intermittent feeding is used in hot climates, then broilers should be fed for a period of 1.5 h, after which the lights go off, allowing the broilers to rest and digest their meal. After 3 h of darkness, the lights are turned on again for another period of feeding. When intermittent feeding is used, 20–30% more feeder and drinker space is provided so that all birds can eat and drink during the 1.5 h when the lights are on. This method should have a possible application in hot climates since the hours of darkness provide minimum activity on the part of the birds and therefore reduced heat production.

Fast growth rate in the modern broiler has not been without its problems. It has contributed to increased mortality due to heart attacks, increased leg problems and increased incidence of ascites. By restricting feed and/or nutrient intake early in life, it has been possible to reduce some of these problems. Body fat, an undesirable component in the modern broiler, can also be reduced by feed and/or nutrient restriction. Several methods have been used to reduce very rapid growth rate in the early life of the broiler. One method is by altering the lighting programme; another is by actual feed restriction; and the third is by reducing energy and/or protein content of the diet by adding to it extra fibre.

Sex-separate feeding of broilers has become quite popular in many countries. In areas where uniformity in the final product is important and where uniformity has been hard to attain, this practice is highly recommended. Furthermore, this practice can save in feed costs since rations for the females can be reduced by 1–2% in protein. This can be done by shifting from the starter to the grower and from the grower to the finisher at earlier ages (see Table 9.2). Males and females have the same nutrient requirements during the first 2 weeks of life. After this period females tend to respond differently to certain nutrients in the feed. This is why reducing the plane of nutrition for females may save in feed costs. Sex-separate feeding also allows for marketing the males earlier than the females and thus provides for greater yearly output from the same farm.

Feed withdrawal

Fasted animals produce less heat than fed animals. Thus the practice of feed withdrawal to reduce heat stress has been tested by several researchers. Feed withdrawal has been shown to reduce heat production, leading to a decrease in body temperature and mortality of broilers (Francis et al., 1991; Yalcin et al., 2001). McCormick et al. (1979) reported that fasting of 5-week-old broilers for 24, 48 or 72 h resulted in progressively increased survival time when exposed to heat stress.

Feed withdrawal during the hottest hours of the day has become a common management practice in many broiler-producing areas. Short-term feed withdrawal can lower the bird's body temperature and increase its ability to survive acute heat stress. Smith and Teeter (1987, 1988) studied the influence of fasting duration on broiler gain and survival. They showed that fasting intervals beginning 3–6 h prior to heat-stress initiation and totalling up to 12 h daily during significant heat stress (up to 37°C) reduce mortality significantly. One suggested practice when a heatwave is expected is to remove feed at 8.0 a.m. and return it at 8.0 p.m. Fasting will probably result in reduced weight gains, a longer growing period and thus a delay in marketing age. Therefore the producer will have to weigh the importance of a more rapid growth rate versus a greater mortality risk. Another concern about feed withdrawal was recently raised by Thompson and Applegate (2006), who showed that during a short-term feed withdrawal, there are alterations in intestinal morphology and depletion of intestinal mucus, which may reduce the integrity of the intestine.

Dual feeding programme

Since dietary proteins have a higher thermogenic effect than carbohydrates or fats, a dual-feed regime has been proposed by some workers, using a low-protein diet during the warmer times of the day and a high-protein diet to be offered during the cool night. De Basilio et al. (2001) found that growth and feed efficiency were slightly reduced (–4%) by dual feeding. Dual feeding, however, reduced mortality compared with regular feeding. Further research is needed on the value of this practice.

Early feed restriction

Early feed restriction was shown to induce enhanced thermotolerance in White Plymouth Rocks (Zulkifli et al., 1994). Realizing the importance of this practice in commercial broiler production in hot regions, Zulkifli *et al.* (2002) studied the effect of early-age feed restriction on heat tolerance in female commercial broilers. These workers observed that chicks subjected to 60% feed restriction at 4, 5 and 6 days of age have improved growth and survivability in response to the subsequent heat treatment at marketing age (35–41 days of age). This improvement in heat tolerance was explained by these authors as being due to enhanced ability to express heat-shock protein 70 in the brain.

Khajavi *et al.* (2003) investigated the effect of feed restriction early in life on the humoral and cellular immunity of broiler chickens under high-temperature conditions. These workers concluded that feed restriction early in life reduced some of the negative effects of the heat stress on the immune system of broilers when exposed to high environmental temperature later in life. These authors also suggested that the differences between strains for some measured variables may imply that there may be genetic differences in response as well as sex differences, since males did not respond equally to females.

Liew *et al.* (2003) investigated the effect of early-age feed restriction and heat conditioning on heat-shock protein 70 expression, antibody production, resistance to infectious bursal disease and growth of heat-stressed male broiler chickens. They concluded that the combination of feed restriction and heat conditioning could improve weight gain and resistance to IBD in male broilers under heat-stress conditions. They suggested that the improved heat tolerance and disease resistance in these birds could be attributed to better heat-shock protein 70 response.

Drug administration

Drug activity under heat stress may not be the same as in a temperate environment. Furthermore, drug administration through the drinking water needs much more careful calibration than under normal temperature conditions.

Producers tend to avoid the use of nicarbazine as an anticoccidial drug during heat stress periods. Reports of nicarbazine-induced mortality during heat stress vary greatly. Keshavarz and McDougal (1981) reported that the incorporation of nicarbazine at the regular dose rate of 125 p.p.m. can result in excessive broiler mortality following heat stress. This relationship between nicarbazine and high environmental temperature had been reported earlier by McDougal and McQuinston (1980). These workers showed that environmental temperatures exceeding 38°C for 11 days during a 50-day broiler cycle caused a mortality of 91% in broilers fed nicarbazine compared with 27% in non-medicated controls. This effect of nicarbazine on broiler mortality at high temperature had also been previously reported in South Africa (Buys and Rasmussen, 1978). At temperatures of 40°C, broilers aged 3–5 weeks and receiving nicarbazine showed from 60 to 75% mortality, compared with only 10–25% in the corresponding age groups which were medicated with 125 p.p.m. amprolium. Smith and Teeter (1988) reported on the adverse effects of nicarbazine, which caused a 16% mortality in broilers. In a second study, birds were exposed to a hot climate prior to the start of the experiment. Nicarbazine-treated birds and controls had the same mortality (20%). The same workers showed that the high mortality noted with nicarbazine was reduced when potassium chloride was added to the drinking water. Macy *et al.* (1990) studied the effect of Lasalocid on broilers raised in a hot, cyclic (26.7–37.8°C) versus a moderate (21°C) environment. In the hot, cyclic environment, broilers fed Lasalocid gained more weight and had better feed conversion than those fed the basal diet. Inclusion of amprolium in feed or water at very high levels may cause thiamine deficiency since this drug functions as a specific antagonist of thiamine (Polin *et al.*, 1962). It is important, therefore, to check that adequate levels of this vitamin are present in the feed whenever amprolium is used.

Vaccination

Vaccination is the main method for the control of infectious diseases in poultry and, when properly used and administered, can reduce the overall weight of infection on a farm and result in great benefits from improved performance. Vaccination programmes used in several countries in the hot regions of the world have been described by Aini (1990). This section will present some suggestions and recommendations designed to aid in reducing vaccine failures in hot climates.

Water vaccination is one of the most widely used methods of vaccine administration in broiler production. Vaccines that are given in the drinking water are live vaccines and are sensitive to high temperatures. Mutalib (1990) recommended the following for handling vaccines:

1. Use ice or a thermos to transport vaccines from source to farm.
2. Avoid exposure of vaccine to heat and direct sunlight (do not leave vaccine bottles in car).
3. Store vaccines refrigerated until use.
4. Once reconstituted, the vaccine should be used within the hour.
5. It is advisable to vaccinate early in the morning to avoid the heat of the day.
6. Do not use outdated vaccines.
7. Dilute vaccines only in clean, non-chlorinated water.

In hot climates, it is important to determine the amount of water consumed by the flock and to use that figure for diluting the vaccine rather than a book figure. Another precaution to take in hot climates is not to deprive the birds of water for a prolonged period of time, usually not more than 1 h prior to vaccination. Every effort should be made to reduce stress in birds before and after vaccination. Therefore avoid vaccination on days when there are exceptionally high temperatures and try to provide as much comfort for the birds as possible. All other practices used in temperate regions for vaccination should also be applied in hot regions.

Beak trimming

Beak trimming is a common practice with the young commercial laying pullet but is not as commonly used for broilers. However, in hot climates, where feather picking and cannibalism may be frequent, producers resort to beak trimming as a means of reducing bird losses. Furthermore, when the broiler house is open-sided in hot climates, the light intensity cannot be controlled. This high light intensity necessitates beak trimming to prevent feather picking and cannibalism.

Results obtained by different researchers have varied, but in general no effect on final body weight has been observed when broiler chicks were beak-trimmed at 1 day of age (Harter-Dennis and Pescatore, 1986; Trout et al., 1988). Even an improvement in feed efficiency was observed by Harter-Dennis and Pescatore (1986) when birds were beak-trimmed at 1 day of age. Andrews (1977), however, reported that, in broilers that had been beak-trimmed at 1 or 10 days of age, both procedures resulted in reduced 8-week body weights when compared with controls that had not been beak-trimmed. Christmas (1993) compared the performance of spring- and summer-reared broilers as affected by beak trimming at 7 days of age. He observed that beak trimming resulted in significantly lower body weights and feed intake in summer-reared broilers but not in spring-reared broilers. Broiler producers in hot climates should therefore weigh the relative importance of a higher growth rate versus higher mortality due to cannibalism. Although beak trimming at 6–7 days of age may produce better results than at 1 day of age, there is the added labour involved with catching the birds at 6–7 days. If broilers must have their beaks trimmed, then it is most appropriate to trim them at the hatchery.

Broiler-house management

Housing for optimum performance in hot climates has been covered in detail in Chapter 5. In this section, therefore, selected aspects of broiler-house management that have not been covered earlier will be presented. It cannot be overemphasized that, in high-temperature conditions, all practices that aid in reducing heat production by the bird and those that aid in the elimination of heat already produced in the house need to be implemented. There is, of course, heat produced by lighting and by machinery present in the broiler house, but these are minor sources of heat in comparison with that coming from the birds. From the standpoint of heat production, the easiest practice to implement is to reduce the density of birds in the broiler house. Navahari and Jayaprasad (1992) evaluated space requirements of broilers on three floor types (deep litter, wire floor and slatted floor) during three seasons (summer, monsoon and winter). They concluded that, for a broiler weighing 1.1 kg at market, a litter floor space of 400 cm^2 or a wire-slatted floor area of 280 cm^2 was sufficient for that region of India. North and Bell (1990) recommended

500 cm^2, for a broiler weighing 1.3 kg at market, raised on a litter floor during the summer. As for the elimination of heat already produced, the most economical method is to increase ventilation rates in the house.

Relative humidity in the poultry house has a significant bearing on heat loss by evaporative heat dissipation. The effect of humidity on thermal regulation response of broilers depends on age and air temperature (Lin et al., 2005a). Humidity is of extreme importance to broiler performance when temperature exceeds 28°C. High humidity alone (60%) impaired heat transmission from body core to peripheral parts at 35°C, but helped its transmission at 30°C in 4-week-old broilers (Lin et al., 2005b).

The optimum temperature and humidity within a poultry house depend on the type of stock and age of the flock. In broiler houses, the climate has to be properly adjusted to the age of the flock. Special attention should be paid not only to air temperature and humidity but also to purity and draught-free flow. Afsari (1983) suggested maximum rates, in percentage by volume, for the three important chemical factors, which should not be exceeded. These are carbon dioxide (3.5), ammonia (0.05) and hydrogen sulfide (0.01).

Litter management

The control of litter quality is important to the successful management of broiler flocks. High-moisture litter due to improper operation of drinkers or poor ventilation will lead to accumulation of ammonia, which increases the incidence of respiratory problems and ocular lesions, as well as favouring intestinal infections.

It is common practice to reduce litter depth in hot weather for more bird comfort. This is a good practice, provided litter depth is not reduced to less than 5 cm. Litter should be kept in good crumbly condition. Both quantity and quality of litter affect breast blister incidence and thus broiler condemnation rate.

Excess of sodium in the diet for broilers increases water intake and moisture content of the droppings (Damron and Johnson, 1985). However, too low a level can have a negative effect on performance. Therefore, it is important to meet the exact sodium requirement in broiler diets. Vahl and Stappers (1992) reported that it was possible to decrease the dietary sodium content without negative effects on performance during the latter part of the broiler cycle. They recommended a sodium level of 0.9 g/kg diet after 5 weeks of age.

Broiler-house cooling devices

Cooling systems for both naturally ventilated and closed, fan-ventilated houses have been described in Chapter 5. This section covers a few selected systems that have been tested and used for broiler houses.

In hot, dry climates, broiler producers have for many years used desert-type air coolers after the 2–3-week brooding period. These coolers have been found to be the cheapest and most efficient means of lowering air temperature and increasing humidity in hot, dry climates (see Fig. 9.2). They are still used for small operations and for naturally ventilated houses, while large

Fig. 9.2. Naturally ventilated house equipped with desert-type air cooler.

integrated operations in hot climates use closed, environmentally controlled housing with different sophisticated evaporative-cooling systems.

Several studies have been conducted to show the effects of using these desert air coolers on broiler performance and economic efficiency. Al-Zujajy et al. (1978) studied the effect of air coolers on broilers housed at different densities in two naturally ventilated houses in Iraq. Differences in weight gain between birds in cooled and uncooled houses reached 385 g per bird at market at a density of 16 birds/m^2. Feed conversion efficiency, meat yield and carcass conformation were much better in the cooled house than in the uncooled house and financial returns per square metre were 40% higher in the cooled house.

Another method of reducing heat stress in naturally ventilated broiler houses is by roof insulation. Several workers have shown that roof insulation can considerably reduce the death losses from heat prostration in broilers. Reece et al. (1976) indicated that the radiation from a hot roof alone has no effect on body-weight gain or feed conversion of broilers. However, when the radiation was combined with a 3°C increase in ambient temperature, which was the increase attributed to absence of insulation in a minimally ventilated broiler house, male body-weight gain and feed conversion were adversely affected.

Reilly et al. (1991) conducted a study to determine whether water-cooled floor perches could be used by broilers exposed to a constant hot ambient environment and whether utilization of these perches would improve performance beyond those provided with uncooled floor perches. Their results indicate that water-cooled perches are beneficial in improving broiler performance during periods of high environmental temperatures.

Wet versus dry mashes

Behavioural responses and growth of broilers were investigated by Abasiekong (1989) in three seasons using two wet mashes with feed to water ratios of 2:1 and 3:1. Wet mashes given in the hot, dry season caused significantly lower rates of panting. In the hot, dry season, 60% of chickens preferred wet mashes, whereas 80% preferred dry meal in the cool, dry season. Kutlu (2001) tested a diet mixed with water in a ratio of 1.5:1 and found that it significantly increased body-weight gain, dry matter intake, carcass weight and carcass lipid content in broilers.

Whitewashing house roof

Heat load can be reduced by whitewashing the broiler-house roof, and house temperatures can be reduced by several degrees in uninsulated houses and to a lesser extent in roof-insulated buildings. The temperature differences can range from 3 to 9°C on hot days. Since the effectiveness of whitewash is due to its ability to reflect radiant heat off a surface, this reflective property is greatly reduced by dust accumulation. The dust may need to be removed periodically by sprinkling the roof. Table 9.3 gives the formula proposed by Casey (1983) to be used for whitewashing poultry-house roofs.

Disinfectants used in hot climates

Biosecurity is a very important consideration in hot climates. The all-in, all-out system is highly recommended and preferred, and for thorough decontamination special disinfectants may be needed for use between batches. Effective poultry-house disinfection is not always achieved in hot climates. High ambient temperatures usually lead to rapid evaporation of the disinfectant solution, hence not allowing sufficient contact time for the disinfectant to be effective. Disinfectants used in developing countries are often traditional formulations, which are not always designed to be effective against the specific disease problem at hand. In some areas, the poultry industry still uses outdated housing with dirt or sand floors. These can harbour infection several centimetres below the floor surface, which ordinarily disinfectants cannot reach.

Therefore, in order to overcome some of these problems, a disinfectant with the following properties is recommended for use:

1. A disinfectant that has the ability to mix with both diesel oil and water. A stable emulsion can be formed of the oil, water and disinfectant. This emulsion will dry slowly and allow disinfectant time to be effective.
2. The disinfectant should be broad-spectrum – capable of killing bacteria, viruses, fungi and insects.

Table 9.3. Formula for whitewashing poultry house roofs (from Casey, 1983).

Formula	Area covered (sq. ft)	Years effective
20 lbs hydrated lime + 5 gals water	600	1
20 lbs hydrated lime + 6 gals water + 20 lbs white cement + 0.5 cup blueing	700	3
20 lbs hydrated lime + 5 gals water + 1 qt polyvinyl acetate	700	4

3. The disinfectant must have detergent properties to enable it to penetrate and emulsify the organic matter which harbours infection in the sand or dirt floors.

Dewinged broilers

Feathers of birds are an effective barrier to heat loss from the skin. The wings cover a wide body surface area and their presence reduces the efficiency of heat loss during elevated ambient temperatures. Al-Hassani and Abrahim (1992) conducted two experiments to evaluate the effect of dewinging on performance of broilers reared under cyclic temperatures of 25 to 35°C. Although body temperature and mortality were lower in the dewinged groups than in controls, the body weight of controls was higher. There were no differences in feed intake, feed conversion, dressing percentage and boneless meat percentage between the two groups. The authors concluded that the dewinged birds were more heat-tolerant than controls.

Water Consumption

Water is an important coolant. Drinking water plays an important role in cooling broilers. The cooler the water, the better the birds can tolerate high environmental temperatures. The warmer the water, the more they need to drink. To help them meet this increased need for water, growers provide broilers with approximately 25% more drinker space than the standard temperate climate recommendation. Where possible, wide and deep drinkers, permitting not only the beak but all the face to be immersed, are used. Vo and Boone (1977) observed that birds consuming water from drinkers large enough to allow head dunking survived longer under heat stress than those not provided with such drinkers. Morrison *et al.* (1988), in a survey on broiler farms, reported that drinker type had a significant effect on percentage increase in mortality during the summer months. These workers found that, among four types of drinkers, the trough drinker that allowed head dunking had the lowest mortality rate.

Ambient temperature is probably the most important factor affecting water intake in broilers. Figure 9.3, adapted from data presented by North and Bell (1990), shows the effect of house temperature on water consumption of straight-run broilers. Broilers at a house temperature of 38°C consume four times as much water as those at 21°C. Harris *et al.* (1975) demonstrated the adverse effect of providing drinking water at 35–40°C compared with 17–23°C. Water at low temperature functions as a heat sink in the intestinal tract. Insulation of header tanks and supply piping is necessary if the temperature of the water at the point of consumption exceeds 25°C. If possible, bury water pipes at least 60 cm underground and shade the area where the pipes run. The water:food intake ratio increases from approximately 2:1 at moderate temperatures to about 5:1 at 35°C (Balnave, 1989). This can have drastic effects on the intake of nutrients, drugs or any other substance

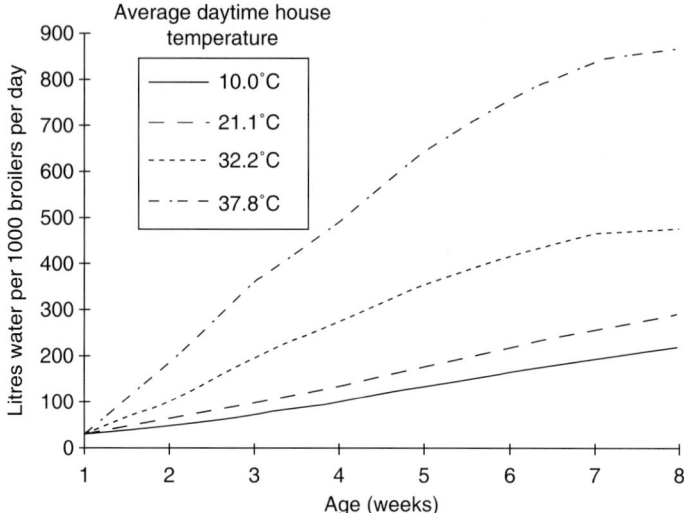

Fig. 9.3. Effect of house temperature on water consumption of straight-run broilers.

present in the drinking water. The most important of these are the minerals, and many producers believe that the mineral content of drinking water can influence broiler performance, especially when well water is used. Barton (1989) examined the effect of well water in a survey on broiler performance in Arkansas. The only mineral ion to show a significant effect on performance was nitrate, with lower nitrate concentrations in well water being associated with better performance. Increased concentrations of calcium, on the other hand, were associated with increased growth and improved feed conversion, but with increased mortality and downgrading. The overall results of the study were not very conclusive and more controlled tests on this subject need to be conducted.

Although ambient temperature is the major factor affecting water intake, other factors have also been shown to be important. Austic (1985) summarized these factors and stated that electrolyte content of both diet and water is a major determinant of water intake. Since the chicken has a limited capacity to concentrate urine, excessive dietary levels of soluble mineral elements can increase water consumption. In many hot regions of the world, the water used on poultry farms is brackish water with high levels of total dissolved solids. These contribute to the solute load which must be excreted by the bird. The intake of excessive levels of minerals therefore increases water requirements of broilers and contributes to wet litter problems. High-protein diets also cause increased water intake because of the need for excretion of by-products of protein catabolism. If chlorinated water is being used on the farm, it is recommended to discontinue chlorination on extremely hot days (Murphy, 1988) and to clean drinkers at least once daily using a disinfectant rinse to help reduce slime build-up. Not all studies on the use of brackish water available for poultry in hot regions have been negative. Ilian et al. (1982) showed that brackish water containing up to 3000 p.p.m. of total dissolved solids did not adversely affect the overall performance of broilers.

The common practice of walking the birds during heat stress has always been shown to increase water consumption by about 10%. Walk birds early, before temperatures reach critical levels. One of the best ways of increasing water consumption is by lowering its temperature. Header tanks and water piping should be insulated to ensure that water temperature does not exceed 25°C. One suspended drinker should be provided per 100 broilers. Water lines should be flushed every few hours during the high-heat period of the day to keep water cool. Add extra waterers to the pens and, if emergency conditions occur, consider adding ice or mechanically cooled water. Do not disturb birds when air temperatures are high. Burger (1988) emphasized that, during heat stress, anything that can be done to keep the broilers quiet will reduce mortality. This has also been emphasized by Van Kampen (1976), who cautioned against disturbing birds during heat stress.

May and Lott (1992) studied feed and water consumption patterns of broilers at a constant 24°C or at a cyclic 24, 35 and 24°C. Their data show that the increased water intake and decreased feed intake observed due to high, cyclic temperatures arise from changes that occur during certain times of the day and no changes occur at other times. The increase in water intake precedes the reduction in feed intake. Maintenance of both carbon dioxide and blood pH is critical to the heat-stressed broiler, and the beneficial effects of adding ammonium chloride and potassium chloride to the drinking water are well documented (see Chapter 6). Increased consumption of cool drinking water is crucial to survival of heat-stressed broilers.

Acclimatization to Heat Stress

It has been known for some time that chickens can adapt to climatic changes. Van Kampen (1981) classified responses to climatic changes into three phases: neuronal, hormonal and morphological. The neuronal phase is the first and is seen immediately, while the morphological response is observed much later. Changes in metabolic rate occur fairly quickly since they involve neuronal and hormonal aspects. Morphological changes, such as enlarged combs and wattles, less body fat and less feather cover, which have been known to occur in chickens reared under high temperature, require a much longer time to appear.

Based on these earlier studies, attempts have been made at reducing heat-stress mortality in broilers by acclimatization. Raising house temperature prior to the onset of a heatwave has been shown to reduce mortality. This is partly attributed to a reduction in feed intake in response to the stress. Reece et al. (1972) demonstrated that broilers can acclimatize in 3 days and resist extremely high mortality from heat prostration. The treatment used by these workers consisted of birds taken from 21°C and exposed for 3 days to a 24-h cycle of 24, 35 and 24°C before a temperature stress of 40.6°C was imposed. This treatment reduced mortality from 33 to 0%. Ernst et al. (1984), however, showed that exposing chicks at 1 day of age for 2 h to 43°C reduced performance up to 16 days of age. Arjona *et al.* (1990) observed that exposure for 24 h at 35–38°C at 5 days of age reduced mortality when these birds were heat-stressed for 8 h at 44 days of age. Teeter and Smith (1986) reported that at least 50% of the hypothermic effect of acclimatization immediately prior to heat stress may be attributed to a reduction in feed intake in response to the stress.

Another metabolic adaptation of broilers exposed to high temperature has been reported by Hayashi *et al.* (1992). These workers studied the effects of increasing ambient temperature on protein turnover and oxygen consumption in skeletal muscle. They found that rates of protein synthesis and breakdown and oxygen consumption were higher in birds raised at 20°C than in those raised at 30°C. They suggested that skeletal muscle may function as a regulator of body temperature by changing the rate of skeletal muscle protein turnover in response to ambient temperature changes.

Several adaptations involving the cardiovascular and respiratory systems take place when birds are exposed to high temperatures. These changes have previously been described in Chapter 4. Furthermore, birds undergo certain metabolic adaptations (Chwalibog, 1990). Heat acclimatization involves certain specific adaptations that regulate dehydration and hypovolaemia. These adaptations tend to maximize body water reserves needed for evaporative cooling and blood volume maintenance (Van Kampen, 1981). Belay and Teeter (1993) showed that water consumption increases to match increased respiratory water loss during heat stress.

Early-age thermal conditioning has been used to reduce losses caused by heat stress (Arjona *et al.*, 1988; Yahav and Hurwitz, 1996; Yahav and Plavnik, 1999). These birds have lower body temperature at normal or high temperature, indicating a change in metabolic status (De Basilio *et al.*, 2003). Yahav *et al.* (1997) reported that the long-term development of heat tolerance does not appear to be associated with the induction of heat-shock proteins (see Chapter 4). It has been suggested that a temperature of 36–37.5°C at three days of age is optimum for early heat conditioning of broilers (Yahav and McMurty, 2001).

Arjona *et al.* (1988) showed that exposing broilers to elevated environmental temperature at 5 days of age improved their tolerance to heat stress without reducing subsequent performance. Lott (1991) exposed broilers to high temperature for 3 consecutive days at 21 days of age and was able to decrease mortality resulting from high ambient temperature later in life. This worker observed that acutely heat-acclimatized birds consume more water than non-acclimatized controls during heat stress. Realizing that heat acclimatization may include adaptations in kidney function, Wideman *et al.* (1994) conducted a study on renal function in broilers at normal and high temperature and found that heat-acclimatized birds have significantly lower glomerular filtration rates, filtered loads of sodium and tubular sodium reabsorption rates than the respective control groups. These changes in kidney function, according to these workers, would minimize urinary fluids and solute loss when heat-acclimatized broilers consume large quantities of water to support evaporative cooling. Although this practice of acclimatization is still in the experimental stage, it has definite potential for the broiler industry in hot climates.

Lighting Programmes

The majority of broilers the world over are grown on a light programme of 23 h of light each day. At some time during the night, the lights are turned off for 1 h. This is done to allow the birds to become accustomed to the lights suddenly being turned off. Thus, in the event of a power failure, the flock will accept the loss of light without panic, and crowding and possible suffocation is avoided. This practice is very important because power failures are more common in the developing, hot regions than in the developed, temperate areas of the world.

The use of intermittent light has been shown to increase feed consumption during the cooler part of the day. Intermittent lighting has been shown to improve feed efficiency in broilers, and this improvement can be due to the lower heat production during the dark period (Aerts et al., 2000). Buyse et al. (1994) found that broilers under 1 L:3 D light schedule produce less heat during the early and later growth stage. Francis et al. (1991) reported that the induction of a 4-h period of darkness in the midst of a 14-h day when the temperature was raised from 25 to 35°C reduced the rise in rectal temperature of 28–35-day female broilers. Very few studies have been conducted on broiler lighting programmes in hot, tropical climates. A study on the integration of lighting programmes and operation of feeders has been reported from Singapore (Ngian, 1982), using convection-ventilated housing, in which improved performance was obtained by illumination and feeding during the night. This programme produced the highest live weight at 56 days, but feed conversion and mortality were lower than other combinations examined. Diab et al. (1981) at the Kuwait Institute of Scientific Research showed that sequential 7-h periods of darkness and illumination provided better growth and feed conversion in broilers when compared with continuous feeding. Renden et al. (1992) compared performance of light-restricted broilers (16 h light, 8 h darkness (16 L:8 D) with those reared under an extended lighting programme (23 L:1 D) during the summer. Although relative growth was greater from 28 to 48 days of age for light-restricted broilers, final body weights were similar. Smith (1994) studied the effects of electrolytes and lighting regimens (23 L:1 D versus 16 L:8 D) on broilers grown at elevated temperatures. This worker found that photoschedule had no effect on body-weight gain, feed consumption or carcass characteristics.

Hulan and Proudfoot (1987) studied the effects of light source, ambient temperature (20–34°C) and dietary energy source on the general performance and incidence of leg abnormalities of roaster chickens. They found that light source had no significant effect on mortality, body weight or feed conversion. Incidence of angular deformity (AD) and total leg abnormalities (TLA) was lower under fluorescent as opposed to incandescent light. Cooler ambient rearing temperature increased linearly the incidence of mortality, curly toes (CT), AD, enlarged hocks (EH) and TLA. Mean financial returns were not significantly better for roasters reared under fluorescent versus incandescent light and were better for birds reared under warm versus cool ambient temperature.

Contamination from the digestive tract is a persistent problem in broiler processing. May *et al.* (1990) studied the effects of light and environmental temperature on quantity of crop, gizzard and small intestine contents during feed withdrawal. Results indicated that crop clearance is improved by lighting before and after cooping. No differences were observed in broilers maintained at 18 or 27°C.

Very few studies have been conducted on the effects of light quality on broiler performance in hot climates. In the USA, incandescent lamps have been the standard for many years and are still used in many broiler houses, especially those with side-wall curtains, which can use daylight as a supplement to artificial lighting (Weaver, 1992). Boshouwers and Nicaise (1992), in a study on the effects of light quality on broiler performance, reported that light source significantly affected physical activity, energy expenditure and body weight. Physical activity was lowest under 100 Hz fluorescent light and highest under incandescent. Since in hot climates we may want to reduce the activity of birds, fluorescent light may have an advantage over incandescent lights.

Conclusions

Several studies have been conducted during the past two decades on various techniques to improve broiler performance in hot climates. Nutritional manipulation, such as the addition of fat and the reduction of excess protein and amino acids, has been widely adopted. Birds should also be fed during the cool hours of the day. Feed withdrawal prior to heat-stress initiation has become a viable management tool. Fasting intervals beginning 3–6 h prior to maximum heat stress and totalling up to 12 h daily during significant heat stress reduces broiler mortality. Fasting will probably reduce weight gains. Therefore, the producer will have to weigh the importance of a more rapid growth rate versus a greater mortality risk. Maintenance of both carbon dioxide and blood pH is critical to the heat-stressed broiler, and the beneficial effects of adding ammonium chloride and potassium chloride to the drinking water are well documented. Acclimatization to heat stress by exposing birds to high temperature before the onset of the heatwave has potential, but more research and testing are needed before large-scale field application becomes a reality. The addition of extra vitamins and electrolytes to the drinking water has been helpful in most tests and under most situations. The use of ascorbic acid in the feed or in the drinking water has become a common practice in many hot regions of the world.

Microingredient premixes should be purchased from reputable sources and vitamins should be in gelatine-encapsulated and stabilized forms. Antioxidants should be incorporated into vitamin premixes and into rations to prevent production of toxic free radicals as a result of oxidation. This is particularly important in rations with added fats.

Besides ambient temperature, season per se has an effect on broiler performance, and the best performance is usually attained in winter, while the worst is in summer. High temperatures not only reduce growth but also cause a reduction in efficiency in the utilization of feed energy for productive purposes.

Environmental temperature has been shown to influence body composition in broilers. Broilers reared at normal temperatures (21°C) retain more energy in their carcasses as fat than those reared at high temperature (30°C). High ambient temperatures tend to reduce the proportion of polyunsaturated fatty acids in abdominal fat. Polyunsaturated to saturated fatty acid ratio in broiler carcass declines with age regardless of temperature. Reports on the effect of temperature on carcass proteins are not conclusive.

Biosecurity is a very important consideration in hot climates. The all-in, all-out system is highly recommended and for thorough decontamination special disinfectants may be needed for use between batches. Properties of such disinfectants and adequate methods for their application have been proposed.

Intermittent lighting is potentially a sound practice for hot climates, particularly when used to increase feed consumption during the cooler parts of the day. The provision of cool drinking water plays an important role in reducing heat stress. Increased consumption of cool drinking water is crucial to the survival of heat-stressed broilers. Electrolyte content of both diet and water increases water consumption. High-protein diets also cause increased water intake because of the need for excretion of by-products of protein catabolism. Be sure to discontinue chlorination of water on extremely hot days if this practice is used. The above practices can improve growth and feed efficiency of broilers in hot climates. They can also help in reducing mortality due to heat stress.

References

Abasiekong, S.F. (1989) Behavioural and growth responses of broiler chickens to dietary water content and climatic variables. *British Poultry Science* 29, 563–570.

Aerts, J.M., Berckmans, D., Saevels, P., Decuypere, E. and Buyse, J. (2000) Modelling the static and dynamic responses of total heat production of broiler chickens to step changes in air temperature and light intensity. *British Poultry Science* 41, 651–659.

Afsari, N. (1983) Air conditioning of houses in hot climates. *Poultry International* 22, 64–73.

Aini, I. (1990) Control of poultry diseases in Asia by vaccination. *World's Poultry Science Journal* 46, 125–132.

Ait-Boulahsen, A., Garlich, J.D. and Edens, F.W. (1992) Relationship between blood ionized calcium and body temperature of chickens during acute heat stress. *Proceedings of the 19th World's Poultry Congress*, Vol. 2, pp. 87–92.

Ait-Tahar, N. and Picard, M. (1987) Influence of ambient temperature on protein requirements of broilers. Research Note, INRA Laboratoire de Recherches Avicoles, Nouzilly, France, pp. 1–12.

Akit, M., Yalcin, S., Ozkan, S., Metin, K. and Ozdemin, D. (2005) Effects of temperature during rearing and crating on stress parameters and meat quality of broilers. *Poultry Science* 85, 1867–1874.

Al-Hassani, D.H. and Abrahim, D.K. (1992) Performance of normal and dewinged broiler chickens under high ambient temperatures. *Proceedings of the 19th World's Poultry Congress*, Vol. 2, pp. 709–711.

Al-Zujajy, R.J., El-Hammady, H. and Abdulla, M.A. (1978) The use of air-coolers in broiler houses under subtropical conditions in Iraq. *British Poultry Science* 19, 731–735.

Andrews, L.D. (1977) Performance of broilers with different methods of debeaking. *Poultry Science* 56, 1689–1690.

Anonymous (1982) Energy and protein requirements for broilers in the tropics. *Poultry International*, August, pp. 62–63.

Arjona, A.A., Denbow, D.M. and Weaver, W.D. (1988) Effect of heat stress early in life on mortality of broilers exposed to high environmental temperatures just prior to marketing. *Poultry Science* 67, 226–231.

Arjona, A.A., Denbow, D.M. and Weaver, W.D. (1990) Neonatally induced thermotolerance: physiological responses. *Comparative Biochemistry and Physiology* 95A, 393–399.

Austic, R.E. (1985) Feeding poultry in hot and cold climates. In: Yousef, M.K. (ed.) *Stress Physiology in Livestock*, Vol. 3, Poultry. CRC Press, Boca Raton, Florida, pp. 123–136.

Baghel, R.P.S. and Pradhan, K. (1989a) Energy, protein and limiting amino acid requirement of broilers in their different phases of growth during hot-humid season. *Indian Journal of Animal Science* 59, 1467–1473.

Baghel, R.P.S. and Pradhan, K. (1989b) Studies on energy and protein requirement of broilers during hot-humid season at fixed level of limiting amino acids. *Indian Journal of Poultry Science* 24, 127–132.

Balnave, D. (1989) Mineral in drinking water and poultry production. *Monsanto Nutrition Update* 7(3), 1–8.

Barton, T.L. (1989) Effects of water quality on broiler performance. *Zootecnica International*, March, 44–46.

Belay, T. and Teeter, R.G. (1993) Broiler water balance and thermobalance during thermoneutral and high ambient temperature exposure. *Poultry Science* 72, 116–124.

Bertechini, A.G., Rostagno, H.S., Fonseca, J.B. and Oliveira, A.I.G. (1991) Effects of environmental temperature and physical form of diet on performance and carcass quality of broiler fowls. *Revista da Sociedade Brasileira de Zootecnia* 20, 257–256.

Boshouwers, F.M.G. and Nicaise, E. (1992) Light quality affects physical activity, energy expenditure and growth of broilers. *Proceedings of the 19th World's Poultry Congress*, Vol. 3, pp. 178–181.

Bottje, W.G. and Harrison, P.C. (1985) Effect of carbonated water on growth performance of cockerels subjected to constant and cyclic heat stress temperatures. *Poultry Science* 64, 1285–1292.

Burger, R.E. (1988) Bird death at high temperatures: is there anything we can do? *California Poultry Newsletter*, April, 6–7.

Buys, S.B. and Rasmussen, R.M. (1978) Heat stress mortality in nicarbazine-fed chickens. *Journal of South African Veterinary Association* 49, 127–131.

Buyse, J., Decuypere, E. and Michels, H. (1994) Intermittent lighting and broiler production. 2. Effect on energy and on nitrogen metabolism. *Archiv für Geflügelkunde* 58, 78–83.

Cahaner, A. and Leenstra, F.R. (1992) Effects of high temperature on growth and efficiency of male and female broilers from lines selected for high weight gain, favourable feed conversion and high or low fat content. *Poultry Science* 71, 1237–1250.

Casey, J.M. (1983) White wash formula for the poultry house roof. *Poultry Tips*, P.S. 1. Cooperative Extension Service, University of Georgia, Athens, Georgia.

Christmas, R.B. (1993) The performance of spring- and summer-reared broilers as affected by precision beak trimming at seven days of age. *Poultry Science* 72, 2358–2360.

Chwalibog, A. (1990) Heat production, performance and body composition in chickens exposed to short time high temperature. *Archiv fur Geflugelkunde* 54, 167–172.

Chwalibog, A. and Eggum, B.O. (1989) Effect of temperature on performance, heat production, evaporative heat loss, and body composition in chickens. *Archiv fur Geflugelkunde* 53, 179–184.

Cier, D., Rimsky, Y., Rand, N., Polishuk, O., Gur, N., Benshoshan, A., Frisch, Y. and BenMoshe, A. (1992a) The effects of supplementing ascorbic acid on broiler performance under summer conditions. *Proceedings of the 19th World's Poultry Congress*, Vol. 1, pp. 586–589.

Cier, D., Rimsky, Y., Rand, N., Polishuk, O., Frisch, Y., Gur, N., Benshoshan, A. and BenMoshe, A. (1992b) The effects of different dietary levels of available phosphorus on broiler performance. *Proceedings of the 19th World's Poultry Congress*, Vol. 2, pp. 264–265.

Coelho, M.B. (1991) Effects of processing and storage on vitamin stability. *Feed International*, December, 39–45.

Dale, N.M. and Fuller, H.L. (1979) Effects of diet composition on feed intake and growth of chicks under heat stress. I. Dietary fat levels. *Poultry Science* 58, 1529–1534.

Dale, N.M. and Fuller, H.L. (1980) Effect of diet composition on feed intake and growth of chicks under heat stress. II. Constant versus cycling temperatures. *Poultry Science* 59, 1434–1440.

Damron, B.L. and Johnson, W.L. (1985) Relation of dietary sodium chloride to chick performance and water intake. *Nutrition Reports International* 31, 805–810.

De Basilio, V., Vilarino, M., Yahav, S. and Picard, M. (2001) Early age thermal conditioning and a dual feeding program for male broilers challenged by heat stress. *Poultry Science* 80, 29–36.

De Basilio, V., Requena, F., Leon, A., Vilarino, M. and Picard, M. (2003) Early age thermal conditioning immediately reduces body temperature of broiler chicks in a tropical environment. *Poultry Science* 82, 1235–1241.

Deaton, J.W., Reece, F.N. and Lott, B.D. (1984) Effect of differing temperature cycles on broiler performance. *Poultry Science* 63, 612–615.

Decuypere, E., Buyse, J., Vanlsterdael, J., Michels, H. and Hermans, A. (1992) Growth, feed conversion and carcass quality in broiler chickens in hot and humid tropical conditions. *Proceedings of the 19th World's Poultry Congress*, Vol. 2, pp. 97–100.

Diab, M.F., Husseini, M.D. and Salman, A.J. (1981) Effect of thermal stress, lighting and feeding regimens on performance of broilers. *Poultry Science* 60, 1464 (Abstract).

El-Husseiny, O. and Creger, C.R. (1980) The effect of ambient temperature on carcass energy gain in chickens. *Poultry Science* 59, 2307–2311.

Ernst, R.A., Weathers, W.W. and Smith, J. (1984) Effects of heat stress on day-old broiler chicks. *Poultry Science* 63, 1719–1721.

Faber, H.V. (1964) Stress and general adaptation syndrome in poultry. *World's Poultry Science Journal* 20, 175–182.

Farrell, D.J. and Swain, S. (1977a) Effects of temperature treatments on the heat production of starving chickens. *British Poultry Science* 18, 725–734.

Farrell, D.J. and Swain, S. (1977b) Effects of temperature treatments on the energy and nitrogen metabolism of fed chickens. *British Poultry Science* 18, 735–743.

Fattori, T.R., Mather, F.B. and Hilderbrand, P.E. (1990) Methodology for partitioning poultry producers into recommendation domains. *Agricultural Systems*, 32, 197–205.

Ferket, P.R. and Qureshi, M.K. (1992) Performance and immunity of heat-stressed broilers fed vitamin and electrolyte-supplemented drinking water. *Poultry Science* 71, 88–97.

Filho, D.E., Rosa, P.S., Figneiredo, D.F., Dahlke, F., Macari, M. and Furlan, R.L. (2004) Low protein diet impairs broiler performance under hot environmental temperature. *Proceedings of the 22nd World's Poultry Congress*, Istanbul, Turkey, p. 547.

Francis, C.A., Macleod, M.G. and Anderson, J.E.M. (1991) Alleviation of acute heat stress by food withdrawal or darkness. *British Poultry Science* 32, 219–225.

Fuller, H.L. (1978) The extra value of fat and reduced heat increment. *Proceedings of the Florida Nutrition Conference*, pp. 21–35.

Fuller, H.L. and Dale, N.M. (1979) Effect of diet on heat stress in broilers. *Proceedings Georgia Nutrition Conference*, University of Georgia, Athens, pp. 56–60.

Fuller, H.L. and Rendon, M. (1977) Energetic efficiency of different dietary fats for growth of young chicks. *Poultry Science* 56, 549–557.

Garlich, J.D. and McCormick, C.C. (1981) Interrelationship between environmental temperature and nutritional status of chicks. *Proceedings of the Federation of American Societies for Experimental Biology* 40, 73–76.

Geraert, P.A., Guillaumin, S. and Leclercq, B. (1992) Effect of high ambient temperature on growth, body composition and energy metabolism of genetically lean and fat male chickens. *Proceedings of the 19th World Poultry Congress*, Vol. 2, pp. 109–110.

Gorman, I. (1992) Dietary mineral supplementation of broilers at high temperature. *Proceedings of the 19th World's Poultry Congress*, Vol. 3, p. 651.

Harris, G.C., Nelson, G.S., Seay, R.L. and Dodgen, W.H. (1975) Effects of drinking water temperature on broiler performance. *Poultry Science* 54, 775–779.

Harter-Dennis, J.M. and Pescatore, A.J. (1986) Effect of beak trimming regimen on broiler performance. *Poultry Science* 65, 1510–1515.

Hayashi, K., Kaneda, S., Otsuka, A. and Tomita, Y. (1992) Effects of ambient temperature and thyroxine on protein turnover and oxygen consumption in chicken skeletal muscle. *Proceedings of the 19th World's Poultry Congress*, Vol. 2, pp. 93–96.

Hayat, J., Balnave, D. and Brake, J. (1999) Sodium bicarbonate and potassium bicarbonate supplements for broilers can cause poor performance at high temperatures. *British Poultry Science* 40, 411–418.

Hoffman, L. (1991) Energy metabolism of growing broiler chickens kept in groups in relation to environmental temperature. 1. Feed intake, heat production, and energy utilization. *Archives of Animal Nutrition* 41, 245–255.

Hoffman, L., Schiemann, R. and Klein, M. (1991) Energy metabolism of growing broiler chickens kept in groups in relation to environmental temperature. *Archives of Animal Nutrition* 41, 167–181.

Howlider, M.A.R. and Rose, S.P. (1989) Rearing temperature and the meat yield of broilers. *British Poultry Science* 30, 61–67.

Hulan, H.W. and Proudfoot, F.G. (1987) Effects of light source, ambient temperature and dietary energy source on the general performance and incidence of leg abnormalities of roaster chickens. *Poultry Science* 66, 645–651.

Hurwitz, S., Weiselberg, N., Eisner, U., Bartov, I., Riesenfield, G., Shareit, M., Nir, A. and Bornstein, S. (1980) The energy requirements and performance of growing chickens and turkeys as affected by environmental temperature. *Poultry Science* 59, 2290–2299.

Ilian, M.A., Diab, M.F., Husseini, M.D. and Salman, A.J. (1982) Effects of brackish water utilization by broilers and growing pullets on performance. *Poultry Science* 60, 2374–2379.

Keshavarz, K. and McDougal, L.R. (1981) Influence of anticoccidial drugs on losses of broiler chickens from heat stress and coccidiosis. *Poultry Science* 60, 2423–2426.

Khajavi, M., Rahimi, S., Hassan, Z.M., Kamali, M.A. and Mousavi, T. (2003) Effect of feed restriction early in life on humoral and cellular immunity of two commercial broiler strains under heat stress conditions. *British Poultry Science* 44, 490–497.

Kleiber, M. and Dougherty, J.E. (1934) The influence of environmental temperature on the utilization of food energy in baby chicks. *Journal of General Physiology* 17, 701–726.

Koh, M.T., Wei, H.W. and Shen, T.F. (1989) The effects of environmental temperature on protein and energy requirements of broilers. *Journal of Taiwan Livestock Research* 22, 23–41.

Kutlu, H.R. (2001) Influences of wet feeding and supplementation with ascorbic acid on performance and carcass composition of broiler chicks exposed to a high ambient temperature. *Archiv fur Tierernahrung* 54, 127–139.

Leenstra, F. and Cahaner, A. (1992) Effects of low and high temperature on slaughter yield of broilers from lines selected for high weight gain, favourable feed conversion and high or low fat content. *Poultry Science* 71, 1994–2006.

Liew, P.K., Zulkifli, I., Hair-Bejo, M., Omar, A.R. and Israf, D.A. (2003) Effects of early age feed restriction and heat conditioning on heat shock protein 70 expression, resistance to Infectious Bursal Disease and growth in male broiler chickens subjected to heat stress. *Poultry Science* 82, 1879–1885.

Lin, H., Zhang, H.F., Jiao, H.C., Zhao, T., Sui, S.J., Gu, X.H., Zhang, Z.Y., Buyse, J. and Decuypere, E. (2005a) The thermoregulation response of broiler chickens to humidity at different ambient temperatures. I. One week age. *Poultry Science* 84, 1166–1172.

Lin, H., Zhang, H.F., Du, R., Gu, X.H., Zhang, Z-Y., Buyse, J. and Decuypere, E. (2005b) The thermoregulation responses of broiler chickens to humidity at different ambient temperatures. II. Four-week age. *Poultry Science* 84, 1173–1178.

Lott, B.D. (1991) The effect of feed intake on body temperature and water consumption of male broilers during heat exposure. *Poultry Science* 70, 756–759.

Lu, Q., Wen, J. and Zhang, H. (2007) Effect of chronic heat exposure on fat deposition and meat quality in two genetic types of chicken. *Poultry Science* 86, 1059–1064.

Lyle, G.R. and Moreng, R.E. (1968) Elevated environmental temperature and duration of post-exposure ascorbic acid administration. *Poultry Science* 47, 410–417.

Macy, L.B., Harris, G.C., Delee, J.A., Waldroup, P.W., Izat, A.L., Gwyther, M.J. and Eoff, H.J. (1990) Effects of feeding Lasalocid on performance of broilers in moderate and hot temperature regimens. *Poultry Science* 69, 1265–1270.

Manfueda, P. (2004) Feeding broilers in hot climates, dietary amino acid balance and protein level. *Proceedings of the 22nd World's Poultry Congress*, Istanbul, Turkey, p. 561.

May, J.D. and Lott, B.D. (1992) Feed and water consumption patterns of broilers at high environmental temperature. *Poultry Science* 71, 331–336.

May, J.D., Lott, B.D. and Deaton, J.W. (1990) The effect of light and environmental temperature on broiler digestive tract contents after feed withdrawal. *Poultry Science* 69, 1681–1684.

McCormick, C.C., Garlich, J.D. and Edens, F.W. (1979) Fasting and diet affect the tolerance of young chickens exposed to acute heat stress. *Journal of Nutrition* 109, 1089–1097.

McCormick, C.C., Garlich, J.D. and Edens, F.W. (1980) Phosphorus nutrition and fasting: interrelated factors which affect survival of young chickens exposed to high ambient temperatures. *Journal of Nutrition* 110, 784–790.

McDougal, L.R. and McQuinston, T.E. (1980) Mortality in heat stress in broiler chickens influenced by anticoccidial drugs. *Poultry Science* 59, 2421–2425.

McDowell, R.E. (1972) *Improvement of Livestock Production in Warm Climates*. W.H. Freeman, San Francisco.

Miraei-Ashtiani, S.R., Zamani, P., Shirazad, M. and Zare-shahned, A. (2004) Comparison of the effect of different diets on acute heat stressed broilers. *Proceedings of the 22nd World's Poultry Congress*, Istanbul, Turkey, p. 552.

Morrison, W.D., Braithwaite, L.A. and Leeson, S. (1988) Report of a survey of poultry heat stress losses during the summer of 1988. Unpublished report from the Department of Animal and Poultry Science, University of Guelph, Guelph, Ontario, Canada.

Murphy, D.W. (1988) Non-nutritional nutrition effects. *Proceedings, Monsanto Technical Symposium*, Fresno, California, pp. 49–55.

Mutalib, A. (1990) How to reduce water vaccination failures. *Poultry Digest*, March, 14–16.

Nakamura, Y., Aoyagi, Y. and Nakaya, T. (1992) Effect of ascorbic acid on growth and ascorbic acid levels of chicks exposed to high ambient temperature. *Japanese Poultry Science* 29, 41–46.

Navahari, D. and Jayaprasad, L.A. (1992) Influence of season, floor type and space on broiler performance in humid tropics. *Proceedings of the 19th World's Poultry Congress*, Vol. 2, p. 138.

Ngian, M.F. (1982) Night feeding of broilers optimizes feed efficiency. *Poultry International*, November, 48.

Nir, I. (1992) Optimization of poultry diets in hot climates. *Proceedings of the 19th World's Poultry Congress*, Vol. 2, pp. 71–76.

Njokn, P.C. (1984) The effect of ascorbic acid supplementation on broiler performance in a tropical environment. *Poultry Science* 63 (Suppl. 1), 156.

North, M.O. and Bell, D.D. (1990) *Commercial Chicken Production Manual*, 4th edn. Van Nostrand Reinhold, New York.

Olson, D.M., Sunde, M.L. and Bird, H.R. (1972) The effect of temperature on ME determination and utilization by the growing chick. *Poultry Science* 51, 1915–1922.

Orban, J.I. and Roland, D.A. (1990) Response of four broiler strains to dietary phosphorus above and below the requirement when brooded at two temperatures. *Poultry Science* 69, 449–455.

Osman, A.M.K., Tawfik, E.S., Klein, F.W. and Hebeler, W. (1989) Effect of environmental temperature on growth, carcass traits, and meat quality of broilers of both sexes and different ages. *Archiv fur Geflugelkunde* 53, 163–175.

Polin, D., Wynosky, E.R. and Porter, C.C. (1962) Amprolium: influence of egg yolk thiamine concentration on chick embryo mortality. *Proceedings of the Society of Experimental Biology and Medicine*, 110, 844–848.

Pope, D.L. (1960) Nutrition and environmental studies with broilers. *Proceedings of the University of Maryland Nutrition Conference for Feed Manufacturers*, pp. 48–51.

Rajmane, B.V. and Ranade, A.S. (1992) Remedial measures to control high mortality during summer season in tropical countries. *Proceedings of the 19th World's Poultry Congress*, Vol. 1, pp. 343–345.

Reece, F.N., Deaton, J.W. and Kubena, L.F. (1972) Effects of high temperature and humidity on heat prostration of broiler chickens. *Poultry Science* 51, 2021–2025.

Reece, F.N., Deaton, J.W. and Harwood, F.W. (1976) Effect of roof insulation on the performance of broiler chickens reared under high temperature conditions. *Poultry Science* 55, 395–398.

Reilly, W.M., Koelkebeck, K.W. and Harrison, P.C. (1991) Performance evaluation of heat stressed commercial broilers provided water cooled floor perches. *Poultry Science* 70, 1699–1703.

Renden, J.A., Bilgili, S.F. and Kincaid, S.A. (1992) Live performance and carcass yield of broiler strain crosses provided either 16 or 23 hours of light per day. *Poultry Science* 71, 1427–1435.

Smith, M.O. (1993) Nutrient content of carcass parts from broilers reared under cycling high temperatures. *Poultry Science* 72, 2166–2171.

Smith, M.O. (1994) Effects of electrolytes and lighting regimen on growth of heat-distressed broilers. *Poultry Science* 73, 350–353.

Smith, M.O. and Teeter, R.G. (1987) Potassium balance of the 5 to 8 week-old broiler exposed to constant heat or cycling high temperature stress and the effects of supplemental potassium chloride on body weight gain and feed efficiency. *Poultry Science* 66, 487–492.

Smith, M.O. and Teeter, R.G. (1988) Practical application of potassium chloride and fasting during naturally occurring summer heat stress. *Poultry Science* 61 (Suppl. 1), 36.

Sonaiya, E.B. (1988) Fatty acid composition of broiler abdominal fat as influenced by temperature, diet, age and sex. *British Poultry Science* 29, 589–595.

Sonaiya, E.B. (1989) Effects of environmental temperature, dietary energy, sex, and age on nitrogen and energy retention on the edible carcass of broilers. *British Poultry Science* 30, 735–745.

Sonaiya, E.B., Ristic, M. and Klein, F.W. (1990) Effect of environmental temperature, dietary energy, age and sex on broiler carcass portions and palatability. *British Poultry Science* 31, 121–128.

Tawfik, E.S., Osman, A.M.A., Ristic, M., Hebeler, W. and Klein, F.W. (1989) Einflus der Stalltemperatur auf Mastleistung, Schlachtkorperwert und Fleischbeschaffenheit von Broiler Unterschiedlichen alters und geschlechts. 2. Mitteilung Schlachtkorperwert. *Archiv fur Geflugelkunde* 53, 235–244.

Tawfik, E.S., Osman, A.M.A., Hebeler, W., Ristic, M. and Freudenreich, P. (1992) Effect of environmental temperature, sex and fattening period on amino acid composition of breast meat of broilers. *Archiv fur Geflugelkunde* 56, 201–205.

Teeter, R.G. and Smith, M.O. (1986) High chronic ambient temperature stress effects on broiler acid-base balance and their response to supplemental NH_4Cl, KCl and K_2CO_3. *Poultry Science* 65, 1777–1781.

Teeter, R.G., Smith, M.O., Owens, F.N. and Arp, S.C. (1985) Chronic heat stress and respiratory alkalosis: occurrence and treatment in broiler chicks. *Poultry Science* 64, 1060–1064.

Thompson, K.L. and Applegate, T.J. (2006) Feed withdrawal alters small-intestinal morphology and mucus of broilers. *Poultry Science* 85, 1535–1540.

Thornton, P.A. (1961) Increased environmental temperature influences on ascorbic acid activity in the domestic fowl. *Proceedings of the Federation of American Societies for Experimental Biology* 20, 210A.

Trout, J.M., Bierlmaier, S.J. and Mashaly, M.M. (1988) Effect of beak trimming on performance of broiler chicks. *Poultry Science* 61 (Suppl. 1), 166.

Vahl, H.A. and Stappers, H.P. (1992) Effect of lower sodium levels in broiler diets. *Proceedings of the 19th World's Poultry Congress*, Vol. 1, pp. 598–602.

Van Kampen, M. (1976) Activity and energy expenditure in laying hens: the energy cost of eating and posture. *Journal of Agricultural Science* (Cambridge) 87, 85–88.

Van Kampen, M. (1981) Water balance of colostomized hens at different ambient temperatures. *British Poultry Science* 22, 17–23.

Vo, K.V. and Boone, M.A. (1977) Effect of water availability on hen survival time under high temperature stress. *Poultry Science* 56, 375–377.

Waldroup, P.W. (1982) Influence of environmental temperature on protein and amino acid needs of poultry. *Federation Proceedings* 41, 2821–2823.

Waldroup, P.W., Mitchell, R.J., Payne, J.R. and Hazen, K.R. (1976) Performance of chicks fed diets formulated to minimize excess levels of essential amino acids. *Poultry Science* 55, 243–253.

Weaver, W.D. (1992) Broiler housing in the USA. *Proceedings of the 19th World's Poultry Congress*, Vol. 3, pp. 161–163.

Wideman, R.F., Ford, B.C., May, J.D. and Lott, B.D. (1994) Acute heat acclimatization and kidney function in broilers. *Poultry Science* 73, 75–88.

Yaghi, A. and Daghir, N.J. (1985) Protein requirement for broiler starter, grower and finisher rations. *Poultry Science* 64 (Suppl.), 201.

Yahav, S. and Hurwitz, S. (1996) Induction of thermotolerance in male broiler chickens by temperature conditioning and early age. *Poultry Science* 75, 402–406.

Yahav, S. and McMurty, J.P. (2001) Thermotolerance acquisition in broiler chicken by temperature conditioning early in life. The effect of timing and ambient temperature. *Poultry Science* 80, 1662–1666.

Yahav, S. and Plavnik, I. (1999) Effect of early-age thermal conditioning and food restrictions on performance and thermotolerance of male broiler chickens. *British Poultry Science* 40, 120–126.

Yahav, S., Shamai, A., Horev, G., Bar-Ilan, D., Geuina, O. and Freidman-Einat, M. (1997) Effect of acquisition of improved thermotolerance on the induction of heat shock proteins in broiler chickens. *Poultry Science* 76, 1428–1434.

Yalcin, S., Settar, P., Ozkan, S. and Cahaner, A. (1997) Comparative evaluation of three commercial broiler stocks in hot versus temperate climates. *Poultry Science* 76, 921–929.

Yalcin, S., Ozkan, S., Acikgoz, Z. and Ozkan, K. (1999) Effect of dietary methionine on performance, carcass characteristics and breast meat composition of heterozygous naked neck (Na/na+) birds under spring and summer conditions. *British Poultry Science* 40, 688–694.

Yalcin, S., Ozkan, S., Türkmut, L. and Siegel, P.B. (2001) Responses to heat stress in commercial and local broiler stocks. 1. Performance traits. *British Poultry Science* 42, 149–152.

Yo, T., Siegel, P.B., Guerin, H. and Picard, M. (1997) Self-selection of dietary protein and energy by broilers grown under a tropical climate: effect of feed particles size on feed choice. *Poultry Science* 76, 1467–1473.

Zulkifli, I., Dunnington, A., Gross, W.B. and Siegel, P.B. (1994) Food restriction early or later in life and its effect on adaptability, disease resistance and immunocompetence of heat stressed dwarf and non dwarf chickens. *British Poultry Science* 35, 203–214.

Zulkifli, I., Che Noma, M.T., Israf, D.A. and Omar, A.R. (2002) The effect of early-age feed restriction on heat shock protein 70 response in heat-stressed female broiler chickens. *British Poultry Science* 43, 141–145.

10 Replacement Pullet and Layer Feeding and Management in Hot Climates

N.J. Daghir

Faculty of Agricultural and Food Sciences, American University of Beirut, Lebanon

Introduction	261
Replacement pullets	262
Body weights of replacement pullets	262
Feeding the replacement pullet	263
Water consumption	266
Acclimatization	267
Replacement pullet management	267
Layers	269
Feeding the laying hen	269
Water quality and quantity for layers	278
Acclimatization	281
Effects of temperature on egg quality	281
Layer management practices	282
Conclusions and recommendations	285
References	287

Introduction

This chapter covers selected aspects of feeding and management of the replacement pullet as well as the laying hen in hot climates. It presents a combination of nutrition and management strategies because the author believes that this is the most adequate approach to overcome problems of heat stress in laying stock. The chapter starts with a section on pullet body weights at housing, since this is probably the most challenging pullet production practice in hot climates. Feeding the young pullet and the laying hen, acclimatization of both growing pullets and layers, lighting programmes, water quality and quantity, and several other management tips for hot climates are covered.

A section has been included in this chapter on the detrimental effects of high temperature on egg weight, shell quality and interior egg quality, because it is fairly well documented that these effects on egg quality are somewhat independent of the effect of reduced feed intake at high temperature. This has previously been discussed in Chapter 6, where it was pointed out that the detrimental effects of heat stress on performance can be divided into those that are due to high temperature per se and those that are due to reduced feed intake.

Replacement Pullets

Body weights of replacement pullets

The success of a table egg production enterprise depends, to a very large extent, on the quality of pullets at housing. A quality pullet can be defined as one of optimum body weight and condition required for optimum performance in the laying house. Jensen (1977) reported that a major problem in rearing pullets in the southern USA was obtaining acceptable body weights at sexual maturity for pullets reared in the hot months of the year. A study by Bell (1987a) on over 100 commercial flocks in the USA showed that April, May and June hatches had the lowest production per hen housed. He indicated that this was probably due to lighter pullets grown in hot weather. Payne (1966) reared pullets from 6 to 21 weeks of age at mean temperatures of either 20 or 33°C. Birds reared at 33°C were 118 g lighter at 21 weeks of age and their eggs were consistently smaller throughout the laying period. Stockland and Blaylock (1974), in a similar study, observed a difference of 130 g in body weight between birds reared at 29.4°C and those reared at 18.3°C. Vo *et al.* (1978) reared Leghorn pullets at constant temperatures of 21, 29 and 35°C. They showed that birds reared at 35°C weighed 20–30% less than those reared at 21°C. Vo *et al.* (1980) also demonstrated that sexual maturity was significantly delayed for pullets reared at 35°C as compared with those reared at 21 or 29°C. Escalante *et al.* (1988) studied the effect of body weight at 18 weeks of age on performance of White Leghorn pullets in Cuba. Egg production from 21 to 66 weeks was higher for the heavier-weight birds. Body weight at 18 weeks, however, had no effect on age at sexual maturity or age at 50% production.

Considering the above studies, it is therefore recommended that pullets should be weighed frequently, starting as early as 4 weeks of age, and that weight and uniformity be watched closely during hot weather. Flock uniformity is very important in obtaining optimum performance and the greatest profitability. In situations where uniformity is a problem, growers should sort out all the small birds and pen them separately at about 5 weeks. One practice used when it is known that pullets are going to be grown during warm weather is to start about 10% fewer pullets for a given space as compared with normal temperature conditions. The result will mean increased floor space per pullet along with more water and feeder space. It is important to take body-weight measurements every 2 weeks, from 6 weeks of age and on to housing, in order to determine if the pullets are growing satisfactorily during this critical period.

In hot weather, it is desirable to get pullets as heavy as possible before the onset of egg production, because the larger the weight of the pullet at maturity, the larger is her weight throughout the laying period and the greater is her feed intake. It is therefore recommended that pullets at housing be above the breeders' recommended target weight. Since adequate feed consumption is of primary concern in hot climates, heavier hens will consume more feed, which will result in higher peaks and better persistency in production.

Feeding the replacement pullet

It has been shown for many years that house temperature is one of the most important factors affecting feed consumption. There is a change in feed consumption as house temperatures increase or decrease, but the relationship is not constant at various house temperatures. Table 10.1 shows that the percentage change in feed consumption is much larger during hot weather than during cold weather.

The influence of temperature on the nutrient requirements of replacement pullets has not been widely investigated. Stockland and Blaylock (1974) in their study on rearing pullets at 29.4°C and 18.3°C concluded that protein requirement as percentage of diet was increased in a hot environment. McNaughton et al. (1977), on the other hand, reported that neither dietary protein nor energy influenced body weights at 20 weeks of age under high temperature conditions. These workers, however, used a cyclic temperature of 24 to 35°C between 12 and 20 weeks of age for the duration of the study.

Table 10.1. Temperature and feed consumption for growing pullets (from North and Bell, 1990).

Average daytime house temperature		% Change in feed consumption for each 1 °F (0.6°C) change in temperature
°F	°C	
90–100	32.2–37.8	3.14
80–90	26.7–32.2	1.99
70–80	21.1–26.7	1.32
60–70	15.6–21.1	0.87
50–60	10.0–15.6	0.55
40–50	4.4–10.0	0.30

They also reported that increasing lysine and methionine plus cystine levels above National Research Council (NRC) recommendations did not influence body weights under high temperature conditions. Henken *et al.* (1983) confirmed this by showing in controlled metabolic studies that protein deposition in the growing pullet was not influenced by temperature and that protein anabolism was relatively independent of environmental temperature.

In general, however, pullet growth can be improved at high temperatures by increasing nutrient density. Leeson and Summers (1981) found that pullets reared at 26°C after brooding did respond to increased nutrient density up to 8 weeks of age in terms of improved weight gains. The same workers (Leeson and Summers, 1989) tested pullets reared at a constant 22°C or at cyclic temperatures of 22–32°C and given diets ranging from 2650 to 3150 kcal/kg with 15–19% crude protein. Increasing the metabolizable energy (ME) of the diet increased body weight and the effect was most pronounced in the hot environment. Increasing dietary protein increased body weight initially, but at 140 days of age body weights were not affected by dietary protein. They concluded that, given adequate protein, pullet growth is most responsive to energy intake. Rose and Michie (1986) concluded that high-protein or nutrient-dense rearing feeds increased body weights of pullets reared under high temperatures and decreased the time taken by the pullets to reach sexual maturity.

Researchers at the University of Florida have studied feeding programmes for raising pullets in hot climates for many years. Douglas and Harms (1982) showed that, when low-protein diets are fed to pullets to be housed in hot weather, birds did not consume sufficient layer feed to gain optimum body weight and maintain maximum egg weight. Douglas *et al.* (1985) conducted three experiments using a step-down protein programme for commercial pullets. Maximum body weight was obtained by these workers at 20 weeks when a 20% protein diet was fed from 0 to 8 weeks and 17% from 8 to 12 weeks, followed by reducing the protein level by one percentage point at biweekly intervals. Douglas and Harms (1990) published a similar study on dietary amino acid levels for commercial replacement pullets. On the control diet, winter-reared pullets averaged 1443 g and summer-reared pullets 1322 g at 20 weeks of age, both groups receiving 21% protein starter and 18% protein from 8 to 20 weeks. These authors recommended the use of a step-down amino acid programme with linear reductions at biweekly intervals of the following ranges: 0.54–0.42% total sulfur amino acids (TSAA) (0.28–0.21% methionine), 0.64–0.48% lysine, 0.18–0.14% tryptophan and 0.96–0.78% arginine. Such a programme would allow the protein source to be gradually lowered and the energy source to be raised, which is needed in most hot-weather situations. El-Zubeir and Mohammed (1993) fed commercial egg-type pullets from 10 to 19 weeks of age diets containing 13, 15 or 17% protein and ME of 10, 11 or 12 MJ/kg, followed by a layer diet containing 16% protein and 11.7 MJ/kg. At summer temperatures in Sudan ranging between 26 and 45°C, the diet containing 15% protein and 10 MJ/kg was best for optimal growth, minimum number of days required to reach point of lay and age at 25 and 50% egg production.

Under hot weather conditions, optimum pullet growth cannot be achieved with low-energy diets and, in many cases, high-energy diets containing fat are essential. Although growing rations of 2750–2900 kcal/kg are adequate under most conditions, during hot weather birds will not eat enough feed and body weights will be low. Metabolizable energy levels of 3000–3100 kcal/kg of diet may be necessary to get enough energy into the bird and increased protein is needed to maintain the same calorie to protein ratio, particularly early in the rearing period. The use of a 'broiler type' starter (0–3 weeks) can help in getting early growth. If birds are doing well, then one can change to a normal starter at 21 days. Table 10.2 presents suggested ration specifications for the starter, grower and prelay, while Table 10.3 presents recommended levels for vitamins and trace minerals per tonne of complete feed. Normally, a starter is fed up to 6 weeks of age, and the change from starter to the first grower takes place during the 42–49-day period. If body weights are not up to those specified by the breeder, then feed change should be postponed and the starter is continued until standard weight for age is reached. Usually, a starter is not fed beyond the tenth week of age. If house temperatures exceed 30°C, an increase in these suggested

Table 10.2. Ration specifications for chick starter, grower and prelay.

Nutrients	Starter	Grower I	Grower II	Prelay
Protein (%)	19–20	16–17	15–16	16–17
Metabolizable energy				
(kcal/kg)	2850–2950	2750–2850	2700–2800	2750–2850
(MJ/kg)	11.9–12.3	11.5–11.9	11.3–11.7	11.5–11.9
Calcium (%)	0.90	1.00	1.00	2.00
Available phosphorus (%)	0.45	0.40	0.35	0.40
Sodium (%)	0.18	0.18	0.18	0.18
Chloride (%)	0.17	0.17	0.17	0.17
Potassium (%)	0.60	0.60	0.60	0.60
Lysine (%)	1.00	0.75	0.70	0.75
Methionine (%)	0.42	0.36	0.34	0.36
Methionine + cysteine (%)	0.72	0.62	0.58	0.65
Tryptophan (%)	0.20	0.17	0.16	0.18
Arginine (%)	1.05	0.90	0.84	0.94
Histidine (%)	0.40	0.34	0.32	0.36
Phenylalanine (%)	0.70	0.65	0.56	0.65
Threonine (%)	0.70	0.60	0.56	0.63
Leucine (%)	1.44	1.22	1.15	1.30
Isoleucine (%)	0.84	0.72	0.67	0.76
Valine (%)	0.86	0.73	0.69	0.77

Table 10.3. Recommended vitamin–trace mineral levels per tonne of complete feed.

		Starter	Grower	Layer
Vitamin A	(IU)	9,500,000	8,500,000	10,000,000
Vitamin D_3	(ICU)	2,000,000	1,000,000	2,200,000
Vitamin E	(IU)	15,000	15,000	15,000
Vitamin K_3	(g)	2	1.5	2
Thiamine	(g)	2.2	1.5	2.2
Riboflavin	(g)	5.0	5.0	6.5
Panthotenic acid	(g)	12.0	10.0	15.0
Niacin	(g)	40.0	30.0	40.0
Pyridoxine	(g)	4.5	3.5	4.5
Biotin	(g)	0.2	0.15	0.2
Folic acid	(g)	0.8	0.5	1
Vitamin B_{12}	(g)	0.012	0.010	0.014
Choline	(g)	1,300	1,000	200
Iron	(g)	96	96	96
Copper	(g)	10	10	10
Iodine	(g)	0.4	0.4	0.4
Manganese	(g)	66	66	66
Zinc	(g)	70	70	70
Selenium	(g)	0.15	0.15	0.15

Note: antioxidants should be added at levels recommended by the manufacturer.
Antioxidants are especially important in hot climates and where fats are added to the ration.

specifications may be needed. The extent of this increase would depend on the relative decrease in feed intake.

Several methods have been suggested for increasing feed consumption in growing pullets. Feeding crumbled feeds has been shown to help. These feeds are eaten faster and digested more easily than mash. However, the quality of the crumble should be checked for too much dust. Feed consumption can be encouraged by increasing the frequency of feeding and by stirring the feed between feedings. Spraying the feed with water can help encourage eating, but care needs to be taken to avoid mould growth.

Water consumption

Water consumption in growing pullets varies with age, breed, ambient temperature, humidity, density of the feed and several nutritional factors. Ambient temperature is by far the most important factor affecting water intake. Figure 10.1 shows weekly water consumption of growing Leghorn pullets at four house temperatures. It is seen that Leghorn pullets drink about twice as much water per day at 38°C as at 21°C. House temperatures below 21°C, however, do not significantly reduce water intake.

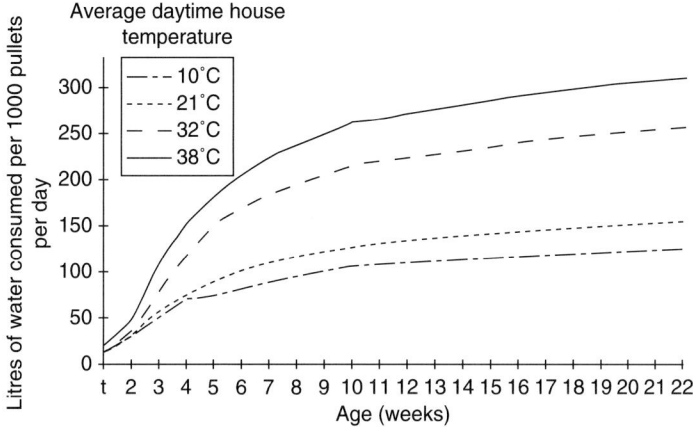

Fig. 10.1. Water consumption of growing Leghorn pullets.

Acclimatization

There are indications that rearing pullets at a relatively high temperature (26–29°C) will help in getting them acclimatized to subsequent high-temperature exposure in the laying house. Although there has been a lot of work on acclimatization of broilers (May et al., 1987) and laying hens (Sykes and Fataftah, 1980; Bohren et al., 1982), very little has been done in this area on replacement pullets. Sykes and Fataftah (1986a) studied the effect of acclimatization of chicks to a hot, dry climate by evaluating changes in rectal temperature during regular daily exposure to an ambient temperature of 42°C and 26% relative humidity (RH). They reported that young chicks (5–47 days of age) were able to acclimatize and showed a progressive reduction in the rate of increase in rectal temperature over the period of exposure and the ability to survive conditions that initially would have been fatal. The heat tolerance of the youngest chickens was high and these acclimatized rapidly to an ambient temperature of 42°C. This initial heat tolerance was reduced at 19 days of age and more so at 47 days. Both of these age groups were able to acclimatize to this very hot climate within 5 and 8 days respectively. The question to be raised here is how long will these birds in practice retain this ability to withstand heat stress and what effect will this acclimatization have on body weight at housing and subsequent laying performance. This area needs further investigation.

Replacement pullet management

Managing the growing pullet is one of the most important activities of a table egg production enterprise. If pullets are grown well, then the prerequisite for success of the enterprise is satisfied. A great deal has been written on the management of growing Leghorn pullets. This section, however, will cover only a few selected aspects that have some bearing on hot climates and that can help in reducing the detrimental effects of heat stress. High environmental temperatures have been shown to create a severe stress on young growing pullets. Pullets will eat less and drink more water than they do at normal temperatures. It is important, therefore, that adequate feed and water space be provided in hot climates and additional waterers be used during the very hot periods of the year. This increased water consumption will aggravate litter problems, and ventilation and air movement through the house become more critical. In severe wet-litter cases it is recommended to add 2.5 kg of superphosphate to each 10 m² of floor space and mix the phosphate with the litter by stirring.

Growing pullets do not perform well at high temperatures. Their growth is reduced; they feather poorly; flocks are not uniform; and feed conversion is poor. Therefore, every effort should be made to reduce house temperature and thus reduce heat stress on the bird.

Beak trimming

Beak-trimming stress is a likely cause of underweight pullets, and its detrimental effects are more severe on birds reared in hot climates than on those in temperate areas. Research has suggested that beak trimming at an early age will reduce this problem (Bell, 1987b). Precision beak trimming at 7–10 days of age is highly recommended in hot regions because it is less stressful than at an older age (6–9 weeks). Retrimming, if necessary, should be done no later than 12 weeks.

Light

The rearing of pullets between latitudes 30°N and 30°S requires special light consideration. These areas present a special problem for the producer. Producers need not only to take account of the amount of natural daylight and the amount of light needed for maximum production but also to consider adding light during the coolest part of the night to stimulate feed consumption. Intermittent lighting of pullets from 2 to 20 weeks of age has resulted in improved weights (Ernst *et al.*, 1987). Biomittent lighting for pullets has also been recommended by Purina Mills Inc. (1987) in hot weather. They claim that pullets perform better during heat stress on this programme than on standard lighting. They are less underweight owing to lessened activity and have better feed utilization. The programme consists of 24 h light during the first week. At 2 weeks of age, light is reduced to 8 h daily. From 3 to 18 weeks, lights are maintained at 8 h daily. During these 8 h, the lights are usually on 15 min/h (15 min light and 45 min darkness) (15 L:45 D). The exception is the last hour of each day, when the pattern is 15L:30 D:15 L. One recommended programme for pullets in light-tight houses is shown in Table 10.4.

The light phase of short-day programmes should be started early in the day to encourage feed consumption. One hour of light in the dark period for open-sided houses can be beneficial to feed consumption.

Handling pullets

Pullets should not be serviced (vaccinated, beak-trimmed, etc.) or moved on hot days. Always move pullets on cool days whenever possible or, better

Table 10.4. Programme for pullets in light-tight housing.

Age (weeks)	Programme
1	24 h constant
2	8 h constant
3–18	8 h (15L:45D)

still, during the night. Moving is less stressful if it is done before pullets reach sexual maturity. Put 30% fewer birds per crate for moving and move birds at least 2 weeks before the first egg is expected. Stress can increase vitamin needs in birds. Nockels (1988) pointed out that vitamin levels, particularly vitamins A, C and E, required for stressed birds are greater than those needed for birds under normal environmental conditions. Furthermore, in hot and humid areas, vitamin stability in feeds is considerably reduced (McGinnis, 1989). It is a good practice, therefore, to give vitamins in the drinking water for 3 days prior to moving and electrolytes for another 3 days after moving.

Housing

When the time comes to house pullets, many growers house by body weight and maturity. This makes it possible to provide the lighter and more immature birds with a separate lighting and feeding programme to speed up their development and maturity.

The watering system in the laying house should be similar to that in the rearing house. This is very critical in hot climates, because if pullets fail to locate the water immediately after housing, they can be severely stressed and damaged for a long time.

Very little work has been done on methods of housing pullets in hot climates. Owoade and Oduye (1992) tested the use of a slatted floor in an open-sided house for pullet rearing in Nigeria. They were able to significantly reduce rearing mortality and suggested that this was a good option of housing for the prevention of disease during rearing in a hot environment. The environmentally controlled house, however, offers a better means of creating a good environment for rearing pullets in hot climates for large, integrated operations. The open-sided house suggested by the above workers may be feasible for small-scale operations in a hot and humid environment.

Layers

Feeding the laying hen

Just as with the growing pullet, house temperature is one of the most important factors affecting feed consumption in laying hens. Sterling *et al.* (2003)

summarized the performance of 14.7 million commercial layers in the USA, representing 11 different White Leghorn strains, and found that environmental temperature was highly correlated with several measures of performance, including feed and water intake, body weight, egg production, feed conversion and egg weight.

Sykes (1976) reported that the average decrease in energy intake was 1.6% per 1°C as environmental temperature increased above 20°C. Reid (1979) indicated that ME intake declined 2.3% per 1°C as environmental temperature increased from 20 to 30°C. Above 30°C, both feed intake and egg production were markedly reduced. Smith and Oliver (1972) found that feed intake declines an average of 5% per 1°C between 32 and 38°C. Therefore, the value of a 1.5% drop in food intake per degree rise in temperature reported by Austic (1985) is only valid up to 30°C. It is very important, therefore, to monitor daily feed consumption in hot weather to ensure an adequate intake of nutrients on a per bird basis. This is particularly important early in the production cycle. Petersen *et al.* (1988), in a study on the effect of heat stress on performance of hens with different body weights, reported that a permanent laying stop is observed in heavy birds and in hens with a low food intake during the first months of laying. Data presented by North and Bell (1990) show that feed consumption is reduced by half when house temperatures increase from 21.1 to 37.8°C. Most of this reduction in feed intake is due to reduced maintenance requirement. Figure 10.2 shows that the maintenance requirement of White Leghorns as well as brown layers is reduced by 30 kcal/day when ambient temperatures rise from 21 to 38°C.

Energy and protein

ME requirements decrease with increasing temperature above 21°C. This reduced requirement is mainly due to a reduction in the requirements for

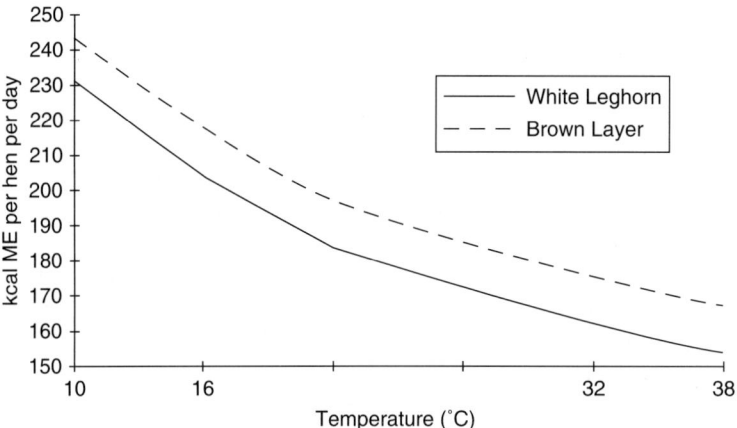

Fig. 10.2. Effect of ambient temperature on the ME requirements for maintenance of laying hens.

maintenance and continues to reach a low at 28°C, followed by an increase in requirement as temperature increases. It has been known for a long time that hens eat mainly to satisfy their energy requirements and therefore energy requirements for maintenance decrease as environmental temperature increases. This relationship holds true only within the zone of thermoneutrality. At very low temperature birds overeat and at high temperature they under-eat. Above 28°C, feed intake decreases more rapidly and the hen's energy requirements begin to increase. This increase enables the bird to get rid of the extra heat burden caused by high temperature. The use of high-energy rations during the early production phase has therefore become quite common in warm regions. This practice should be accompanied by increased density of all nutrients, particularly critical amino acids.

It has long been suggested that feed for laying hens should contain more protein in hot weather than in cold weather (Heywang, 1947). Reid and Weber (1973) studied different methionine levels for laying hens at 21 or 32°C and did not detect any limitations in egg production due to sulfur amino acids. They used regression equations to estimate the requirement, which was found to be 498 and 514 mg per hen per day, respectively, at 21 and 32°C. Valencia et al. (1980) varied protein levels from 12 to 20% at 21 or 32°C. They concluded that the protein requirement was not affected by temperature. Since the work of Heywang (1947), there has been no clear-cut evidence that the protein and/or amino acid requirements of laying hens are higher under high temperature conditions. Several nutritionists advocate, however, protein- and amino acid-rich programmes for layers in hot climates. The important practice to follow is to adjust density of amino acids in the feed to ensure the same daily intake of these nutrients as that normally consumed at 21°C. DeAndrade et al. (1977) fed laying hens under heat stress a diet containing 25% more of all nutrients except energy, which was increased by only 10%. They found that these dietary adjustments overcame most of the detrimental effects of high temperature on percentage egg production and led to a limited improvement in egg weight. Eggshell quality was not improved by dietary adjustments. Amino acid imbalances can be critical at high temperature because these imbalances increase the reduction in feed intake. Protein levels, therefore, could be reduced by about 1–2% at high temperature, provided amino acid levels are kept constant and well balanced (Strighini et al. 2002).

Some workers have raised the hypothesis of a harmful effect of high-protein diets under high temperature conditions (Waldroup et al., 1976). The explanation here is twofold. One is that excess amino acids in the bloodstream may depress food intake because of their effect on the hypothalamus. The second reason is the high heat increment of protein: thus a reduction in protein catabolism would result in a decrease in heat production and help the bird to maintain its energy balance under high temperature conditions. What is being advocated now is a low-protein diet balanced with commercial feed-grade methionine and lysine.

Scott and Balnave (1989) studied the influence of hot and cold temperatures and diet regimen (complete versus self-select) on feed and nutrient intake and selection and egg mass output. Although pullets maintained under high temperatures consumed less food and nutrients and produced less egg mass, there were no differences between the protein:ME ratio selected by self-select-fed pullets under either temperature treatment. The authors point out that, for those concerned with feeding pullets at high temperatures, their work shows that pullets fed by self-selection are able to consume up to 20 g protein/day by the third week following sexual maturity, whereas pullets fed the complete diets were only consuming 12 g of protein/day. The National Research Council (NRC, 1994) estimates the daily protein requirement for egg-laying hens to be between 16 and 17 g/hen/day, while Scott and Balnave (1989) showed actual protein intakes by pullets at hot and cold temperatures to be 19 and 30 g/hen/day, respectively, by 3 weeks after laying the first egg. Austic (1985) proposed that protein and amino acid levels as percentage of diet be increased linearly as environmental temperatures increase from 20 to 30°C. However, beyond that temperature, no further increase is needed since rate of egg production begins to decline.

It has also been observed that energy consumption during the summer months drops significantly in comparison with winter or spring (Daghir, 1973). Energy intake during the summer can be 10–15% lower than during the winter according to work conducted by the author over a period of 3 years in Lebanon. The data of Chawla *et al.* (1976) show a difference in energy consumption of 10–25% between summer and winter in India (Punjab Agriculture University). Results of work at the University of California on 100 commercial Leghorn flocks are illustrated in Fig. 10.3, which shows the effect of season and age on feed and ME consumption. Feed consumption varied over the whole year, with an average of 98 and 106 g per bird per day in summer and winter respectively, while ME consumption varied from 274 to 297 kcal per bird per day. This is expected because, as stated earlier, hens eat mainly to satisfy their energy requirements, and thus energy requirements for maintenance decrease as environmental temperatures rise. Figure 10.4 shows the relationship between temperature, energy content of the diet and feed consumed per 100 hens per day. At high temperature, the difference in feed consumption is greater for each degree change in temperature than at low ambient temperatures. Thus hens could become energy-deficient when subjected to high temperature. In practice, such a deficiency could be aggravated by increasing protein at the expense of energy in the diet. Protein has a relatively high heat increment and thus increases the heat burden of the laying hen. Realizing this energy shortage under high-temperature conditions, nutritionists have tested high-energy diets during the summer and found that the addition of fat stimulates feed and ME consumption (Reid, 1979). The use of high-energy layer rations is now fairly common practice in hot areas, particularly during the early production period (20–30 weeks) when feed consumption is still low. We have shown that the addition of 5% fat not only improves feed intake at high temperature (30°C) but also improves egg weight and shell thickness (Daghir, 1987).

Ramlah and Sarinah (1992), in experiments with laying hens in Malaysia, found that hens had higher performance in terms of egg production, egg

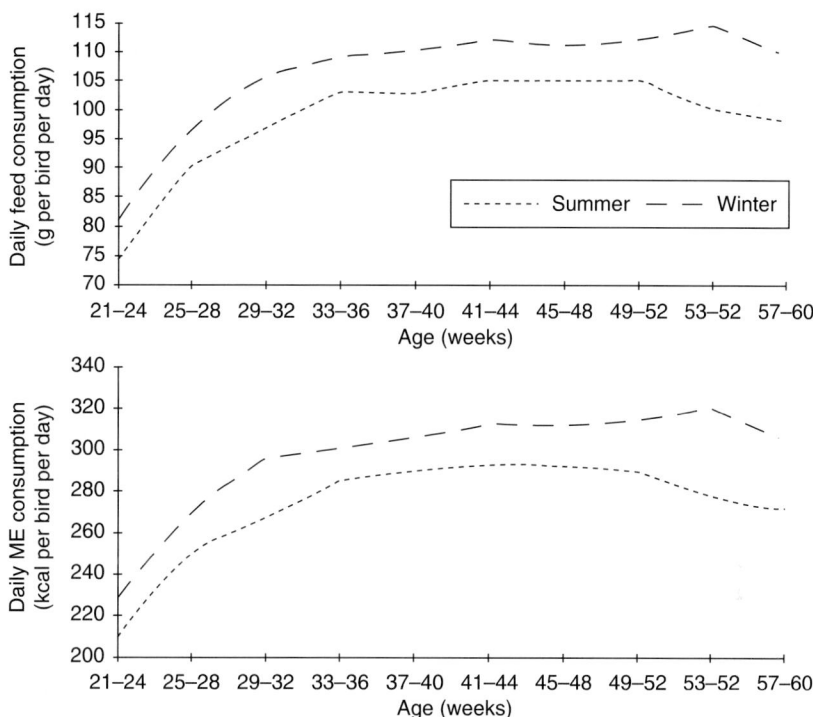

Fig. 10.3. Feed and ME consumption of Leghorn layers in relation to age and season.

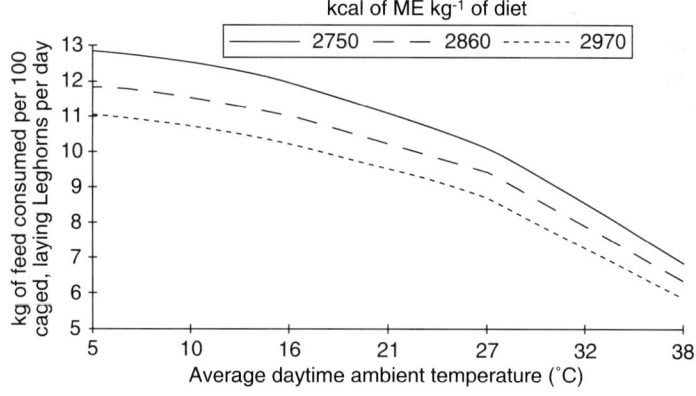

Fig. 10.4. Relationship between ambient temperature, energy content of the diet, and kg feed consumed per 100 caged, laying Leghorns per day.

mass and better feed efficiency when offered choices of feeds that included diets with supplemented fat and with a higher level of both protein and energy, when compared with choice-fed hens with lower levels of both protein and energy in their diets. They observed that hens under their conditions tended to select more of the feed with supplemental fat. Dietary fat not only increases palatability of the feed but also reduces the amount of heat increment produced during its utilization in the body. Nir (1992), however, cautioned that, with the use of dietary fat in hot climates, special care needs to be taken to prevent its oxidation. He suggested the use of saturated rather than polyunsaturated fats under those conditions.

Marsden and Morris (1987) summarized data of 30 published experiments in an attempt to examine the relationship between environmental temperature and ME intake, egg output and body-weight change in laying pullets. They concluded that the relationship between temperature and ME intake is curvilinear, with food intake declining more steeply as ambient temperature approaches body temperature. They calculated that the energy available for production is at a maximum at 23°C for brown birds and at 24°C for White Leghorns. Although gross energetic efficiency is at a maximum at 30°C, egg output is reduced at this temperature. The authors concluded that the optimum operating temperature for laying houses will depend upon the local cost of modifying ambient temperature and on the cost of supplying diets of appropriate protein content. Marsden et al. (1987), in two experiments with laying hens, found that it was not possible to maintain egg weight or egg output at 30°C by feeding a high-energy, high-protein diet. Peguri and Coon (1991), in a study on the effect of temperature and dietary energy on layer performance, found that feed intake was 5–9 g lower when dietary ME was increased from 2645 to 2976 kcal/kg and was 21.7 g lower when temperatures were increased from 16.1 to 31.1°C. Egg production was not affected by either temperature or dietary energy density. Egg weight increased 0.78 g with increases in dietary energy from 2675 to 2976 kcal/kg and decreased 3.18 g when temperatures were increased from 16.1 to 31.1°C.

Sohail et al. (2002b) studied the economics of using fat and protein on egg size under summer conditions. They observed that maximum profits are attained where fat is added to the highest-protein diet (19.8%) from 21 to 37 weeks of age. Increasing either fat or protein in a maize–soybean diet increased egg weight. The same workers (Sohail et al., 2002a), in working with older hens (60 weeks old) under summer conditions, reported that increasing dietary energy levels above 2783 kcal/kg feed had no beneficial effect on egg weight. Pirzadeh (2002) studied the effect of energy levels at different protein levels on egg weight of older hens under hot environmental conditions and concluded that increasing dietary energy beyond 2700 kcal/kg feed had no effect on egg weight.

Researchers are looking at other means of increasing energy intake under high-temperature conditions. Picard et al. (1987) suggested using a low-calcium diet along with a marine shell source fed separately and with free choice. At 33°C there was a clear-cut improvement in energy intake and in calcium intake and a significant increase in egg output and shell weight. Uzu (1989) also recommended separate calcium feeding in the form of oyster shell, offered in the afternoon, along with a low calcium level in the diet. There is a tendency in hot regions of the world to give excessively high levels of calcium and this has a negative effect on feed intake, not only because of a physiological effect on appetite but also because of reduced palatability of the diet. This has been confirmed by Mohammed and Mohammed (1991) in a study on commercial layers in a hot, tropical environment; they reported that feed intake was significantly reduced when dietary calcium increased from 2.5 to 3.9%. Devegowda (1992) reported that, in India, separate calcium feeding, along with reducing the calcium level in the diet to about 2%, improved feed intake and increased egg production and shell quality. Another suggestion for improving feed intake and thus energy intake in hot climates is pelleting of the diet. Since low-energy, high-fibre feedstuffs are used in many hot regions, Picard (1985) suggested that feeding pellets rather than mash should be considered for laying hens. Devegowda (1992) also recommended feeding pellets where low-energy and high-fibre diets are used.

Feeding a wet diet increases the dry matter intake of layers at high temperature. Okan et al. (1996) found that the dry matter intake, egg production and egg weight in laying quail were all increased by wet feeding. They added tap water to the diet in ratios ranging from 1:0.5 to 1:3. Feeding layers a pelleted ration during the summer months increases egg production, feed efficiency and water intake (Almirall et al., 1997). Feeding during the cooler times of the day has been shown to be useful for increasing nutrient intake. Midnight feeding has also been shown to help under extreme temperature conditions.

Minerals and vitamins

Some aspects of laying-hen mineral nutrition have been covered in Chapter 6. As for vitamin nutrition, apart from the work on ascorbic acid there has been very little research done on laying hens in hot climates. Heywang (1952), in a study on the level of vitamin A in the diet of laying and breeding chickens during hot weather, indicated that the absolute requirement of laying hens for vitamin A is not altered by high temperature. Thornton and Moreng (1959) were among the first to report the improvement in performance of laying hens receiving vitamin C in a hot environment. Although there is no evidence that vitamin C is an essential nutrient for poultry maintained under moderate environmental temperatures, numerous early reports have shown beneficial effects on laying hens receiving this vitamin in hot climates (Perek and Kendler, 1963; Kechik and Sykes, 1974). Other more recent studies have also shown that vitamin C helps laying hens maintain adequate performance under high-temperature conditions. Cheng et al. (1990) report that laying-hen mortality due to heat stress can be significantly reduced by using as little as 100 p.p.m. ascorbic acid. Manner et al. (1991) studied the influence of different sources of vitamin C on performance of layers at varying environmental temperatures. Three forms of vitamin C were used, namely crystalline ascorbic acid, protected ascorbic acid (Cuxavit C 50) or phosphate–ascorbic acid ester. Performance and eggshell quality of treated hens improved only at 34°C and not at 20°C. The best results were obtained with the protected ascorbic acid and the phosphate ester. The authors suggested that the wide variations in the recommendations for supplementing layer diets with ascorbic acid might be due to higher losses during storage.

Morrison *et al.* (1988) reported the results of a survey conducted on 162 layer flocks that either received or did not receive a vitamin and electrolyte water additive during heat stress. Flocks receiving water additives experienced a smaller drop in egg production during heat stress and this drop in production was shorter in duration than that of those with no additives. Balnave and Zhang (1992b) showed that dietary supplementation with ascorbic acid is effective in preventing a decline in eggshell quality when laying hens are given saline drinking water. Shell-quality characteristics were maintained at control values with a dietary ascorbic acid supplement of 2 g/kg diet. Results of this study indicate that supplementation of the diet with ascorbic acid prevents the decline in eggshell quality caused by saline drinking water. The effects of dietary sodium zeolite (Ethacal) on poultry have been investigated extensively. Because of its high ion selectivity for calcium, a great deal of research has been done on the effects of this product on calcium and phosphorus utilization, eggshell quality and bone development. Sodium zeolite has been shown to be beneficial in reducing the effects of heat stress on laying hens (Ingram and Kling, 1987). The exact way in which this product works is unknown but it may be acting as a buffer in the gut and reducing the alkalosis associated with panting.

Feeding programmes

Since a major problem in rearing pullets in hot climates is obtaining acceptable body weights at housing, the use of a prelay ration is recommended. During the 2–3 weeks prior to the first egg, the liver and reproductive system increase in size in preparation for egg production. At the same time, calcium reserves are being built up to meet the future demands for shell formation. Table 10.2 gives the specifications for a prelay ration, which is usually similar to a layer ration except for a 2.0–2.5% level of total calcium. Such a ration is usually fed until 5% production is reached and helps pullets to attain the desired body weight at this early stage of production.

Phase feeding has become common practice for layer operations all over the world. Besides reduction in feed cost, phase feeding contributes to reduced nitrogen excretion, which in turn lowers ammonia emissions. When phase feeding is used, the dietary nutrient profile is adjusted over time to better meet the needs of the layer. A layer ration is usually fed when a flock reaches 5% production. The first layer feed is a high-nutrient-density ration to ensure that birds receive the required nutrients for sustained production and early egg size. Table 10.5 gives recommended specifications for a phase-feeding programme consisting of four formulae. Changes from a high-nutrient-density ration to a lower-density ration should be made on the basis of daily egg mass. Daily egg mass output is calculated by multiplying the actual hen-day rate of egg production by the average egg weight in grams. For example, a flock laying at 90% with an average egg weight of 55.6 g has a

Table 10.5. Nutrient recommendations for laying hens in the Middle East region.

Age	ME – kcal/kg	Protein g/b/d	Ca (%)
Phase I 18–40 wks	2800–2850	20	3.50
Phase II 41–54 wks	2750–2800	19	3.75
Phase III > 54 wks	2700–2750	18	4.00

daily egg mass output of 50 g per bird. Normally, a shift from the first layer ration to the second layer ration is made when daily egg mass reaches a peak and starts declining. The change from the second to the third layer ration is not usually made before daily egg mass is down to 50 g, and the change from the third to the fourth layer ration is made when daily egg mass goes below 47 g.

Energy intake of the laying hen is often more limiting than protein or amino acid intake, and this is especially true in warm climates and at onset of production, when feed intake is low. The energy level, as well as density of all other nutrients in the ration, should be adjusted in accordance with actual intake of feed. Energy densities between 2700 kcal/kg (11.3 MJ/kg) and 2950 kcal/kg (12.3 MJ/kg) are suitable for the different phases of production. Table 10.5 has been compiled by the author based on field data in the Middle East region. Our recommendation for energy levels in that region range from 2700 to 2850 kcal/kg, depending on stages of production, since producers in the region have benefited from using higher levels of energy during the early states of production, but available research reports do not justify levels above 2850 kcal/kg.

The formulae suggested in Table 10.6 differ in the level of protein and other nutrients. Protein and amino acid requirements are greatest from the onset of production up to peak egg mass. This is the period when body weight, egg weight and egg numbers are all increasing. The attainment of adequate egg size is one of the problems of the egg industry in hot climates. If satisfactory egg size is not attained with 19% protein in the ration, the levels of the most critical amino acids should be checked, particularly that of methionine. The best way of correcting a methionine limitation is by adding a feed-grade form of methionine. Daily intake of about 360 mg of methionine should be maintained. Small egg size can be due to low energy intake as well as low protein and amino acid intake. The use of fat in layer rations has been shown to be helpful, not only because of its energy contribution but also because it can increase the linoleic acid level, which should be over 1.2% in the ration. The importance of calcium has previously been discussed in this chapter as well as in Chapter 6. If separate feeding of calcium, as recommended earlier, is not feasible, then at least 50% of the calcium in the feed should be in granular form rather than all in powder form. Table 10.5 shows that available phosphorus levels vary from 0.45% to 0.37%. It is important not to overfeed phosphorus since it has been shown that excessive levels are detrimental to eggshell quality, particularly in hot climates.

Table 10.6. Ration specifications for the laying period.

Nutrients	Layer I	Layer II	Layer III	Layer IV
Crude protein (%)	18–19	17–18	16–17	15–16
Metabolizable energy				
(kcal/kg)	2850–2950	2800–2900	2750–2850	2700–1800
(MJ/kg)	11.93–12.34	11.72–12.34	11.50–11.39	11.30–11.72
Calcium (%)	3.25	3.50	3.75	4.00
Available phosphorus (%)	0.45	0.43	0.40	0.37
Sodium (%)	0.18	0.18	0.17	0.16
Chloride (%)	0.17	0.17	0.17	0.17
Potassium (%)	0.60	0.60	0.60	0.60
Methionine (%)	0.40	0.38	0.36	0.34
Methionine + cysteine (%)	0.70	0.67	0.63	0.60
Lysine (%)	0.84	0.80	0.75	0.70
Tryptophan (%)	0.20	0.19	0.18	0.17
Threonine (%)	0.70	0.65	0.63	0.59
Leucine (%)	1.40	1.32	1.25	1.18
Isoleucine (%)	0.80	0.76	0.71	0.67
Valine (%)	0.82	0.78	0.73	0.69
Arginine (%)	1.00	0.95	0.89	0.84
Phenylalanine (%)	0.85	0.80	0.76	0.72
Histidine (%)	0.40	0.38	0.36	0.34

Water quality and quantity for layers

Underground water supplies, which are high in total dissolved solids, are an important source of drinking water for poultry in many countries of the hot regions. Information on the effects that these types of water have on the performance of laying hens is limited. Koelkebeck *et al.* (1999) reported that well waters are inferior in quality to city water and negatively affect performance of laying hens. Balnave (1993) reviewed the literature on the influence of saline drinking water on eggshell quality and formation. He indicated that saline drinking water supplied to mature laying hens with contents similar to those found in underground well water has an adverse effect on eggshell quality. Furthermore, these effects on shell quality can occur without adverse effect on egg production, feed intake or egg weight. Sensitivity of hens to saline drinking water has been shown to increase with age of the hen and with increases in egg weight (Yoselewitz and Balnave, 1989a). Yoselewitz and Balnave (1990) also showed strain differences in sensitivity to saline drinking water and considerable variation in response of hens within a strain. The incidence of poor shell quality also increases with higher concentrations of sodium chloride in the water (Balnave and Yoselewitz, 1987).

Balnave *et al.* (1989) suggested that one of the major causes of poor eggshell quality in laying hens receiving sodium chloride in the drinking water may be a reduced supply of bicarbonate ions to the lumen of the shell gland. Furthermore, Yoselewitz and Balnave (1989b) showed that specific activity of carbonic anhydrase was significantly lower in hens receiving saline drinking water than in hens receiving regular water. These studies brought Balnave (1993) to conclude that the primary metabolic lesion associated with the poor eggshell quality resulting from intake of saline drinking water is related to the supply of bicarbonate rather than calcium to the lumen of the shell gland for shell formation.

Balnave (1993) recommends two treatments for this problem, besides using desalination of the drinking water. One is the use of ascorbic acid supplements in the diet or in the drinking water and the other is the use of zinc–methionine supplements in the diet. These recommendations are based on studies conducted by Balnave *et al.* (1991) and Balnave and Zhang (1992a) on ascorbic acid and Moreng *et al.* (1992) with zinc–methionine. Balnave (1993) cautions, however, that these treatments are preventive rather than remedial in nature and should be applied from the first time sexually mature hens are exposed to saline drinking water. With the present economically feasible systems of desalination of water, the installation of desalination units on poultry farms with this problem may be the best solution.

The early work on the effects of water temperature on laying hen performance has been reviewed by Adams (1973), who reported that providing drinking water at 35–40°C has a detrimental effect on performance. Cooling the drinking water has been shown to improve performance of layers in many tests. North and Bell (1990) presented data from work at the University of Guelph which shows that water temperature at 35°C compared with water at 3°C reduced egg production by 12% and daily feed intake by 12 g per hen. Leeson and Summers (1991) showed the advantage of providing cool water to laying hens at high temperature to maintain peak egg production. Xin *et al.* (2002) evaluated the effects of drinking-water temperature on laying hens subjected to warm cyclic air temperature conditions and found that 23°C is the optimum water temperature for heat-challenged laying hens. It is always helpful to raise pullets on the same type of watering system that is going to be used in the laying house. In a study conducted by Odom *et al.* (1985) it was shown that, during periods of high environmental temperature (35°C), birds given carbonated water to drink had a significant relief from the reduction in eggshell quality as a result of a delay in time for the decline to occur.

Laying hens drink twice as much water per day when the temperature is 32°C compared with 24°C. Figure 10.5 shows the effect of house temperature on water consumption of laying hens in cages. It also shows the variation in water consumption throughout the laying period, which is the result of changes in body weight as well as rate of egg production. Maximum water consumption at all temperatures is shown to occur at 6–7 weeks of production, which coincides with peak production. For cage operations, one cup per cage of up to five commercial layers is recommended for hot climates. Egg production drops when hens are not able to drink enough water. The amount of production loss

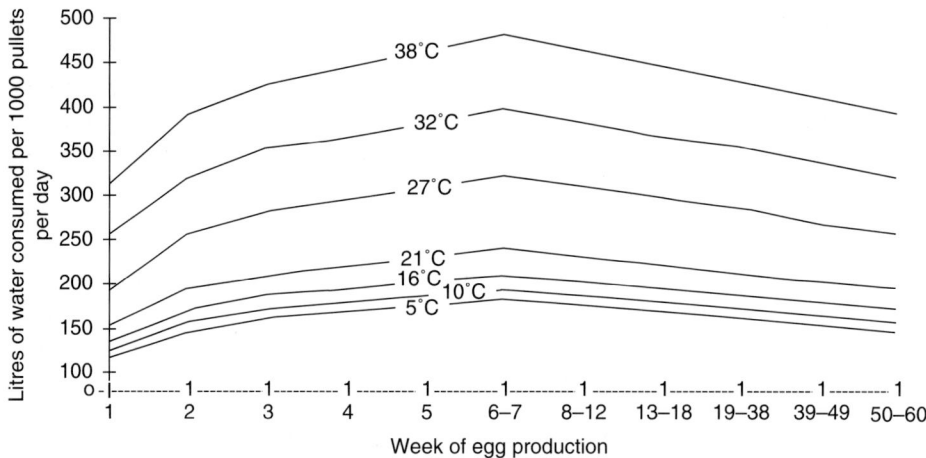

Fig. 10.5. Water consumption of standard laying Leghorn pullets in cages.

is proportional to the amount of water not consumed. Savage (1983) proposed certain practices for a closed watering system as well as for a flow-through type waterer to ensure that hens receive enough water during hot weather.

1. For a closed watering system (cup or nipple), in-line water filters should be checked and cleaned and may have to be replaced often. Some wells pull more sediment than others and require that the filters be changed much more often. It is good to have functional water pressure gauges on both sides of the water filter. A 3–8 lb differential between incoming and outgoing water pressure should be maintained. Line pressure gauges should be checked often to see that water pressure is maintained. During the hottest part of the day, at least one cup or nipple per line should be checked for pressure at the far end of the laying house. The drinker should be triggered and held in open position for a few seconds to observe the rate of water flow. A cup should fill in 2–3 s and a nipple should flow in a steady stream.

2. For a flow-through-type waterer, water troughs should be checked for flow rate at both ends of the house. The amount of waste water at the far end of the water trough should be checked closely during the hot part of the day to ensure that all birds have access to water when they need it most. To ensure that hens have good access to the water, check the spacing above the water trough. Sometimes, sagging feeders or cage fronts restrict the space available for the hens to reach the water. At least monthly, walk slowly down each aisle and observe the hens drinking. You can often find several cages where the hens must struggle to get their heads through a tight space to the drinking trough. This is more likely to occur in older and moulted flocks since their combs are larger and they therefore require a greater space above the water troughs. Water troughs should be cleaned more often in hot weather to improve water flow and reduce obstructing feed or algae growth in the troughs.

Acclimatization

Hutchinson and Sykes (1953) were the first to study heat acclimatization in inbred Brown Leghorn hens exposed to a hot, humid climate by measuring changes in body temperatures. Smith and Oliver (1971) reviewed the early literature on laying-hen acclimatization to high temperature and suggested that the process is mainly associated with a low basal metabolic rate at high temperatures. Several studies have been published since that time showing that laying hens are able to survive hot, lethal conditions if they are previously exposed to a daily, intermittent heat-stress situation (Hillerman and Wilson, 1955; Deaton et al., 1982; Arad and Marder, 1984; Sykes and Fataftah, 1986b). The increased heat tolerance in those studies was reflected in the lower body temperatures, higher panting rates and decreased evaporative water loss. Strain differences in the response to heat stress were reported by both Arad and Marder (1984) and Sykes and Fataftah (1986b), but it was not clear whether these differences are a reflection of body size and metabolic rate or some other genetically determined character. White laying strains were not always more heat-tolerant than brown strains. Khan (1992) confirmed that a laying hen's acclimatization to heat stress varies with strain of bird and that deep body temperature has a significant influence on egg production. There are variations in laying hens in response to summer stress with respect to deep body temperature. Fataftah (1980) has shown that heat tolerance is somewhat labile and can be increased or decreased considerably by reducing or increasing energy intake. Another factor that has been shown to affect acclimatization is the presence or absence of water. Arad (1983) showed that water depletion reduces heat tolerance. Sykes and Fataftah (1986b), on the other hand, could not demonstrate that hens which remained in positive water balance were any more heat-tolerant than those that did not.

The value of acclimatization should be considered in relation to survival and production during acute heat stress. It may be desirable to allow laying birds to be exposed to temperatures of 29–33°C before a very hot day is expected even though it may have been possible to keep maximum house temperature below this level.

Effects of temperature on egg quality

There is general agreement among researchers that high ambient temperatures have a negative effect on egg quality. Sauveur and Picard (1987) reviewed the literature on the effects of high ambient temperature on egg quality. Balnave (1988) reviewed factors affecting eggshell calcification and methods of optimizing calcium supply. The adverse effects of high ambient temperature, shell gland lesions, inadequate mineral supplies and bicarbonate ions on eggshell quality were all considered in this review. Many researchers have reported a reduction in egg weight associated with increases in environmental temperature (Payne, 1966; Stockland and Blaylock, 1974; DeAndrade et al., 1977; Vohra *et al.*, 1979). These workers have reported decreases in egg weight ranging from 0.07 to 0.98 g per egg for every 1°C rise in temperature. Mueller (1961) found that maintaining laying hens in a cycling temperature of 13–32–13°C resulted in the production of smaller eggs than from hens kept at a constant 13°C.

Shell quality has also been shown to be reduced as environmental temperatures rise. Harrison and Biellier (1969) found very quick reductions in shell weight as temperatures increase. This was confirmed later by Wilson *et al.* (1972), DeAndrade *et al.* (1977) and Wolfenson *et al.* (1979). Data from the University of California (North and Bell, 1990) show that the decrease in shell thickness during the summer is greater in eggs produced by older birds. The difference in shell thickness between winter and summer was reported to be 10 µm in 50-week-old hens and 17 µm in 60-week-old hens. Shells less than 356 µm in thickness amounted to 47% in summer versus 30% in winter. This reduced shell thickness at high temperature has been attributed to reduced calcium intake as a result of the reduced feed intake. High temperatures are known to increase respiratory rate, resulting in respiratory alkalosis, which alters the acid–base balance and blood pH. Attempts to improve shell thickness through modification of the acid–base balance by adding sodium bicarbonate to feed or giving carbonated water have been found to be helpful (Odom *et al.*, 1985). Grizzle *et al.* (1992), in a study of the nutritional and environmental factors involved in eggshell quality, suggested that midnight lighting programmes provide a means of supporting eggshell quality in older laying hens during the summer months without a significant reduction in egg production. Eggs in a hot environment should be collected more often and cooled quickly in a properly equipped egg-storage room to maintain their internal quality. More care should be taken in handling eggs in hot areas because of the reduced shell quality.

Layer management practices

Several management tips have been recommended by different workers that help to reduce the detrimental effects of high temperature on laying hens. This section will present some of these recommendations.

Ernst (1989) recommends getting the flock up early in the summer to encourage feed consumption before temperatures begin to rise. Feed consumption should be measured weekly and ration compositions adjusted to match intake. Tadtiyanant *et al.* (1991) recommended the use of wet feed to increase feed consumption in laying hens under heat stress. They showed that the use of wet feed gave a 38% increase in dry matter intake when compared with the use of dry feed at 33.3°C. The removal of comb or wattles from commercial layers is not recommended in high temperature regions because these organs, with their good surface blood supply, help in cooling. It is helpful during hot weather to plan work schedules so that hens are not disturbed during the hot part of the day. Furthermore, the use of low light intensities can help to reduce bird activity and thus heat production (Ernst *et al.*, 1987).

Lighting effects

Nishibri et al. (1989) studied diurnal variation in heat production of laying hens at temperatures of 23 and 35°C. Heat production was higher during the light period than during darkness. At 23°C, the differences between the light and dark periods were greater than at 35°C. Body temperature was higher during the light period than during darkness. Li et al. (1992) observed that both heat production and abdominal temperature in laying hens declined with decreasing light intensity and this was considered by these authors to result from changes in physical activity. These workers also observed that, above 28°C, abdominal temperature increased with both environmental temperature and feed intake, indicating that the heat production associated with feed intake adds to the heat load of high environmental temperature.

Al-Hassani and Al-Naib (1992) tested two lighting regimes (night versus day) and three nutritional treatments (pellet, mash and 4-h withdrawal of mash) on egg quality in brown layers. Their results indicated that in countries like Iraq, characterized by great diurnal variation in ambient temperature, night lights coupled with pelleted feed and 4-h feed withdrawal during the day can help to alleviate heat stress in laying hens. Oluyemi and Adebanjo (1979) tested various combinations of lighting and feeding patterns in Nigeria on medium-strain commercial layers. They found that night feeding under a reversed lighting programme (6 p.m. to 6 a.m.) produced a significantly higher level of egg production than daytime feeding. Nishibri et al. (1989), in a study on diurnal variation in heat production of laying hens, reported that body temperature was higher during the light period than during the dark periods of the day. It is therefore recommended that laying hens are not provided with lights during the hot periods of the day, since this will help hold down body temperature. Another practice that helps in holding down body temperature is reducing traffic through the laying house and keeping it at a minimum so that birds will not be unnecessarily excited.

Cage space and shape

Cage space or density significantly affects performance in hot climates. This is because higher densities make ventilation more difficult in those areas. Extensive studies by North Carolina State University have shown an advantage of about 10 eggs per hen when density is reduced from 350 cm^2 to 460 cm^2 per bird in group cages located in fan-ventilated houses. The literature on this subject has been reviewed by Adams and Craig (1985), who found that increasing the density of hens in cages significantly reduced the number of eggs per hen housed, decreased feed intake, increased feed required per dozen eggs and increased mortality. Reducing floor space per hen from an average of 387 cm^2 to 310 cm^2 reduced eggs per hen housed by 16.6, increased mortality by 4.8% and decreased feed consumption by 1.9 g/hen/day. Reducing space from 516 cm^2 to 387 cm^2 reduced egg production by 7.8 eggs/hen housed, increased mortality by 2.8% and decreased feed consumption by 4.3 g/hen/day. Teng et al. (1990) tested stocking densities of layers in cages with two or three birds per cage in Singapore. Egg production and feed consumption decreased with decreasing floor space per bird. There was a difference of nine eggs per bird during a period of 350 days of production. Their results indicated that birds can be housed at 387 cm^2 per bird in a hot climate, provided sufficient feeding space is available. Gomez Pichardo (1983) evaluated performance of laying hens in Mexico when housed three or four in cages measuring 1350 cm^2, thus allowing 450 or 337.5 cm^2 per bird. They observed that, at the higher space allowance, the weight of eggs produced per m^2 was 28% greater than at the lower space allowance. Rojas Olaiz (1988), also in Mexico, looked at the effect of housing semi-heavy and light breeds housed three or four birds to a cage of 1800 cm^2 during the second production cycle. Birds housed three to a cage had lower mortality, higher egg production and body weight and higher Haugh unit scores than those housed four to a cage, but the latter produced a higher total egg weight per cage. There were no significant differences between the two groups in egg weight or feed consumption. Egg production was significantly affected by cage shape. Hens in shallow cages produced 5.8 more eggs/hen housed than those in deep cages. Research has demonstrated that performance of hens in shallow cages (reverse-type cage) is better than in deep cages. The greater eating space in these shallow cages may be helpful in maintaining feed consumption in hot regions. Ramos *et al.* (1990) did not find a difference in performance between birds housed in deep versus shallow cages. Ulusan and Yildiz (1986) studied the effects of climatic factors in Turkey on hens caged singly versus those caged in groups of four with the same space per hen. For group-caged birds, egg production was significantly correlated with ambient temperature and with relative humidity. For singly caged birds, egg production was significantly correlated with relative humidity. For both groups, egg production was significantly correlated with atmospheric pressure.

External and internal parasites

High ambient temperatures usually increase the population of insects responsible for the transmission of disease. Houseflies (*Musca domestica*) and related species are very active in hot climates and are involved in the transmission of several poultry diseases (Shane, 1988a). Droppings under cages should be allowed to cone and dry out, or be cleaned up completely at intervals of less than 7 days. This is because the fly's life cycle in the heat is about 7 days. Chemical fly treatments are an aid but do not replace good management. Therefore premises should be kept clean, dry and tidy. In hot climates, both endo- and ectoparasites can be a problem all year round, particularly in extensive management systems. Warm and humid conditions favour the propagation of endoparasites, including round worms (*Ascaridia* spp. and *Hetarakis* spp.) and hair worms (*Capillaria* spp.), as reported by Shane (1988b). Commercial layer operations located in hot climates which use cages and apply good management and hygiene are less prone to parasitism. A sample of each flock should be inspected for mites and lice each month. Chemical sprays for flocks in cages and powders for floor-housed flocks are usually recommended. Worms can also present a problem and should be brought under control as soon as recognized.

Wet droppings

Layers drink more water when they are on wire than when kept on a litter floor. Furthermore, they drink more when temperatures rise and therefore they eliminate more water through the droppings. Water consumption also increases with increase in production. At 70% production, 1000 pullets in cages drink 201 l/day at 21°C while at 90% production they drink 239 l/day at the same temperature.

Wet droppings are affected by several factors, such as relative humidity and temperature of the outside air, relative humidity and temperature inside the house and the amount of air moving through the house. High levels of protein and salt in the ration have been shown to increase moisture in the droppings. Wet droppings rarely occur because of the amount of salt in the feed, unless a mixing error has occurred at the feed mill. Sodium levels in the diet have to be very high before one begins seeing significant increases in manure moisture. Sodium levels should be kept in the range of 0.15–0.20% of the diet. The early introduction of the high calcium layer diet has in many cases caused an increase in manure moisture. The effects have been transitory but in some flocks have persisted for several weeks past peak. The use of a prelay ration has been helpful in these situations. The use of high levels of barley in the ration has been shown to increase water consumption and wet droppings. Using crumbled feeds has also been shown to increase wet droppings.

Some feed additives have been shown to increase moisture in the droppings. Keshavarz and McCormick (1991) reported that the use of sodium aluminium silicate at 0.75% of the diet increased the dropping moisture with or without sodium correction and when tested in both summer and winter. Leenstra *et al.* (1992), in a study on the inheritance of water content and drying characteristics of droppings of laying hens, showed that both of these characteristics can be improved by selection without negative consequences on production traits. Leaky watering devices are a major contributor to excessive water in the droppings' collection area. Wet droppings also increase obnoxious odours coming from ammonia and bacterial action in the droppings. Figure 10.6 shows normal manure accumulation underneath laying cages compared with watery manure accumulation.

Conclusions and Recommendations

1. A major problem of rearing pullets in hot climates is obtaining acceptable body weights at sexual maturity.
2. Pullets should be weighed frequently, starting as early as 4 weeks of age, and weight and uniformity should be watched closely. When uniformity is a problem, small birds should be sorted out and penned separately at about 5 weeks of age.
3. During the hot months of the year, about 10% fewer pullets should be started for a given space as compared with the cool months. This will mean increased floor space per pullet, along with more water and feeder space.

Fig. 10.6. Left: normal manure accumulation underneath laying hens in cages. Right: watery manure underneath laying hens in cages.

4. Pullets at housing should be above the breeder's recommended target weight, because heavier hens will consume more feed and this will result in higher peaks and better persistency in production.

5. High environmental temperatures depress feed intake in growing pullets and the percentage decrease in feed consumption varies from 1.3% per 1°C rise in temperature at 21°C to over 3% decrease at 38°C. Although results of studies on the influence of temperature on nutrient requirements of replacement pullets are conflicting, pullet growth can be improved at high temperature by increasing nutrient density of the diet.

6. A feeding programme for growing pullets that would allow the protein level to be gradually lowered and the energy level to be raised in the diet is needed under most hot-weather situations. Extending the feeding period of a starter beyond 6 weeks of age may be necessary to attain the desired body weights.

7. Several methods have been suggested for increasing feed consumption in growing pullets. Feeding crumbled feeds has been shown to help. Feed consumption can also be encouraged by increasing frequency of feeding and stirring feed between feedings.

8. Ambient temperature is by far the most important factor affecting water intake. Growing Leghorn pullets drink about twice as much water per day at 38°C as they do at 21°C.

9. Acclimatization of replacement pullets to high temperature is an area that has not been researched adequately. The questions to be raised are how long replacement pullets will retain their ability to withstand heat stress and what effect this acclimatization will have on subsequent performance.

10. Beak trimming of growing pullets at an early age (7–10 days) is recommended for hot regions because it is less stressful and not detrimental to body weight at housing.

11. Intermittent and biomittent lighting for growing pullets has been recommended because pullets tend to perform better in hot regions on this programme than on standard lighting.
12. The average decrease in feed intake of laying hens is about 1.6% per 1°C as environmental temperatures increase from 20 to 30°C. Food intake, however, declines an average of 5% per 1°C between 32 and 38°C. Therefore, feed consumption should be monitored daily in hot weather to ensure adequate intake of nutrients, i.e. those normally consumed at 21°C. Dietary adjustments can overcome most of the detrimental effects of high temperature on percentage egg production.
13. The use of high-energy layer rations is recommended in hot regions during the early production period (20–30 weeks), because feed consumption during this period is still low and hens could easily become energy deficient.
14. Every means of increasing energy intake under high temperature conditions should be used, particularly early in the production cycle. Feeding the calcium source separately, along with the use of low-calcium diets, helps to improve feed intake.
15. Several studies have shown that vitamin C supplements in the diet or water help laying hens maintain adequate performance under high temperature conditions.
16. Ration specifications have been presented for starter, grower and prelay rations, as well as for four stages of production for laying hens.
17. Water available for use on poultry farms in many hot regions is high in total dissolved solids. Some studies have shown adverse effects of such water on eggshell quality. Two treatments have been suggested for this problem, besides desalination of the drinking water. One is the use of ascorbic acid supplements in the diet or in the drinking water and the other is the use of zinc–methionine supplements in the diet.
18. Many tests have shown that cooling the drinking water improves the performance of layers in hot weather.
19. Laying hens are able to survive hot, lethal conditions if they have been previously exposed to a daily, intermittent heat-stress situation.
20. High ambient temperatures have a negative effect on egg quality. Decreases in egg weight range from 0.07 to 0.98 g per egg for every 1°C rise in temperature.
21. Several management practices, such as lighting adjustments, cage space and shape modifications, egg handling and good hygienic practices, have been described to improve performance in hot climates.

References

Adams, A.W. (1973) Consequences of depriving laying hens of water a short time. *Poultry Science* 52, 1221–1225.

Adams, A.W. and Craig, J.V. (1985) Effect of crowding and cage shape on productivity and profitability of caged layers: a survey. *Poultry Science* 64, 238–242.

Al-Hassani, D.H. and Al-Naib, A.Y. (1992) Egg quality as influenced by lighting and feeding regimes of laying hens during hot summer in Iraq. *Proceedings of the 19th World's Poultry Congress*, Vol. 2, p. 106.

Almirall, M., Cos, R., Esteve-Garcia, E. and Brufau, J. (1997) Effect of inclusion of sugar beet pulp, pelleting and season on laying hen performance. *British Poultry Science* 38, 530–536.

Arad, Z. (1983) Thermoregulation and acid-base status in the panting dehydrated fowl. *Journal of Applied Physiology* 54, 234–243.

Arad, Z. and Marder, J. (1984) Strain differences in heat resistance to acute heat stress, between the Bedouin desert fowl, the white Leghorn and their crossbreeds. *Comparative Biochemistry and Physiology* 72A, 191–193.

Austic, R.E. (1985) Feeding poultry in hot and cold climates. In: M. Youssef (ed.) *Stress Physiology in Livestock*, Vol. 3, *Poultry*. CRC Press, Boca Raton, Florida, pp.123–136.

Balnave, D. (1988) Egg shell calcification in the domestic hen. *Proceedings of the Nutrition Society of Australia* 13, 41–48.

Balnave, D. (1993) Influence of saline drinking water on egg shell quality and formation. *World's Poultry Science Journal* 49, 109–119.

Balnave, D. and Yoselewitz, I. (1987) The relation between sodium chloride concentration in drinking water and egg shell damage. *British Journal of Nutrition* 58, 503–509.

Balnave, D. and Zhang, D. (1992a) Responses in egg shell quality from dietary ascorbic acid supplementation of hens receiving saline drinking water. *Australian Journal of Agricultural Research* 43, 1259–1264.

Balnave, D. and Zhang, D. (1992b) Dietary ascorbic acid supplementation improves egg shell quality of hens receiving saline drinking water. *Proceedings of the 19th World's Poultry Congress*, Vol. 1, pp. 590–593.

Balnave, D., Yoselewitz, I. and Dixon, R.J. (1989) Physiological changes associated with the production of defective egg shells by hens receiving sodium chloride in the drinking water. *British Journal of Nutrition* 61, 35–43.

Balnave, D., Zhang, D. and Moreng, R.E. (1991) Use of ascorbic acid to prevent the decline in egg shell quality observed with saline drinking water. *Poultry Science* 70, 848–852.

Bell, D. (1987a) Flock management-quality pullets. *California Poultry Letter*, March, p. 4.

Bell, D. (1987b) Age of beak trimming and high fibre diets. *Proceedings of the University of California Symposium for Success*, pp. 1–5.

Bohren, B.B., Rogler, J.C. and Carson, J.R. (1982) Performance at two rearing temperatures of White Leghorn lines selected for increased and decreased survival under heat stress. *Poultry Science* 61, 1939–1943.

Chawla, J.S., Lodhi, G.N. and Ichponani, J.S. (1976) The protein requirement of laying pullets with changing season in the tropics. *British Poultry Science* 17, 275–283.

Cheng, T.K., Cook, C. and Hamre, M.L. (1990) Effect of environmental stress on the ascorbic acid requirement of laying hens. *Poultry Science* 69, 774–780.

Daghir, N.J. (1973) Energy requirements of laying hens in a semi-arid continental climate. *British Poultry Science* 14, 451–459.

Daghir, N.J. (1987) Nutrient requirements of laying hens under high temperature conditions. *Zootecnica International*, May, 36–39.

DeAndrade, A.N., Rogler, J.C., Featherston, V.R. and Alliston, C.W. (1977) Interrelationships between diet and elevated temperature on egg production and shell quality. *Poultry Science* 56, 1178–1183.

Deaton, J.W., McNaughton, J.L. and Lott, B.D. (1982) Effect of heat stress on laying hens acclimated to cyclic versus constant temperatures. *Poultry Science* 61, 875–878.

Devegowda, G. (1992) Feeding and feed formulation in hot climates for layers. *Proceedings of the 19th World's Poultry Congress*, Vol. 2, pp. 77–80.

Douglas, C.R. and Harms, R.H. (1982) The influence of low protein grower diets on spring-housed pullets. *Poultry Science* 61, 1885–1890.

Douglas, C.R. and Harms, R.H. (1990) An evaluation of a step-down amino acid feeding programme for commercial pullets to 20 weeks of age. *Poultry Science* 69, 763–767.

Douglas, C.R., Welch, D.M. and Harms, R.H. (1985) A step-down protein programme for commercial pullets. *Poultry Science* 64, 1137–1142.

El-Zubeir, E.A. and Mohammed, O.A. (1993) Dietary protein and energy effects on reproductive characteristics of commercial egg type pullets reared in arid hot climate. *Animal Feed Science and Technology* 41, 161–165.

Ernst, R.A. (1989) *Hot Weather Management Techniques*. California Extension Leaflet, University of California, Davis, California.

Ernst, R.A., Millan, J.R. and Mather, F.B. (1987) Review of life-history lighting programmes for commercial laying fowls. *World's Poultry Science Journal* 43, 43–55.

Escalante, R., Chernova, I., Herrera, J.A. and Exposito, A. (1988) Effect of body weight at 18 weeks of age on the lifetime performance of White Leghorn pullets. *Revista Cubana de Ciencia Avicola* 16, 53–59.

Fataftah, A.R.A. (1980) Physiological acclimatization of the fowl to high temperatures. PhD thesis, University of London.

Gomez Pichardo, G. (1983) Production of laying hens when housed three or four in cages measuring 1350 cm^2. *Veterinaria, Mexico* 14, 268–270.

Grizzle, J., Iheanacho, M., Saxton, A. and Broaden, J. (1992) Nutritional and environmental factors involved in egg shell quality of laying hens. *British Poultry Science* 33, 781–794.

Harrison, P.C. and Biellier, H.V. (1969) Physiological response of domestic fowl to abrupt change of ambient air temperature. *Poultry Science* 48, 1034–1045.

Henken, A.M., Groote, A.M.J. and Vanderhel, W. (1983) The effect of environmental temperature on immune response and metabolism of the young chicken. 4. Effect of environmental temperature on some aspects of energy and protein metabolism. *Poultry Science* 62, 59–67.

Heywang, B.W. (1947) Diets for laying chickens during hot weather. *Poultry Science* 11, 38–43.

Heywang, B.W. (1952) The level of vitamin A in the diet of laying and breeding chickens during hot weather. *Poultry Science* 31, 294–300.

Hillerman, J.P. and Wilson, W.O. (1955) Acclimatization of adult chickens to environmental temperature changes. *American Journal of Physiology* 180, 591–595.

Hutchinson, J.C.D. and Sykes, A.H. (1953) Physiological acclimatization of fowls to a hot, humid environment. *Journal of Agricultural Science, Cambridge* 43, 294–322.

Ingram, D.R. and Kling, C.E. (1987) Influence of Ethacal feed component on performance of heat-stressed White Leghorn hens. *Poultry Science* 66, 22 (Abstract).

Jensen, L.S. (1977) The effect of pullet nutrition and management on subsequent layer performance. *Proceedings of the 37th Semiannual Meeting, American Feed Manufacturers Association Nutrition Council*, pp. 36–39

Kechik, I.T. and Sykes, A.H. (1974) Effect of dietary ascorbic acid on the performance of laying hens under warm environmental conditions. *British Poultry Science* 15, 449–457.

Keshavarz, K. and McCormick, C.C. (1991) Effect of sodium aluminosilicate, oyster shell and their combinations on acid-base balance and egg shell quality. *Poultry Science* 70, 313–325.

Khan, A.G. (1992) Influence of deep body temperature on hen's egg production under cyclic summer temperatures from 72 to 114°F. *Proceedings of the 19th World's Poultry Congress*, Vol. 2, pp. 128–131.

Koelkebeck, K.W., Mckee, J.S., Harrison, P.C. and Parsons, C.M. (1999) Performance of laying hens provided water from two sources. *Journal of Applied Poultry Research* 8, 374–379.

Leenstra, F.R., Flock, D.K., Van den Berge, A.J. and Pit, R. (1992) Inheritance of water content and drying characteristics of droppings of laying hens. *Proceedings of the 19th World's Poultry Congress*, Vol. 2, pp. 201–204.

Leeson, S. and Summers, J.D. (1981) Effect of rearing diet on performance of early maturing pullets. *Canadian Journal of Animal Science* 61, 743–749.

Leeson, S. and Summers, J.D. (1989) Response of Leghorn pullets to protein and energy in the diet when reared in regular or hot cyclic environments. *Poultry Science* 68, 546–557.

Leeson, S and Summers, J.D. (1991) *Commercial Poultry Nutrition*. University Books Publisher, Guelph, Ontario, Canada, pp. 58–59.

Li, Y., Ito, T., Nishibori, M. and Yamamoto, S. (1992) Effects of environmental temperature on heat production associated with food intake and on abdominal temperature in laying hens. *British Poultry Science* 33, 113–122.

Manner, K., Singh, R.A. and Kamphues, J. (1991) Influence of varying vitamin C sources on performance and egg shell quality of layers at varying environmental temperature. *Proceedings of a Symposium: Vitamine und Weitere Zusatzstaffe bei Mensch und Tier*, pp. 266–269.

Marsden, A. and Morris, T.R. (1987) Quantitative review of the effects of environmental temperature on food intake, egg output and energy balance in laying pullets. *British Poultry Science* 28, 693–704.

Marsden, A., Morris, T.R. and Cronarty, A.S. (1987) Effects of constant environmental temperature on the performance of laying pullets. *British Poultry Science* 28, 361–380.

May, J.D., Deaton, J.W. and Branton, S.L. (1987) Body temperature of acclimated broilers during exposure to high temperature. *Poultry Science* 66, 378–380.

McGinnis, C.H., Jr. (1989) Vitamins in pullet nutrition. *Multi-State Poultry Meeting* 16–17 May.

McNaughton, J.L., Kubena, L.F., Deaton, J.W. and Reece, F.N. (1977) Influence of dietary protein and energy on the performance of commercial egg-type pullets reared under summer conditions. *Poultry Science* 56, 1391–1398.

Mohammed, T.A. and Mohammed, S.A. (1991) Effect of dietary calcium level on performance and egg quality of commercial layers reared under tropical environment. *World Review of Animal Production* 26, 17–20.

Moreng, R.E., Balnave, D. and Zhang, D. (1992) Dietary zinc methionine effect on egg shell quality of hens drinking saline water. *Poultry Science* 71, 1163–1167.

Morrison, W.D., Braithwaite, L.A. and Leeson, S. (1988) Report of a survey of poultry heat stress losses during the summer of 1988. Unpublished Report from Department of Animal and Poultry Science, University of Guelph, Ontario, Canada.

Mueller, W.J. (1961) The effect of constant and fluctuating temperature on the biological performance of laying pullets. *Poultry Science* 40, 1562–1571.

National Research Council (NRC) (1994) *Nutrient Requirements of Domestic Animals – Nutrient Requirements of Poultry*, 9th edn. National Academy of Science, Washington, DC.

Nir, I. (1992) Optimization of poultry diets in hot climates. *Proceedings of the 19th World's Poultry Congress*, Vol. 2, pp. 71–76.

Nishibri, M., Li, Y., Fujita, M., Ito, T. and Yamamoto, S. (1989) Diurnal variation in heat production, heart rate, respiration rate and body temperature of laying hens at constant environmental temperature of 23 and 35°C. *Japanese Journal of Zootechnical Science* 60, 529–533.

Nockels, C.F. (1988) *Proceedings of the Georgia Nutrition Conference for the Feed Industry*, Atlanta, Georgia, 16–18 November, p. 9.

North, M.D. and Bell, D. (1990) *Commercial Chicken Production Manual*, 4th edn. Van Nostrand Reinhold, New York.

Odom, T.W., Harrison, P.C. and Darre, M.J. (1985) The effects of drinking carbonated water on the egg shell quality of single comb white Leghorn hens exposed to high environmental temperature. *Poultry Science* 64, 594–596.

Okan, F., Kutlu, H.R., Baykal, L. and Canogullari, S. (1996) Effect of wet feeding on laying performance of Japanese quail maintained under high environmental temperature. *British Poultry Science* 37(Suppl.), 570–571.

Oluyemi, J.A. and Adebanjo, A. (1979) Measures applied to combat thermal stress in poultry under practical tropical environment. *Poultry Science* 58, 767–770.

Owoade, A.A. and Oduye, O.O. (1992) Use of slatted floor house for pullet rearing – a trial in Nigeria. *Proceedings of the 19th World's Poultry Congress*, Vol. 2, p. 136.

Payne, C.G. (1966) Environmental temperature and the performance of light breed pullets. *Proceedings of the 13th World's Poultry Congress*, pp. 480–484.

Peguri, A. and Coon, C. (1991) Effect of temperature and dietary energy on layer performance. *Poultry Science* 70, 126–138.

Perek, M. and Kendler, J. (1963) Ascorbic acid as a supplement for White Leghorn hens under conditions of climatic stress. *British Poultry Science* 4, 191–200.

Petersen, J., Liepert, B.M. and Horst, P. (1988) Sudden laying stop as adaptation reaction to heat stress. *Deutsche Tierarztliche Wochenschrift* 95, 312–317.

Picard, M. (1985) Heat effects on the laying hen, protein nutrition and food intake. *Proceedings of the 5th European Symposium on Poultry Nutrition*, pp. 65–72.

Picard, M., Angulo, I., Antoine, H., Bouchot, C. and Sauveur, B. (1987) Some feeding strategies for poultry in hot and humid environments. *Proceedings of the 10th Annual Conference of Malaysian Society of Animal Production*, pp. 110–116.

Pirzadeh, M. (2002) The effect of energy at different protein levels on egg weight of older hens under hot environmental condition. *Poultry Science, Poscal 80 (Supplement 1)*, p. 90.

Purina Mills Inc. (1987) Bio-mittent lighting saves money, increases egg income. Acculine Leaflet P243F-87A5, pp. 1–3.

Ramlah, H. and Sarinah, A.H. (1992) Performance of layers in the tropics offered diets with and without supplemental fat. *Proceedings of the 19th World's Poultry Congress*, Vol. 2, pp. 107–108.

Ramos, N.C., Germat, A.G. and Adams, A.W. (1990) Effects of cage shape, age at housing and types of rearing and layer waterers on the productivity of layers. *Poultry Science* 69, 217–223.

Reid, B.L. (1979) Nutrition of laying hens. *Proceedings Georgia Nutrition Conference*, University of Georgia, Athens, pp. 15–18.

Reid, B.L. and Weber, C.W. (1973) Dietary protein and sulphur amino acids for laying hens during heat stress. *Poultry Science* 52, 1335–1340.

Rojas Olaiz, L.A. (1988) Performance of laying hens of semi-heavy and light breeds housed three or four birds to a cage measuring 1800 cm^2 during the second production cycle. *Veterinaria, Mexico* 19, 64–66.

Rose, P. and Michie, W. (1986) Effect of temperature and diet during rearing of layer strain pullets. In: Fisher, C. and Boorman, K.N. (eds) *Nutrient Requirements of Poultry and Nutritional Research*. Butterworths, London, pp. 214–216.

Sauveur, B. and Picard, M. (1987) Environmental effects on egg quality. In: Wells, R.G. and Belyavin, C.G. (eds) *Egg Quality: Current Problems and Recent Advances*. Butterworths, London, pp. 219–234.

Savage, S.I. (1983) Water in Hot Weather. Poultry Tips. Cooperative Extension Service, University of Georgia, P.S. 2- 1, Athens, Georgia.

Scott, T.A. and Balnave, D. (1989) Self-selection feeding for pullets – Part 2. *Poultry International*, December, 22–26.

Shane, S.M. (1988a) Factors influencing health and performance of poultry in hot climates. *Critical Reviews in Poultry Biology* 1, 247–269.

Shane, S.M. (1988b) Managing poultry in hot climates. *Zootechnica International*, April, 37–40.

Smith, A J. and Oliver, L. (1971) Some physiological effects of high temperature on the laying hen. *Poultry Science* 50, 912–916.

Smith, A.J. and Oliver, L. (1972) Some nutritional problems associated with egg production at high environmental temperatures. I. The effect of environmental temperature and rationing treatment on the productivity of pullets fed on diets of differing energy content. *Rhodes Journal of Agricultural Science* 10, 3–8.

Sohail, S.S., Bryant, M.M. and Roland, D.A. (2002a) Effect of energy (fat) at different protein levels on egg weight of older hens under summer conditions. *Poultry Science, Poscal 80 (Supplement 1)*, p.140.

Sohail, S.S., Bryant, M.M. and Roland, D.A. (2002b) Economics of using fat and protein on egg size under summer conditions in commercial leghorns (Phase I). *Poultry Science Poscal 80 (Supplement 1)*, p. 120.

Sterling, K.G., Bell, D.D., Pesti, G.M. and Aggrey, S.E. (2003) Relationships among strain, performance and environmental temperature in commercial laying hens. *Journal of Applied Poultry Research* 12, 85–91.

Stockland, W.L. and Blaylock, L.G. (1974) The influence of temperature on the protein requirement of cage-reared replacement pullets. *Poultry Science* 53, 1174–1187.

Strighini, J.H., Jardin Filho, R.M., Leandro, N.S.M., Andrade, L., Café, M.B. and Cunha, W.C.P. (2002) Effects of reducing protein levels in the beginning of the production period of laying hens in hot weather. *Archiv fur Geflugelkunde* 66, 127.

Sykes, A.H. (1976) Nutrition–environment interactions in poultry. *Proceedings of the Nutrition Conference for Feed Manufacturers*, University of Nottingham, UK, pp. 17–21.

Sykes, A.H. and Fataftah, A.R.A. (1980) Heat acclimatization in laying hens. *Proceedings of the 6th European Poultry Congress* 4, 115–119.

Sykes, A.H. and Fataftah, A.R.A. (1986a) Acclimatization of the fowl to intermittent acute heat stress. *British Poultry Science* 27, 289–300.

Sykes, A.H. and Fataftah, A.R.A. (1986b) Effect of a change in environmental temperature on heat tolerance in laying fowl. *British Poultry Science* 27, 307–316.

Tadtiyanant, C., Lyons, J.J. and Vandepopuliere, J.M. (1991) Influence of wet and dry feed on laying hens under heat stress. *Poultry Science* 70, 44–52.

Teng, M.F., Soh, G.L. and Chew, S.H. (1990) Effects of stocking densities on the productivity of commercial layers in the tropics. *Singapore Journal of Primary Industries* 18, 123–128.

Thornton, P.A. and Moreng, R.E. (1959) Further evidence on the value of ascorbic acid for maintenance of shell quality in warm environmental temperature. *Poultry Science* 38, 594–599.

Ulusan, H.O.K. and Yildiz, N. (1986) Effects of climatic factors in poultry houses on egg production, egg weight and food consumption of commercial hybrid white Leghorn. *Ankara Universitsi Vetriner Fakultesi Dergisi* 33, 122–133.

Uzu, G. (1989) Some aspects of feeding laying hens in hot climates. *Rhone-Poulenc Animal Nutrition Report*, 03600 Commentery, France.

Valencia, M.E., Maiorino, P.M. and Keyd, B.L. (1980) Energy utilization in laying hens. III. Effect of dietary protein level at 21 and 32°C. *Poultry Science* 59, 2508–2511.

Vo, K.V., Boone, M.A. and Johnston, W.E. (1978) Effect of three lifetime temperatures on growth, feed and water consumption and various blood components in male and female Leghorn chickens. *Poultry Science* 57, 798–801.

Vo, K.V., Boone, M.A., Hughes, B.L. and Knetshges, K.F. (1980) Effects of ambient temperature on sexual maturity in chickens. *Poultry Science* 59, 2532–2536.

Vohra, P., Wilson, W.O. and Siopes, T.D. (1979) Egg production, feed consumption, and maintenance energy requirement of Leghorn hens at temperatures of 15.6 and 26.7°C. *Poultry Science* 58, 674–680.

Waldroup, P.W., Mitchell, R.J., Payne, J.R. and Hazen, K.R. (1976) Performance of chicks fed diets formulated to minimize excess levels of essential amino acids. *Poultry Science* 55, 243–253.

Wilson, W.O., Siopes, T.D., Ingkasuwan, P.H. and Mather, F.B. (1972) The interaction of temperature of 21 and 32°C and photoperiod of eight and 14 hours on white Leghorn hen production. *Archiv fur Geflugelkunde* 2, 41–44.

Wolfenson, D., Frei, F.E., Snapir, N. and Nerman, A. (1979) Effect of diurnal or nocturnal heat stress on egg formation. *Poultry Science* 58, 167–174.

Yoselewitz, I. and Balnave, D. (1989a) Effect of egg weight on the incidence of egg shell defects resulting from the use of saline drinking water. *Proceedings of Australian Poultry Science Symposium* p. 98.

Yoselewitz, I. and Balnave, D. (1989b) The influence of saline drinking water on the activity of carbonic anhydrase in the shell gland of laying hens. *Australian Journal of Agricultural Research* 40, 1111–1115.

Yoselewitz, I. and Balnave, D. (1990) Strain responses in egg shell quality to saline drinking water. *Proceedings of Australian Poultry Science Symposium*, p. 102.

Xin, H., Gates, R.S., Puma, M.C. and Ahn, D.U. (2002) Drinking water temperature effects on laying hens subjected to warm cyclic environments. *Poultry Science* 8, 608–617.

11 Breeder and Hatchery Management in Hot Climates

N.J. Daghir[1] and R. Jones[2]

[1]Faculty of Agricultural and Food Sciences, American University of Beirut, Lebanon; [2]9 Alison Avenue, Cambridge, Ontario, Canada

Introduction	295
Effects of high temperature on reproductive performance	295
Feeding breeders in hot climates	297
Nutritional experiments	298
Broiler pullet feeding	298
Broiler breeder hen feeding	300
Male broiler breeder feeding	304
Feeding programmes	305
Breeder house management	308
Egg gathering	308
Floor- and slat-laid eggs	309
Nest hygiene	310
Nest type	310
Egg hygiene	311
Water supply	311
Cage versus floor	312
Water-cooled perches	313
Disease control and prevention	313
Storage of hatching eggs	316
Hatchery design	317
Hatchery hygiene	317
Incubation in a tropical climate	318
Incubator conditions	318
Incubator problems in hot climates	319
Incubation time	320
Prewarming eggs for setting	320

© CAB International 2008. *Poultry Production in Hot Climates* 2nd edition (ed. N.J. Daghir)

Water supply and cleanliness	320
Altitude	321
Chick processing and delivery	322
Conclusions and recommendations	323
References	324

Introduction

Commercial incubation was started around 1400 BC in one or more of the warm regions of the world. The native Egyptian hatcheries are still in operation at the present day, producing a sizeable percentage of the baby chicks in the country. Askar (1927) and El-Ibiary (1946) were among the earliest to describe these hatcheries. Ghany *et al.* (1967) gave a detailed description of the structure and operation of the native Egyptian hatchery and reported at that time that these hatcheries were still producing around 95% of the chick output of the country.

This chapter describes some of the important features of breeder management in hot climates as well as hatchery management and operational problems in such an environment. It covers areas of hatching egg production, such as egg gathering, nest types for hot climates, nest and egg hygiene, storage of hatching eggs, water supply for a hatchery, hatchery design, incubator problems and problems of chick processing and delivery.

Effects of High Temperature on Reproductive Performance

The deleterious effects of high environmental temperature on production have been described in previous chapters. The effects of heat stress on the reproductive hormones have also been covered in Chapter 4. This section will therefore consider the effects of high temperature on selected aspects of reproductive performance, mainly fertility and hatchability.

Nature never intended the modern hen to reproduce itself in a hot environment. In the wild, a hot climate that is always above 23°C does not prevent the continuous development of the embryo in the egg. The necessary natural cooling feature normally found in the more moderate climate in the spring is missing in a tropical climate. The retarding of hatching egg temperature, below the embryo development temperature of 23°C, is necessary in order to accumulate sufficient hatching eggs and to have eggs hatch in a predictable time period. In a moderate climate at a certain time of the year, this cooling occurs naturally with night-time temperatures, but in a hot climate and during the summer months in a moderate climate this cooling must be done artificially. The hatching egg, when laid, is at a temperature of 40–41°C. In the spring in a moderate climate, the temperature is around 12–20°C. This permits eggs to cool gradually over a 5–6 h period. Therefore, in a tropical environment, the egg-handling procedure should try to match this criterion.

Broiler chick viability is known to be related to the age of the breeder flock. Chicks produced from young breeder flocks have a lower survival rate than from flocks beyond peak production. Weytjens et al. (1999) determined in two experiments the difference in thermoregulation between chicks originating from young or old broiler breeder flocks. These authors concluded that, post hatch, chicks from old breeder flocks are more resistant to cold, while after a few weeks of age chicks from young breeder flocks are more resistant to heat. Therefore, obtaining chicks from a certain breeder flock age may have practical implications when there is potential for heat or cold stress of these chicks.

Yalcin et al. (2005) studied the effect of pre- and postnatal conditioning-induced thermotolerance on performance of broilers originating from young and old breeder flocks. Their results suggest that this conditioning helps broilers cope with heat stress, and the age of the parent affects ability of broilers to thermoregulate, particularly those originating from younger parents.

Taylor (1949) was among the first to report that the hatchability of eggs decreases with increases in ambient temperature. He observed a drop of 15–20% in hatchability whenever the mean weekly temperature exceeded 27°C. High environmental temperatures have also been associated with lower fertility and poor eggshell quality (Warren and Schnepel, 1940; Heywang, 1944; Clark and Sarakoon, 1967; Miller and Sunde, 1975). Reductions in eggshell quality have also been related to depressed hatchability and weakening of the embryos before eggs were set (Peebles and Brake, 1987). Moving White Leghorn hens from a 21°C to a 32°C environment decreased their egg weight, egg specific gravity and fertility (Pierre, 1989). The detrimental effects of high temperature on egg weight have been discussed in Chapter 10. Since hatching egg weight influences chick weight and early chick performance, early egg size and maintaining adequate hatching egg size in hot climates is of primary importance. Small chicks are often more difficult to start and early mortality is usually high with undersized chicks. Hess et al. (1994) conducted a survey of 17 broiler companies in the southern states of Georgia and North Carolina in the USA, representing placements of approximately 14 million broilers per week. The results of this survey showed that broiler operations routinely setting eggs from young hens (24 weeks old) experienced higher average 7-day mortality than companies setting eggs after 27 weeks of age. Therefore, hatchery operators should avoid setting small eggs by either waiting until a certain percentage production is reached or waiting for a certain prescribed time after breeder housing. Furthermore, methods of improving early egg size in breeders in hot climates are of great economic importance. Nasser et al. (1992) studied the effect of heat stress on egg quality of broiler breeders. Pullets were subjected to cyclic temperatures of 20 to 42°C during a 9-week period starting at 35 weeks of age. There was an inverse relationship between temperature and the specific gravity and shell thickness of eggs. Exposing hens to high temperature induced respiratory alkalosis, as indicated by an elevation in blood pH, accompanied by a lowered haematocrit. These results further illustrate the need for environment-controlled housing for breeders in hot climates.

Seasonal variation in sperm production is well documented, and less sperm is produced during the summer months, which reduces fertility. Part of this reduction in sperm production is associated with reduction in feed intake and thus nutrient intake. Jayarajan (1992) studied the effects of season on fertility and hatchability in three different breeds in Madras, India. Fertility was highest in White Leghorns and White Plymouth Rocks during the cold season (December to February) and for Rhode Island Reds during the summer (March to May) in that country. Hatchability was highest for all breeds during the monsoon season (September to November). In a study of large commercial hatcheries in the USA, hatchability of eggs during July, August and September was 5% lower than during the remainder of the year (North and Bell, 1990). These decreases were shown in that country to be due mainly to two reasons: one was decreased feed consumption by the breeders, which caused embryonic nutritional deficiencies, and the other was deterioration in egg quality during the holding period.

Poor hatchability in hot climates may be partially due to thin-shelled eggs. Bennett (1992) compared hatchability of thin- and thick-shelled eggs for 25 broiler breeder flocks. Shell thickness was determined by using specific gravity measurements. Hatchability of thin-shelled eggs was 3–9% less than thick-shelled eggs in 30–60-week-old breeder flocks. The shell thickness of hatching eggs dropped below the recommended level (specific gravity = 1.080) by 42 weeks of age. The same author (Bennett, 1992) recommends checking shell quality at least once a month for all broiler breeder flocks. Dipping eggs in salt solutions to measure specific gravity had little or no effect on hatchability. This is important in hot climates because the incidence of thin shells is higher on a year-round basis than in temperate regions. Poor hatchability is also closely related to obesity in broiler breeders (Robinson, 1993). Hens that are overweight suffer more from heat stress and lay very erratically. They exhibit low rates of production, oviductal prolapse, poor shell quality and double-yolked eggs. Chaudhuri and Lake (1988) proposed that there would be an advantage in being able to store semen at high ambient temperature in a simple diluent, particularly for genetic selection programmes and commercial breeding in cages in tropical countries. They were able to store semen for up to 17 h at 20 and 40°C in a simple diluent. The diluent was composed of sodium chloride, TES buffer, glucose and antibiotics, and was adjusted to pH 7.4 with sodium hydroxide. For successful storage for 17 h at the high temperature, it was necessary to agitate the diluted semen samples. Samples could be kept in still conditions for 6 h at ambient temperature around 20°C on a bench top. The replacement of chloride with glutamate did not maintain good fertility at high temperature.

Feeding Breeders in Hot Climates

The feeding and feeding programmes of meat-type breeders are very different from those used for egg-type breeders because the former tend to become obese and thus poor performing. Boren (1993) presented some basic rules on broiler breeder nutrition and indicated that nutritional strategies in use today on broiler breeders are to help balance the lower potential for egg production against the economic necessity of maximizing viable hatching egg production and minimizing costs. The management of broiler breeders in the laying period has been reviewed by Dudgeon (1988), who discussed the effects of nutrition, body weight and environmental temperature on different production parameters.

Nutritional experiments

There are very few studies on the feeding of either Leghorn or broiler breeders in hot climates. In the case of Leghorn breeders, it is probably assumed that the feeding of these birds is not too different from that of regular Leghorn layers in hot climates, bearing in mind the extra fortification of the diet in vitamins and trace minerals, particularly those that are known to be critical for fertility and hatchability. Morrison *et al.* (1988) reported the results of a survey conducted on 49 broiler breeder flocks that either received or did not receive a vitamin and electrolyte water additive during heat stress. Results of this survey showed that males were more susceptible to heat stress than females, and younger females experienced a more severe drop in egg production. Both percentage mortality and percentage drop in egg production were reduced in flocks receiving the water additives.

Broiler pullet feeding

Broiler breeder pullets are placed under feed restriction starting at about 14 days of age. An early moderate restriction of growth rate is less stressful than a severe restriction later. This allows for better uniformity and proper fleshing, both of which contribute towards good hatching egg production. There are two methods of restricting growth rate in broiler breeders. There is a qualitative restriction, which utilizes reduced level of nutrients in the feed, and there is a quantitative restriction, which utilizes reduced feed allowances. Although qualitative restriction is simpler and has been shown to maintain average body weight in accordance with breeder recommendations, uniformity using this procedure is poor. This method of restriction has not been used by breeders to any extent. Two methods of quantitative restriction can be used. Birds can be fed either restricted amounts daily or on a skip-a-day programme. The advantage of every-day feeding versus skip-a-day feeding is a feed saving of approximately 1.4 kg per bird to 20 weeks of age (McDaniel, 1991). Leeson and Summers (1991) estimated a saving of up to 10% in feed required to produce a pullet of 2.3 kg body weight. For hot climates, it is important to point out that the heat production of birds on every-day feeding was around 10% lower than those on skip-a-day feeding (Fig. 11.1). Feed allowances for

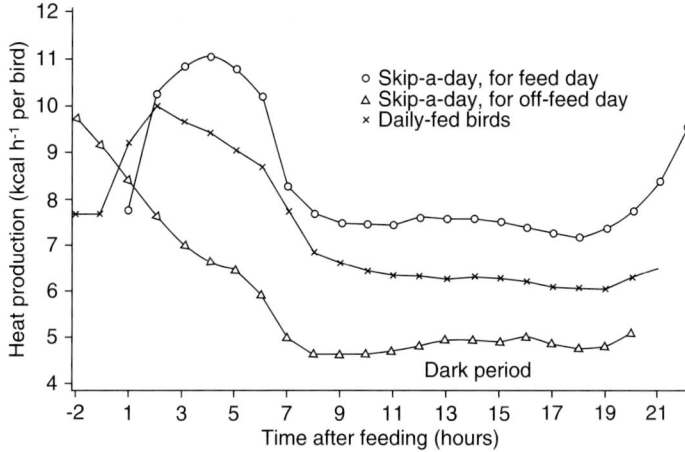

Fig. 11.1. Heat production of broiler pullets (from Leeson and Summers, 1991).

growing broiler breeders should be determined on the basis of kcal metabolizable energy (ME) per bird per day, and these allowances should be adjusted in relation to body weight and the condition and uniformity of the pullets. Feed allowances should be calculated on a flock-by-flock basis, and more feed is required, and probably more feeder space, when uniformity is low or when feed quality is poor and intestinal infections are a problem.

To produce the first egg, a pullet should attain a minimum body weight and age. Body protein content has been more closely linked to onset of production than body fat, and properly fleshed birds can mature and begin laying on time. Several workers have studied the effect of high-protein, and in many cases high-lysine, pre-breeder rations on reproductive performance of broiler breeder hens in a temperate climate. Although many reported improved performance (Cave, 1984; Brake et al., 1985; Lilburn and Myers-Miller, 1990), some doubt that it is necessary (Van Wambeke, 1992). This practice may be useful in certain hot-climate areas where poor-quality protein is used and where breeder pullets may be underweight at onset of production. Furthermore, high-protein pre-breeder diets do not appear to have any positive effect on well-fleshed and adequately developed pullets. Lilburn et al. (1987) studied the relationships between dietary protein, dietary energy, rearing environment and nutrient utilization by broiler breeder pullets. They found that birds exposed to natural daylight and fed 15.5% protein diets had similar caloric efficiencies (kcal/g) at 15 weeks of age, despite dietary density differences of 220 kcal/kg. This, in their opinion, supported the hypothesis that, above some minimal level of protein intake, caloric intake has the greatest control over body-weight gain in restricted pullets. Hagos et al. (1988) studied the effect of different levels of feed restriction on performance of broiler breeders during the growing period in Bangalore, India. They concluded that 30% restriction of a control feed containing 16.6% protein and 2800 kcal ME/kg and 15% restriction of a low-protein, low-energy diet (15% protein and 2500 kcal ME/kg) were both adequate to achieve optimum performance. A survey of 49 breeder flock records from Florida in 1984 concluded that one of the major contributors to poor flock performance to peak was inadequate growth. Many of the flocks exhibited little or no body-weight increase around the time of peak production. Peak weekly egg production was reduced by an average of 5%. Leeson and Spratt (1985) calculated the energy requirements of broiler breeders from 20 to 28 weeks of age and compared these requirements with the feed allowances recommended by the breeder companies. Their results showed that the calculated energy requirements between 20 and 25 weeks of age are much higher than provided for in the recommended feed allowance. They suggested that the broiler breeder is in a very precarious situation with regard to energy at the time of sexual maturity. Energy deficiency at this time may delay sexual maturity and influence ovum development of eggs to be laid in the coming 2–3 weeks. The same workers reported that body weight from 19 to 40 weeks of age was directly related to energy intake. Therefore, weight gain in the pullet until it reaches peak production is critical in hot areas and the use of a high-protein pre-breeder ration may be useful.

The onset of lay in broiler breeders can be advanced if greater body weight is allowed during rearing (Robbins *et al.*, 1988; Yu *et al.*, 1992). Allowing increased body weight during rearing has been suggested as a means of increasing egg weights in pullets stimulated to lay at an early age (Leeson and Summers, 1980; Kling *et al.*, 1985). Therefore, subjecting breeder pullets to photostimulation at an early age, after allowing them to attain greater than recommended body weights during rearing, may reduce rearing time and maximize hatching egg production. Yuan *et al.* (1994) studied effects of rearing-period body weights and early photostimulation on broiler breeder egg production. They observed that the onset of lay by broiler breeders can be advanced by early photostimulation and that increased body weight enhances this. Greater body weight during rearing, however, does not compensate for reduced early egg weights and results in decreased total egg production and mean egg weights when feed is provided at recommended levels during lay.

Broiler breeder hen feeding

The energy requirement of the breeder hen is considered by most nutritionists as the most limiting nutrient. The breeder hen requires energy for body maintenance and activity, growth and egg production. About 50–75% of its energy needs are for maintenance and activity, while requirements for egg production vary from 0 to 35%, depending on the rate of production. Growth requirements can be as high as 30% early in the production cycle (18–24 weeks of age) and as low as 5% later in production. Spratt *et al.* (1990b) in studies on energy metabolism of broiler breeder hens reported that fasting metabolism amounted to 75% of the maintenance energy requirement, while the liver, gut and reproductive tract amounted to 26 and 30% of the total energy expenditure in fed and fasted hens, respectively. The same workers (Spratt et al., 1990a), in studies on the energy requirements of the broiler breeder hen, indicated that individually caged hens between 28 and 36 weeks of age in a thermoneutral environment (21°C) require approximately 1.6 MJ (382 kcal) of apparent metabolizable energy (AME) per bird per day for normal growth (3 g per day) and egg production (85%).

Couto et al. (1990) studied protein requirements of broiler breeders aged 36–52 weeks and fed 18, 20, 22 or 24 g protein per hen per day. Protein intake had no significant effect on fertility or hatchability, but egg weight was significantly lower in hens fed 18 g protein and body weight was significantly higher in hens fed 24 g. The same authors suggested that breeder hens should be fed 20.5 g daily. Brake et al. (1992) fed breeders from 1 day to 18 weeks of age diets containing 2926 kcal ME/kg and 11, 14, 17 or 20% crude protein. All treatments received 17% crude protein from 19 to 25 weeks of age. A high correlation was observed between the level of protein in the rearing ration and skeletal growth. The sternum length varied from 119 mm in the 11% crude protein group to 129 mm in the 17% protein group. There was also more feather retention after onset of production in the high-protein groups. This study further illustrates the need for adequate nutrition during the rearing period and the importance of using good-quality feed when feed intake is restricted for body-weight control purposes.

Feeding standards for broiler breeders in cages are practically non-existent. Pankov and Dogadayeva (1988) conducted a study to determine the protein: energy ratio and develop optimum feeding regimes for caged, meat-type breeders. Caged broiler breeders, according to these workers, may be fed a diet containing 14.5% protein, provided it is supplemented with lysine and methionine to bring it up to the 16% protein level, with an ME of 11.3 MJ/kg at a level of 140 g per bird per day up to 42 weeks of age and 135 g thereafter. Hazan and Yalcin (1988) investigated the effect of different levels of feed intake on egg production and hatchability of caged Hybro broiler breeder hens in Turkey. The authors concluded that the best feed conversion and optimum hatchability and number of chicks can be achieved by feeding 90% of feed consumed by floor breeders. The diet used by these workers had 16.78% protein and 2748 kcal ME/kg. This was confirmed by Glatz (1988) in South Australia, who concluded that breeders in cages fed 10% below the standard levels produced more eggs per bird than hens fed at recommended levels. Spratt and Leeson (1987) studied the effects of dietary protein and energy on individually caged broiler breeders. Feed intake was maintained constant across all treatments and thus nutrient intake was varied by diet formulation. Their results show that 19 g protein and 385 kcal ME per day were sufficient to maintain normal reproductive performance through peak egg production.

Rostagno and Sakamoura (1992) studied the effects of environmental temperature on feed and ME intake of broiler breeder hens. They housed broiler breeders in environmental chambers at 15.5, 21.9 and 26.3°C. Water and feed were provided *ad libitum* for 14 days. Daily feed intake, nitrogen-corrected ME (MEn) and nitrogen-corrected true ME (TMEn) decreased linearly as environmental temperature increased. Hen body weight and rectal temperature were not affected. A 1°C rise in temperature resulted in a decreased intake of 2.43 g feed per hen, 2.10 kcal MEn/kg body weight and 2.20 kcal TMEn/kg body weight. Egg production, egg weight, egg mass and feed conversion were not affected by environmental temperature.

Vitamin A deficiency decreases the number of sperm and increases the number of non-motile and abnormal sperm (Oluyemi and Roberts, 1979). Babu *et al.* (1989) reported that Leghorn breeders require 40 p.p.m. vitamin E and 0.67 p.p.m. selenium for maximum hatchability in two different strains in Madras, India. Surai (1992) reported that increasing the level of vitamin E in the breeder diet raised the concentration of α-tocopherol in both the chicken and turkey spermatozoa and thus increased their membrane stability. It is fairly well established that fertility in hot climates can be improved in breeder flocks by the addition of extra amounts of vitamin E if breeder rations contain the usual level of 15–20 mg/kg of diet. Flores-Garcia and Scholtyssek (1992) studied the riboflavin requirement of the breeding layer and concluded that to obtain optimum hatchability, the breeder diet should contain a minimum of 4.4 mg riboflavin/kg of feed. Biotin requirements of broiler breeders fed diets of different protein content were studied by Whitehead *et al.* (1985). Production of eggs or normal chicks was depressed when practical diets containing 16.8 or 13.7% crude protein were not supplemented with synthetic biotin. Biotin requirement was higher with the diet containing 16.8% protein and was estimated to be about 100 µg of available biotin/kg. The minimum yolk biotin concentration indicative of adequate maternal status was about 550 µg/g.

In recent years, there has been increased interest in the use of ascorbic acid to overcome some of the deleterious effects of heat stress in breeders. The addition of ascorbic acid to breeder rations of chickens and turkeys has in many cases yielded positive responses. Peebles and Brake (1985) demonstrated increased egg production, hatch of fertile eggs and number of chicks per hen with supplementation of broiler breeder rations with 50 or 100 p.p.m. ascorbic acid during the mild summertime stress typical of the southern USA. Supplementation of broiler breeder feeds with 300 p.p.m. ascorbic acid during hot summers in Israel improved offspring performance in both weight gain and feed conversion (Cier *et al.*, 1992a). Mousi and Onitchi (1991) studied the effects of ascorbic acid supplementation on ejaculated semen characteristics of broiler breeder chickens under hot and humid, tropical conditions. They concluded that dietary supplementation of 250 mg ascorbic acid/kg diet is desirable to maintain semen quality during the hot periods (Table 11.1). Dobrescu (1987) reported a 28% increase in semen volume and 31% increase in sperm concentration in turkey toms by the addition of 150 p.p.m. ascorbic acid to breeding tom rations (Table 11.2). Whenever vitamin C is to be added to a pelleted ration, such as in the control of *Salmonella*, it is important to use a stabilized form of the vitamin.

The requirement of calcium of the breeder hen increases with age. Harms (1987) recommended a daily intake of 4.07 g calcium as an average for the whole production period. Therefore, diets should contain about 3.2%

Table 11.1. Effect of dietary ascorbic acid (AA) supplements on semen characteristics of broiler breeder chickens (from Mousi and Onitchi, 1991).

Dietary AA (mg/kg)	Semen vol. (ml)	Sperm concent. ($\times 10^9 ml^{-1}$)	Motile sperm/ ejaculate ($\times 10^9$)	Sperm/ejaculate ($\times 10^9$)	Motility (%)
0	0.47	2.94	0.89	1.42	59.5
125	0.51	2.54	0.79	1.32	58.5
250	0.62	2.86	1.20	1.84	62.6
500	0.77	3.11	1.64	2.40	68.3
SEM	0.07*	0.11	0.19*	0.25*	2.21

*Linear effect of ascorbic acid (significance $P < 0.05$).

Table 11.2. Effect of ascorbic acid on semen parameters in breeding turkey toms (from Dobrescu, 1987).

	Ascorbic acid (p.p.m.) 0	Ascorbic acid (p.p.m.) 150
Semen vol. (ml)	0.32	0.41
Sperm/ejaculate (10^{12})	2.97	3.88

calcium if daily feed intake averages 135 g per hen per day. Birds require slightly more phosphorus at high than at moderate or low temperatures. Breeders need a minimum daily intake of about 700 mg of total phosphorus. Harms (1987) recommended a daily intake of 683 mg. Harms et al. (1984) suggested that, for breeder hens maintained in cages, the requirements of both calcium and phosphorus are significantly higher than for those on litter floors. The sodium requirement of the broiler breeder hen was estimated by Damron et al. (1983) to be not more than 154 mg per hen daily, while Harms (1987) suggested a requirement of 170 mg per hen daily (as shown in Table 11.3).

Gonzales et al. (1991) found that supplementing a maize–soybean meal diet for 50 days with 20% saccharina (dried fermented sugarcane) for broiler breeder hens and cocks significantly increased the percentage of fertile eggs and the hatchability of total eggs incubated.

The effects of pellets, mash, high protein and antibiotics on the performance of broiler breeders in a hot climate were reported by Cier et al. (1992b). No significant differences were observed between crumbles and mash, or from raising protein from 15 to 16.5%. They concluded that, under Israel's climatic conditions, the best biological and economical results are achieved by broiler breeder hens fed daily restricted crumbles, containing 15% protein, and 2700 kcal/kg of ME.

Ubosi and Azubogu (1989) evaluated the effects of terramycin Q and fish meal on heat stress in poultry production. Body weight, feed intake and egg production were increased significantly in the group receiving terramycin Q

Table 11.3. Nutrient requirements of meat-type breeders in units per bird per day.

Nutrient	NRC (1994) Females	NRC (1994) Males	Harms (1987) – females
Protein (g)	19.5	12.0	20.6
Arginine (mg)	1110	680	1379
Lysine (mg)	765	475	938
Methionine (mg)	450	340	400
Methionine + cysteine (mg)	700	490	754
Tryptophan (mg)	190	–	256
Calcium (g)	4.00	0.20	4.07
Non-phytate phosphorus (mg)	350	110	300
Sodium (mg)	150	–	170

and 3% fish meal. Egg weight was decreased in the group receiving terramycin Q only.

McDaniel et al. (1992) fed acetylsalicylic acid (ASA) to White Leghorn breeders for 13 months of production. When fed at 0.4% of the ration, ASA decreased both fertility and hatchability. Chicks from hens given 0.1% ASA weighed more than chicks from hens given no ASA or levels exceeding 0.1%. ASA fed to layer breeders did not improve hatchability of embryos exposed to increased incubation temperature compared with embryos exposed to control incubation temperatures.

Oyawoye and Krueger (1986) reported that the inclusion of 300–400 p.p.m. of monensin in the feed of broiler breeder pullets from 1 to 21 weeks of age will suppress appetite sufficiently to accomplish restriction of body weight. Pullets fed monensin were less uniform in body weight compared with their restricted-feed controls. At low protein levels, the high level of monensin increased mortality.

Male broiler breeder feeding

Several studies have been conducted on feeding programmes for male broiler breeders. Buckner et al. (1986) tested five levels of a diet containing 13% protein and 3170 kcal ME/kg of feed on male adult broiler breeders. The intake of 91 and 102 g feed per day of such a ration reduced the number of males producing semen at 40 weeks of age compared with 136 g. Brown and McCartney (1986) recommended a daily intake of 346 kcal ME per male per day for normal body-weight maintenance and productivity of broiler breeder males grown in individual cages. Buckner and Savage (1986) fed caged broiler breeder males *ad libitum* diets containing 5, 7 or 9% crude protein and 2310 kcal ME/kg from 20 to 65 weeks of age. At 24 weeks of age, body weight, semen volume and sperm counts were reduced for the males fed the

Table 11.4. Percentage fertility of broiler breeder males of different body weights fed diets differing in protein content (from Hocking, 1994).

Body weight (kg)	Dietary protein (%)	
	11.0	16.0
3.0–4.0	93.5	92.0
3.0–4.5	94.4	92.4
3.5–4.5	91.5	93.5

5% protein diet. Average daily protein and energy intake was 10.9, 14.7 and 18.7 g per day and 495, 479 and 473 kcal ME per day for the males fed the 5, 7 and 9% protein diets, respectively. Wilson et al. (1987) reported that broiler breeder males can be fed 12–14% crude protein on a restricted basis after 4 weeks of age with no harmful effects on body weight, sexual maturity or semen quality. More males fed 12% protein continued production of semen beyond 53 weeks than those fed higher protein levels. Hocking (1994), after a series of experiments conducted at the Roslin Institute in the UK, concluded that low-protein diets (11%) increase semen yields in caged broiler breeder males but have little effect on fertility in floor pens. He further recommended that optimum male body weight at the start of the breeding period is 3.0 kg and should increase to 4.5 kg at 60 weeks of age. Table 11.4 shows that high-protein diets (16%) improve fertility in heavy birds, while low-protein diets are more beneficial for lightweight birds.

Feeding programmes

Breeder flocks can be fed a wide range of different feeds as long as these feeds meet the minimum requirements suggested in Table 11.5 and are properly balanced. An 18% protein chick starter with 2850 kcal/kg is usually satisfactory under most conditions. In cases where the 3-week target body weight set by the breeder is not met, then the starter diet may be fed beyond 3 weeks and possibly up to 7 weeks of age. A 15% protein, 2700 kcal/kg growing ration is suggested. Feed allowances for growers are adjusted weekly in order to maintain target body weights. This grower diet is also suitable for feeding males separately throughout the breeding period, provided that the breeder vitamin and trace mineral premix shown in Table 11.6 is used rather than the grower premix. A prelay may be used 2–3 weeks prior to the first egg. Such a ration usually contains 15–16% protein and 2750 kcal/kg with about 2–2.25% calcium. A breeder ration with 16% protein and 2750 kcal/kg is recommended. Such a ration provides 24 g of protein per breeder per day when the flock is fed 150 g of feed per bird per day. Table 11.6 shows the recommended levels of vitamins and trace minerals per tonne of starter, grower and breeder rations. Waldroup et al. (1976) fed diets to broiler breeders in the southern USA that supplied daily protein intakes ranging from 14 to 22 g per day.

Table 11.5. Ration specifications for starter, grower and breeder.

Nutrients	Chick starter (0–3 weeks)	Grower female (3–21 weeks) male (41–64 weeks)	Breeder I (21–40 weeks)	Breeder II (3–64 weeks)
Protein (%)	18	15	16	15
Metabolizable energy				
(kcal/kg)	2850	2700	2750	2700
(MJ/kg)	11.9	11.3	11.5	11.3
Calcium (%)	1.00	1.00	3.20	3.40
Available phosphorus (%)	0.45	0.45	0.45	0.40
Sodium (%)	0.17	0.16	0.16	0.16
Chloride (%)	0.15	0.15	0.15	0.15
Potassium (%)	0.60	0.60	0.60	0.60
Lysine (%)	0.90	0.75	0.75	0.70
Methionine (%)	0.40	0.35	0.36	0.34
Methionine + cysteine (%)	0.70	0.60	0.65	0.60
Tryptophan (%)	0.18	0.16	0.17	0.16
Threonine (%)	0.65	0.58	0.60	0.58
Arginine (%)	0.95	0.80	0.80	0.80

An intake of 20 g per day supported maximum egg production and maximum egg weight. These workers calculated that, with a feed intake of 146 g per day, a dietary level of 13.7% protein would be adequate. Table 11.3 presents daily nutrient requirements of the broiler breeder female as presented by Harms (1987) and by the National Research Council (NRC, 1994).

Feed allowances for breeder hens are usually determined by egg mass output, body weight and changes in time to consume feed by the breeder. Egg mass output usually continues to increase after peak egg production has been reached. Therefore, peak feed, which is usually started from about 40 to 50% of production, should be maintained for 3–4 weeks after maximum egg production has been reached. Changes in consumption time are good indicators of over- or underfeeding. Changes usually precede alteration in body weight by 2–3 days and deviation in egg production by 1–2 weeks. Several stressors have been shown to affect the time required to eat the daily allowance, high environmental temperature being one of the most important. Susceptibility to heat stress has been shown to vary with different broiler breeder lines. McLeod and Hocking (1993) compared two lines of breeder hens divergently selected for fatness and leanness for susceptibility to heat stress. Their results showed that *ad libitum*-fed, fat-line birds were more susceptible to heat stress and this susceptibility was not related to increased heat production but to a decreased ability to lose heat. They suggested that this heat-loss deficiency may be due to elevated blood viscosity and vascular resistance resulting from high plasma triglycerides. Changing the feeding

Table 11.6. Recommended vitamin–trace mineral levels per tonne of complete feed.

Nutrients	Starter	Grower (restricted)	Breeder
Vitamin A (IU)	10,000,000	10,000,000	12,000,000
Vitamin D_3 (ICU)	1,500,000	1,500,000	2,500,000
Vitamin E (IU)	15,000	15,000	25,000
Vitamin K_3 (g)	2	1.5	2.5
Thiamine (g)	2.5	2.5	2.5
Riboflavin (g)	6.0	5.0	8.0
Pantothenic acid (g)	12.0	10.0	16.0
Niacin (g)	40.0	35.0	40.0
Pyridoxine (g)	4.5	3.5	4.5
Biotin (g)	0.20	0.15	0.30
Folic acid (g)	1.30	1.00	1.30
Vitamin B_{12} (g)	0.015	0.010	0.025
Choline (g)	1,300	1,000	1,200
Iron (g)	96	96	96
Copper (g)	10	10	10
Iodine (g)	0.40	0.40	0.40
Manganese (g)	60	60	80
Zinc (g)	70	70	70
Selenium (g)	0.15	0.15	0.30

Note: antioxidants should be added at levels recommended by the manufacturer. Antioxidants are especially important in hot climates and where fats are added to the ration. A coccidiostat is to be used for the starter and grower at a level which would allow the development of immunity from day-old to about 12 weeks of age.

time from once to twice daily has been studied by Samara et al. (1996), who conducted experiments on broiler breeder hens to determine the effect of feeding time and environmental temperature on performance. They fed one daily meal either at 7:00 a.m. or at 6:00 p.m. or one half the daily amount at 7:00 a.m. and the other half at 6:00 p.m. High temperature, as expected, caused significant reduction in egg weight, specific gravity and shell thickness. Significant body-weight loss occurred in hens at high temperature and fed at 7:00 a.m. High temperature and feeding half of daily feed at 7:00 a.m. and the other half at 6:00 p.m. caused a reduction in feed consumption. Changing the time of feeding from 7:00 a.m. to 6:00 p.m. is not an effective method to improve eggshell quality in heat-stressed hens.

Sex-separate rearing for most broiler breeder strains is recommended and especially for 'high-yielding' males. Males, according to Boren (1993), should reach at least 140% of the female weight before mixing them together. The practice of feeding the males separately from females has become widely used in recent years. This is done to avoid overeating and to keep the body weight of the male at the right level. Males that become overweight during the production cycle tend to suffer more from heat stress. Furthermore, the decrease in male fertility can be overcome by the practice of male-separate feeding. Males are usually fed about 120–130 g per bird per day. For maximum benefit of the separate-feeding system, a diet formulated especially for males is essential. McDaniel (1991) recommended 11–12% crude protein and 2800 kcal/kg, the remaining nutrients being the same as that of a pullet grower diet. Daily nutrient requirements of the broiler breeder male as recommended by NRC (1994) are shown in Table 11.3. When it is difficult to formulate a specific male diet, a grower diet can be used by incorporating in it a breeder vitamin and trace mineral premix. Ration specifications for such a grower are shown in Table 11.5 and can be fed to males from 3 to 64 weeks of age.

Wilson *et al.* (1992) compared restricted feeding with an *ad libitum* programme in males aged 22–58 weeks. Full-fed males showed higher body weight at 30 weeks of age (5.03 kg compared with 3.85 kg). By 45 weeks, there was no difference in body weight between the two groups. The full-fed birds produced a high concentration of spermatozoa in their ejaculate during the early production period, but the trend was reversed after 48 weeks of age. The authors concluded, on the basis of the entire cycle, that excess body weight reduces reproductive capacity. The practice of restricted feeding using a separate male system not only gives better reproductive performance but also reduces leg abnormalities, especially during the second half of the production cycle.

Breeder House Management

Proper management procedures can considerably improve performance and reduce mortality in breeders. Whenever ambient temperatures exceed 35°C, mortality can reach 25–30% in a poorly managed house. The greatest losses occur among older and heavier birds, and thus breeders are particularly subject to heat stress.

Egg gathering

When pen temperature approaches 30°C, eggs should be gathered hourly, because embryo development is very rapid at that temperature. Eggs should be moved quickly into a room or building which is cool enough to slowly retard egg temperature. A separate building equipped with mechanical refrigeration is preferred. Failure to provide frequent gathering and cooling can lead to embryos developing to a point where they cannot recover. Eggs should be gathered hourly in such situations and placed in a cool, clean room for 5–6 h. This is because it has been known for many years that hatching eggs require 5–6 h of embryonic development to reach a desired embryo size before they are cooled (Heywang, 1945). More recently, Fasenko *et al.* (1992) conducted a trial on preincubation development of the embryo and subsequent

Table 11.7. Percentage hatchability of two ages and three simulated nest-box temperature treatments (from Meijerhof et al., 1994).

Age (weeks)	'Nest' temp. (°C)	% Hatchability
37	30	91.6[a]
	20	92.3[a]
	10	92.3[a]
59	30	86.1[a]
	20	88.5[b]
	10	86.5[ab]

[a,b]Means within bird age treatment with no common superscript are significantly different ($P < 0.05$).

hatchability of eggs held in nests for periods of up to 7 h under summer conditions in the state of Georgia. Embryonic death, fertility, hatchability and hatch-to-fertile ratio were not affected by duration of holding time in the nest. Embryos from eggs retained in nests for extended periods showed greater development than eggs removed promptly from nests. Meijerhof et al. (1994) studied the effect of differences in nest-box temperature, storage time, storage temperature and presetting temperature on hatchability of eggs produced by broiler breeders of two different ages (37 and 59 weeks of age). A higher temperature in the nest box, longer storage periods and higher storage temperatures, especially at longer storage periods and higher presetting temperatures, significantly reduced the hatchability of fertile eggs from the older birds. For the younger birds, a significant reduction of hatchability was found only for the longest storage period. Table 11.7 shows the effect of nest-box temperature on hatchability of broiler breeder eggs produced at 37 and 59 weeks of age. These authors suggested that the decrease in hatchability of fertile eggs from older birds is related to the increased sensitivity of these eggs to non-optimal preincubation treatments. After cleaning and fumigation, eggs are to be placed in a cold storage room. Failure to cool eggs in this manner can also result in early hatching and a spread in hatching time from first to last chick. To help hatching eggs to cool down, gather eggs on plastic trays. The best tray designs are those with open fabrication. Avoid the use of paper trays until the eggs have cooled. Never pack warm eggs in cartons or cover the eggs until such time as the eggs have lost heat (below 18–10°C).

Floor- and slat-laid eggs

Eggs laid on litter and slat flooring, because of the high faecal contaminants present, are at high risk of being contaminated. For this reason, some companies do not use these eggs, but the vast majority, for financial reasons, set these eggs along with the nest eggs. This practice will not be easy to change;

however, I would suggest that some precautions be taken to perhaps reduce the risks involved. I would suggest the following precautions be taken to handle floor/litter eggs.

Gathering

The longer the egg is in contact with litter/slat, the greater the contamination risk. Litter-/slat-laid eggs should be gathered three times daily and be removed from the breeder house. Soiled eggs should be disposed of and the balance, along with eggs trays, should be sprayed with a suitable disinfectant. Litter-/slat-laid eggs should be kept separate to nest-laid and their hatch results and bacterial contamination monitored. They are high risk and must be carefully watched.

Nest hygiene

In the vast majority of cases, the egg is clean and free from bacteria when first laid. However, if nests and nest litter are contaminated with microorganisms, eggs can become infected. In a tropical climate, it is recommended that nest litter be changed frequently (every 7–14 days) and 25 g of paraformaldehyde added to each nest after a litter change. Nests should be sprayed twice weekly to suppress bacteria. The best time to carry out this treatment is late in the day. A broad-spectrum disinfectant such as 1.5% formalin plus 1.5% of a 20% quaternary ammonium compound in water is recommended. Another good practice to keep nests clean is closure of the nest at night, to prevent birds from roosting and sleeping in them. All nest-laid eggs pass over the tail of table. If it is contaminated, all eggs will be contaminated, so cleaning and disinfecting these tables is very critical. Tables should be cleaned and disinfected each 30–45 min of use or between gatherings.

Nest type

There are many types of nests in use around the world, some of which are more suitable for warm climates and others for cold climates. A colony-type nest, when fully occupied, would be extremely hot and not suited to a hot climate. An automatic nest, which usually holds more than one bird at a time and is open to front, bottom or rear, is suitable for a warm climate. Most automatic nests have the egg roll-away feature, which is good for a warm climate. The conventional open-front nest, when designed with a wire and open construction with an egg roll-away feature, is very suitable for a hot climate. However, when the base is solid and the egg roll-away feature is absent, this nest is the least desirable for a hot climate. The egg roll-away feature is very desirable for hot climates, because it permits the egg to move away from the hen and thus enable cooling to commence.

Subiharta et al. (1985) studied the effect of two types of nests used in rural Indonesia on hatchability and found that for the bamboo cone nest hatchability was 77.37% and for the wooden box nest it was 66.39%.

Egg hygiene

Eggs that come in contact with faecal material run a high risk of being contaminated with bacteria. There are many ways to clean eggs and disinfect them. Generally speaking, if more than 10% of eggs have faecal contamination, a management problem exists and should be corrected. Birds usually consume most of their feed during the cooler parts of the day. Try to adjust the light on–off times and the feeding times to correspond to the coolest times of the day. Make sure that the first egg collection is finished within 3 h of the lights coming on. This is to avoid dirty and nest-soiled eggs. Workers should wash and disinfect hands hourly and just prior to egg gathering. Immediately following gathering, eggs should be fumigated with formaldehyde for a 20 min period. One gram of potassium permanganate to 1.5 ml formalin per cubic foot (approximately 0.03 m^3) of fumigation cabinet space is recommended. Proudfoot and Stewart (1970) evaluated the effects of preincubation fumigation with formaldehyde on hatchability and found that embryo viability was not impaired when the above procedure was followed. Spraying eggs with a disinfectant soon after they are laid is an alternative to fumigation after gathering. The most common disinfectant in use in North America is hydrogen peroxide at a concentration of 5% in water.

Water supply

Additional waterers should be placed in the centre of breeder houses during hot weather. This increased supply of water can be provided by installing additional automatic waterers. It is especially important to provide adequate water in places where birds congregate, usually the centre of the house. Water supplied at the right place will save many birds, since they will not move any great distance to get water when they are overheated.

Since meat-type breeders are feed-restricted during growing, the amount of feed restriction greatly influences the quantity of water consumed by these birds. Figure 11.2, produced from data presented by North and Bell (1990), shows the amount of water that such growing pullets should consume at varying ambient temperatures when feed is restricted. Water consumption is optimal when the average daytime house temperature is 21°C. At 10°C, a bird should drink about 81% of the amount consumed at 21°C, 167% at 32°C and 202% at 38°C. When meat-type breeder pullets are placed on a restricted feeding programme, they consume much more water than usual. In fact, they gorge themselves with water in order to feel full and also out of boredom. This often leads to a wet droppings problem. Limiting the time that

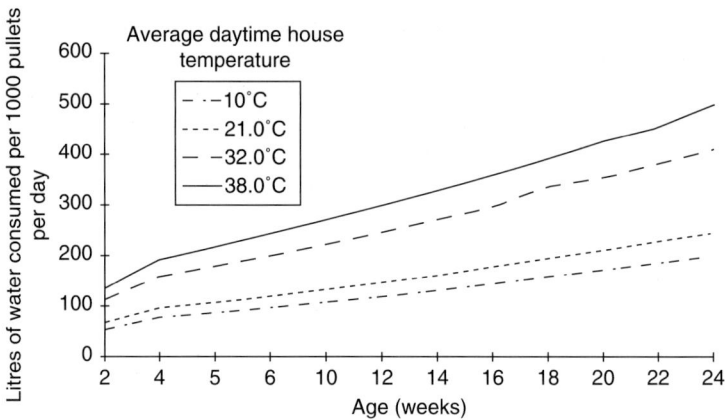

Fig. 11.2. Water consumption by standard meat-type growing pullets for females grown separately, on daily or skip-a-day feed restriction.

water is available to broiler breeders will often help to prevent wet litter problems by reducing the total water usage of the birds (Bennett, 1988). Therefore, these pullets are sometimes placed on a programme of water restriction. In hot climates, however, this is not recommended, and even in temperate regions, when average daytime house temperatures exceed 27°C, we do not restrict water.

Cage versus floor

For breeder houses, the industry has for many years moved from the full-litter floor to a two-thirds slatted floor. The slatted floor has given superior results as compared with the full-litter floor. The move to the slatted floor has helped in allowing for about 15% higher density, fewer dirty-floor eggs, lower disease incidence and considerable saving in cost of litter. Litter is much easier to keep dry and clean when the feed and water systems are located over the slat area. The same advantages have been found to be true in the hot regions of the world. However, this system is not free of problems. High density in breeder houses increases mortality and reduces egg production. Crowding in the centre aisle is a major problem. Some poultry farmers have reduced mortality by cutting the density so that each hen has 0.225 m² of space. The use of cages for broiler breeders was started in some European countries and has moved outside Europe to a limited extent. Its use for breeders in hot climates is not very common. Renden and Pierson (1982) compared reproductive performance of dwarf broiler breeder hens kept in individual cages and artificially inseminated with dwarf breeder hens kept in floor pens and naturally mated. Mean body weights and egg weights were significantly greater for caged hens. There were no differences in feed conversion, hen-day egg production or percentage hatchability between treatments.

Fertility of hens housed in floor pens was significantly better than that of hens housed in cages. Leeson and Summers (1985) studied the effect of cage versus floor rearing on performance of dwarf broiler breeders and their offspring. Cage-reared birds were consistently heavier than floor birds. Floor-reared birds had longer shanks. Rearing treatments had little effect on laying performance of birds in cages from 22 to 44 weeks of age. Broiler chicks from the 40-week hatch were reared to 56 days and no differences in weight were seen at 42 or 56 days of age. Yang and Shan (1992) conducted an experiment on 5000 breeder hens at Beijing University in China. They observed that performance in the cage system was superior to that on the floor. Criteria measured were egg production per hen housed, percentage fertility and viability up to 62 weeks of age. They concluded that the cage system will dominate the broiler breeder industry of China very soon. Yalcin and Hazan (1992) studied the effect of different strains and cage density on performance of caged broiler breeders in Turkey. They concluded that medium-weight breeder strains placed at 677 cm^2 per bird achieve the best performance. In hot climates, where environmentally controlled houses with evaporative cooling devices are used, and where labour is cheap for purposes of artificial insemination, cages for breeders appear to be promising.

Water-cooled perches

Simple and economical methods for alleviating the effects of heat stress on breeder performance are lacking. Muiruri and Harrison (1991) studied the effects of roost temperature on performance in hot ambient environments. They found that water-cooled roosts improved hen-day egg production, daily feed intake and hatchability. The water-cooled roosts minimized deleterious effects of heat stress through conductive heat loss from the birds to the roosts. They improved both the productive and reproductive performance of chicken hens.

Disease control and prevention

The 'all-in, all-out' system is the recommended system for broiler breeder projects in hot regions and is the best for disease control and prevention. In hot climates, birds should be handled only during the cool hours of the day. Vaccinations should be performed in drinking water very early in the morning, because the temperature of the water at that time is usually cool. In extreme high temperature, do not water-fast birds more than 1–2 h. Whenever you administer drugs in the drinking water, remember that, in the case of high environmental temperature, water consumption can be three to four times higher than at a low temperature. Whenever you suspect overdrinking, calculate the dose per bird per day and dissolve the total amount needed per day in a quantity of drinking water which could be consumed in 3–4 h.

The basic principles of disease control, such as cleaning and disinfection of the premises between flocks, control of potential viruses, isolation of brooding and rearing of young stock away from potential sources of infection, etc., are extremely important in hot climates because the environmental conditions of high temperature and humidity are ideal for microbial growth.

One of the major disease problems that has received serious attention in recent years is the problem of salmonellosis. The reason for this increased interest in the disease was the *Salmonella enteritidis* outbreaks that occurred in both the USA and the UK in 1988. These outbreaks focused on the rising worldwide *Salmonella* epidemic and the urgent need for *Salmonella* control in poultry. Today, many countries have developed a fully integrated programme for *Salmonella* control, which includes all sectors of the poultry industry and which has as its ultimate objective the maximum reduction of *Salmonella* transmission through poultry products.

The largest number of *Salmonella* species belong to the paratyphoid group. *Salmonella pullorum* and *Salmonella gallinarum*, which are specific to poultry, have been largely eradicated in many countries but are still causing problems in several developing countries and need to be controlled. *Salmonella enteritidis* is one of several hundred species in this paratyphoid group, but, because of its involvement in food poisoning, it has received most attention. It is believed that the main carrier of *Salmonella* into poultry farms is contaminated feed. However, other natural sources, such as wild birds, insects, rodents and humans, can spread this infection easily within any poultry operation. Once the organism is ingested by the bird, it multiplies very rapidly, creating large colonies in the intestinal tract. These organisms may lie dormant in clinically healthy birds for long periods of time. Transmission of *Salmonella* can take place through faecal contamination of the eggshell during or after the laying process and through subsequent eggshell penetration. Horizontal transmission from numerous live carriers or by other vehicles, such as feed (as pointed out earlier) and water, is a very common route of transmission.

Although *Salmonella* is very difficult to eradicate, it can be reduced by following very strict hygienic practices. A *Salmonella* control programme should include a very strict farm and house sanitation programme, to prevent horizontal transmission of the bacteria and infection of the flock, and a sound monitoring programme, to keep you aware of your disease status and to alert you of any changes that might be taking place. Such a monitoring programme is the most essential for effective *Salmonella* control. Some basic components of a monitoring programme for breeder and hatchery operations are environmental sampling and regular antigen blood testing.

To carry out environmental sampling for breeder flocks, take swab samples of fresh faecal material every 60 days from each flock, one sample per pen or per 500 birds. Each swab should contain a pool of several spots from the pen. The samples are then processed and incubated in the laboratory and the results obtained in 72 h.

A regular plate test (P-T) antigen blood test should be performed every 90 days on 150 birds per house. Tested birds should be leg-banded or individually identified so that you can go back and pick out the reactors for culturing. If *Salmonella* is isolated from these, a sensitivity test should be carried out on isolated bacterial cultures and the proper antibiotic used for treatment for 5–7 days' duration. At the end of the treatment period, nests should be fully cleaned and disinfected and nest litter changed. Floor litter should be removed completely and the floor treated with an application of 5% formalin. During this time, eggs should not be set from the infected flock until the lapse of 10 days from the end of treatment.

Monitoring programmes for hatchery operations should include once-a-month testing of dead-in-shell embryos at transfer time. About six to ten specimens should be collected per source flock (up to 30 embryos depending on number of donor flocks). Samples are sent to the nearest laboratory and results can be known within 72 h. If *Salmonella* shows up, then the follow-up procedure is the same as that for breeder flock sampling. The sensitivity test can be done on the same culture that isolated the *Salmonella* and therefore the idle period in this case can be shorter.

Since feed is considered an important carrier of *Salmonella* into poultry farms, the following steps are recommended.

1. Insist that your feed mill or the mill that you purchase feed from has a *Salmonella* control programme. Although animal by-products are more susceptible to *Salmonella*, plant products can be infected. Pelleted feeds are preferred to mash feeds since pelleting involves some degree of cooking and will expose less surface area to *Salmonella* contamination.
2. Acid treatment of feeds has been found to be effective in reducing *Salmonella*. Propionic acid is most effective against moulds, while formic acid has been found to be effective against *Salmonella*. The use of 0.9% formic acid in feed leads to complete decontamination after 3 days of treatment.
3. Dust has been found to be an important vehicle for *Salmonella*. Dust control contributes to production of uncontaminated feed. All equipment must be cleaned and disinfected with a 25 p.p.m. iodine solution.
4. Samples of feed from each load should be taken and identified by delivery date, flock certification number and type of feed and retained for at least 3 months.
5. Store and transport feed in such a manner so as to prevent possible contamination.

Biosecurity for a breeder and hatchery operation is far more critical than for a regular layer or broiler operation. The progeny of a single parent flock populate many farms, with each breeder hen producing over 150 broiler chicks during her lifetime. As Frazier (1990) points out:

> Biosecurity in a breeder operation must consist of a lot more than a pair of throw-away plastic boots. Dust accumulated on outer clothing and one's hair is just as dangerous as dirty footwear. Unsanitized vehicles accumulate contamination day after day. They don't have to appear grossly dirty to be biologically dirty. These are only a few of the many aspects of a minimum biosecurity programme.

Therefore, it is absolutely essential that breeder and hatchery operators develop their own disease-monitoring and overall biosecurity programmes, based on geographical location and information on disease outbreaks, prevention and control methods known for that area. Biosecurity programmes are often costly, but, to attain the greatest possible success in disease prevention, they have to be instituted and constantly followed.

Storage of Hatching Eggs

As the hatching egg is a highly perishable commodity, it must be stored under specific conditions if its hatching ability and qualities are to be retained. The average egg has between 8000 and 10,000 pores. The pores are the means by which the egg breathes, losing water by evaporation and taking up oxygen, which is transported to the embryo. The egg starts losing water from the time it is laid: the drier the atmosphere, the greater the water loss. With the loss of its moisture, egg vitality is also lost, and this affects not only hatchability but also chick quality. A hatching egg storage room requires a uniform temperature and relative humidity level. The longer the storage period, the more critical these two factors become. For storage up to 7 days, a temperature of 16–17°C and 80% relative humidity are recommended. For a longer period of storage, a lower temperature of around 12°C and 85% relative humidity would be more desirable. For an even longer storage period, enclosing eggs in plastic bags and placing the small end of the egg upwards would be beneficial. The beneficial effects of plastic packaging on hatching eggs have been known for many years. Proudfoot (1964, 1965) reported on several preincubation treatments for hatchability and found that hatchability was maintained at a higher level when eggs were enclosed in plastic film during the preincubation period. Gowe (1965) further illustrated that flushing eggs stored in plastic bags with nitrogen gas improved their hatchability. He showed that hatchability of fertile eggs stored 19–24 and 13–18 days in cryovac bags flushed with nitrogen gas was about 7–8% higher than that of comparable eggs stored for the same periods in cryovac bags and in an atmosphere of air. Obioha et al. (1986) studied the effect of sealing in polythene bags on hatchability of broiler eggs kept at two storage temperatures (15 and 22°C) and held for seven storage periods. Sealing eggs with polythene bags maintained significantly higher hatchability than leaving them untreated under both temperature regimes.

In a recent review on egg storage and embryonic development, Fasenko (2007) reported that one of the methods to reduce the negative effects of long-term storage has been to incubate eggs for short periods before storage. This technique has been successful in improving hatchability of long-term-stored turkey and chicken eggs.

Eggs stored on open incubator trays, even in ideal conditions, are prone to rapid water loss. For storage over 3 days, covering eggs with plastic will extend their shelf-life to about 6 days by reducing water loss. A longer storage time on incubator trays leads to rapid deterioration of egg hatching quality. Baker (1987) studied the effects of storage on weight loss in eggs and found that, irrespective of size, eggs lose an average of 2% in weight when stored at 10°C and 5% at 21°C after 20 days of storage.

Furuta *et al.* (1992) studied the effects of shipping of broiler hatching eggs from a temperate to a subtropical region in Japan on their hatchability. They concluded that, if hatching eggs are properly shipped, there should be no decrease in hatchability. The most critical factor is the time between laying and setting, which should be within 7 days.

Hatchery Design

An incubator is designed to operate within certain environmental limits. If the environmental conditions are not present year-round naturally, they must be provided by artificial means. For incubators to function correctly, the incubator room must provide three elements.

1. A minimum temperature of 20°C and a maximum of 28°C, with an ideal temperature of 25°C.
2. A minimum relative humidity of 50% and maximum of 70%, with an ideal humidity of 60%.
3. A minimum of 14 m^3 of fresh outside air per minute per 100,000 eggs in setters, and a minimum of 14 m^3 of fresh air per minute for each 14,000 eggs in hatchers.

Conditions outside these temperature and humidity parameters will cause an interaction in the incubators which produces a poor environment for incubation. Where the environment is hot and humid, large airy rooms are preferable. The following conditions are recommended.

1. A roof height up to 8 m.
2. A side-wall height of 6 m.
3. A ridge opening at roof of 1.4 m and at side-wall of 2.0 m.

For a hot dry climate, evaporative cooling could be advantageous, with a roof height of 6 m and a wall height of 4.5 m.

Hatchery Hygiene

A hatchery should be isolated and positioned on land at a distance of at least 1 km from other poultry facilities. Prevailing wind direction is a major consideration. When locating the hatchery on the site, the wind should blow from the egg receiving and setting end to the hatching rooms and chick room and never vice versa. The hatchery building must have completely separate rooms, with each room having an independent air supply and temperature and humidity requirement and control. The following room separations are suggested: egg rooms, setting rooms, hatching rooms, chick removal, chick tray wash, box storage, staff lunch, and showers and toilets. Where hatching is to take place twice weekly, one hatch room is satisfactory. Should a hatchery hatch more frequently, two or more hatch rooms are desirable to prevent bacterial contamination from one hatch to the next.

Basic hatchery design should provide a corridor separation of setting and hatching rooms. A separation of hatching and chick rooms is also necessary. Incubator rooms functioning with positive pressure and a strong negative pressure in the corridor prevent air movements from one room to the next.

Incubation in a Tropical Climate

Incubation in a tropical climate is not very different from other climates. The key to good incubation is the hatchery environment and the three elements mentioned above required for incubation. A high level of hygiene is needed inside and outside the incubator. The incubator depends on five elements in its design. Management must ensure that each of these elements is functioning correctly at all times. They are heating, humidity, cooling, ventilation and turning. When any one element malfunctions, problems will result.

Incubator conditions

The chicken embryo absorbs heat from its surroundings up until approximately 288 h of incubation. After this, it commences metabolizing its own energy and producing heat. At 288 h of incubation, embryo temperature is approximately 37°C, while by 480 h its temperature is up to 39.5°C. Embryo body temperature must be held close to 37°C if successful hatching is to occur. Controlling this excess heat has been a major concern in artificial incubation for over 2000 years. Today, it is still the most important factor in an incubator. This is of greater significance in hot climates because more is known at present about the development of thermoregulation in poultry embryos and the effect of incubation temperature on it. Tzschantke (2007) recently discussed the attainment of thermoregulation as affected by environmental factors and indicated that, in poultry, perinatal epigenetic temperature adaptation was developed by changes in the incubation temperature. Further research is needed to determine how long this effect of thermoregulation is on the life of the broiler.

Modern incubators cool by two methods. Some have incorporated both. When purchasing an incubator for a hot climate, the cooling factors must be fully understood before a purchase is made. All modern incubators are designed to use one method of cooling or another, but none of them provides the coolant. This is a separate and usually an expensive item. Two different cooling systems are used.

1. Water cooling. This requires water to be at a temperature of 12°C. In most hot climates, ground water is above 26°C and is of little cooling value. It is therefore necessary to use water-chilling equipment to reduce water temperature and to circulate it around the hatchery to each incubator.
2. Air cooling. When incubator room air exceeds 28°C, some form of air cooling is required. If the outside air is dry, evaporative cooling can reduce hatchery air temperature below the critical level. However, if the environment

is humid, it is necessary to use thermostatically controlled, mechanical air conditioning to cool the air around the incubator air intake.

It is impossible to hatch chicks properly in a hot climate without providing the cooling element. The ideal temperature for incubator and chick rooms is 25°C, while for the egg work room it is 23–25°C. Relative humidity in all live rooms in a hatchery should be about 60%.

Incubator problems in hot climates

One of the problems occurring in incubators in a hot climate is an increase in temperature during midday periods causing room temperature to increase with outside temperature because of insufficient cooling capacity. If incubators are water-cooled, cold water supply temperature at incubator entry should be 12°C. The diameter of the cold water supply line is critical in a hot climate. The pipe must be large enough (minimum 2.5 cm diameter) to supply adequate cold water and insulated to avoid water temperature loss and condensation drip. Incubator cooling coils are usually not large enough for hot climates and therefore doubling the number of coils is normal procedure in this type of environment. If incubators are air-cooled, then air-conditioning is necessary to provide sufficient cold air at the point of use. The air-conditioner duct should be close to the air intake and usually the cool area diameter around the intake should not need to exceed 1 m.

Hatcheries located in very dry climates require mechanical means in order to maintain the desired 60% relative humidity in live rooms, i.e. incubator rooms and chick processing rooms. Failure to maintain this desired humidity level will require humidifiers to work harder. Humidifier systems in incubators are designed mainly to correct small differences in humidity. When they operate for long periods of time and cold water is used, there is a cooling effect, which prompts incubator heaters to operate more frequently. Under such conditions, hatchery managers restrict ventilation, thus limiting fresh air, which produces poor incubation conditions.

Some chicks hatch much earlier than others in the same incubator. If some chicks are hatching much sooner than others, it is most probably because they have received more heat somewhere along the incubation process. In this case, the following steps are suggested.

1. Ensure all eggs receive the same treatment the day they are laid and during storage.
2. Throughout the storage and prewarming period, eggs in the top and bottom trays should be of the same temperature.
3. Follow the incubator manufacturer's recommended set pattern.
4. It is possible to find a different temperature in the incubators with top or middle being somewhat warmer than the bottom. In this case, at the time of transfer, place eggs from the coolest part of the setters to the warmest of the hatchers.
5. When transferring eggs, transfer the eggs nearest the floor first as this is the coolest location.

Incubation time

Normal incubation time for chicken eggs is 21 days or 504 h from the time the eggs are set to chick removal. There are, however, many reasons for chicks to hatch earlier or later.

1. Breed type: differences between breeds are usually no more than 3 h.
2. Broiler breeds in warm climates can hatch up to 12 h early. The important factor here is to remove the chicks when ready. They are considered ready when 5% are still damp on rear of head. Chicks confined to a hatcher can overheat and dehydrate and be exposed to large volumes of bacteria.
3. As flocks advance in age, eggs produced become larger. Large eggs take longer to lose moisture in the same environment as small eggs and thus they need more time. Therefore set older flocks 3–4 h sooner than younger flocks. If you can incubate eggs from older flocks separately, then operate the incubator at a 5% lower humidity level.
4. When breeders first come into production, egg size is very small and eggshells tend to be thick. Eggs laid during the first 4–6 weeks, though quite fertile, usually have lower hatchability than they do when the pullets are fully mature and laying larger eggs consistently. It has been said that eggs from young pullets with thick shells do not lose sufficient water during incubation. Such eggs can benefit from lower setter humidity levels. However, this is not always easy to provide. It is therefore suggested that a minimum egg size of 52 g be used and that eggs from immature breeders be set 2–3 h ahead of what is considered normal.

Prewarming eggs for setting

Some incubators will benefit from several hours of prewarming eggs prior to placing them in the incubator. Others do not benefit from this practice and prewarming can be detrimental. One should check with the incubator manufacturer about the need for prewarming.

For single-stage incubators, prewarming is not necessary. All eggs are set at one time with good air circulation. All eggs warm up at the same time. Jamesway Big Jay and Super Jay incubators, because of their design, do not require egg prewarming. Air passes from warm to cold eggs and is heated prior to recirculation. However, if the eggs could be prewarmed all to the same warm temperature, hatch time would be more punctual and predictable. Placing a thermometer in an egg from the top tray in a buggy and again in an egg from the bottom tray will reveal temperature spread from top to bottom. Usually, top tray eggs are warmer than bottom tray eggs and therefore hatch slightly sooner. The objective is to have hatch time from first to last chick as close as possible.

Water Supply and Cleanliness

A major concern for a hatchery is the cleanliness of its water supply. Conditions in a hot climate are prone to bacterial contamination of water. As incubators are the ideal environment for bacteria to reproduce, a hatchery must take steps to ensure a clean water supply. Chlorine is usually used in the water supply to provide 5 p.p.m. at point of use. Failure to do this will produce bacterial-related diseases in the baby chick.

Much of the water supply contains minerals and different sediments. These elements tend to clog the small orifices of the incubator humidification system, necessitating very frequent cleaning of these orifices. It is therefore recommended that hatcheries use purified water to supply humidity. This is best accomplished with the use of a reverse-osmosis filtration system, which guarantees water purity from virus, bacteria, minerals and sediment. Such systems are not overly expensive and have a short payback period.

A hot climate very often means contaminated water. A high microbial count in the water supply creates problems for the breeders. Contaminated eggshells produce hatchery contamination, which again contaminates eggs and chicks. Egg and incubator rooms should be washed daily, along with all utensils used, with a chlorine-type disinfectant. Floors, the inside of setters and setting rooms should be washed following each and every transfer, using a phenolic disinfectant.

Moulds are a serious problem when they infect chicks. Aspergillosis is a serious problem when it appears in the system. This is a common condition in a hot environment. This mould can infect the litter in the breeder house and be transported to the hatchery in and on eggs and egg carriers. The mould can invade incubators, multiply and infect eggs by penetrating the shell and infecting the lungs of chicks after hatching. In the hatchery, moulds are commonly found in ventilation systems, on evaporative cooling pads and in humidifiers. Moulds should be destroyed from their source, starting with the feed, by the use of a mould inhibitor and in the litter. In hot climates, refuse left following a hatch must be removed from a hatchery site as soon as possible.

Altitude

In a study conducted at an altitude of 5500 feet, Arif *et al.* (1992) reported that hatchability was significantly higher for eggs stored with the small end up (85.8%) than in those with the large end up (74%). Early embryonic mortality did not differ between groups, but late embryonic mortality was lower in eggs stored with the small end up. Christensen and Bagley (1988) examined turkey egg hatchability at high altitudes at two different oxygen tensions and two incubation temperatures. Incubating in a 149 Torr oxygen environment at 37.7°C gave significantly better hatchability than a 109 Torr oxygen environment at 37.5°C. Embryonic mortality data indicated that the higher incubation temperature in combination with increased oxygen tension decreased embryonic mortality during the third and fourth weeks of incubation and resulted in higher hatchability. The data suggest that hatchery managers at high altitudes should supplement with oxygen and incubate turkey eggs at higher temperatures than at lower altitudes.

Chick Processing and Delivery

As mentioned earlier, chicks should be removed from the hatchers when only 5% are left with dampness at the rear of their heads. They should be moved into a well-ventilated room with a temperature of approximately 26°C. At this point, the temperature surrounding the chicks should be lowered from the incubator temperature of 37°C to 33°C for best comfort. A careful monitoring of temperature inside the hatch tray or chick box at this time is important. If the temperature surrounding the chick is too high, quality will be quickly lost. Most people place 100 chicks in the confines of one box. In a hot climate, the number of chicks in a box should be reduced to about 80 (20 per compartment), thus lowering the temperature in the box. Chick boxes need to be well placed while standing in the chick room. If room temperature exceeds 32°C, fans similar in size and power to hatcher fans are needed to blow air through boxes in order to prevent overheating. Chick delivery in a hot climate can be very hazardous, especially if the delivery is for long distances. Chick delivery vans need to be equipped with Thermo-King air-conditioners, especially for daytime delivery. In a hot climate, deliveries before sunrise or after sunset are recommended. High environmental temperatures during transportation of neonatal chicks cause dehydration and death. Neonatal chicks cope with heat by evaporation. Therefore, the initial water content of neonatal chicks may be important. Chicks hatching early lose more body weight in the hatcher than those hatching later. This loss is primarily due to water loss (Thaxton and Parkhurst, 1976). Hamdy *et al.* (1991a) studied the effects of incubation at 45 versus 55% relative humidity and early versus late hatching time on heat tolerance of neonatal male and female chicks. They concluded that chicks that hatched late (i.e. with a short holding period in the hatcher) and coming from eggs incubated at 45% relative humidity had increased heat tolerance in comparison with the other chicks.

It has been shown for some time that chicks exposed to fluctuating higher or lower temperatures have a high incidence of unabsorbed yolks (Leeson *et al.*, 1977). This has recently been confirmed by Mikec *et al.* (2006), who studied the effect of environmental stressors on yolk sac utilization.

Van Der Hel *et al.* (1991) reported that exposing chicks to suboptimal conditions during transportation may lead to impaired production. In order to reduce the negative effects of ambient temperature during and after transport, it is important to assess the thermal limits between which animals can survive without long-lasting negative effect. These workers conducted trials to estimate the upper limit of ambient temperature of neonatal broiler chicks by measuring heat production, dry matter and water loss in the body and yolk sac during 24 h exposure to constant temperature from 30.8 to 38.8°C. Their results showed that this critical temperature was between 36 and 37°C.

Body-weight loss and mortality weights of neonatal chicks are increased when they are transported at high temperatures. After arrival at a farm and placement at normal thermal conditions, chicks that have survived exposure to high temperature will eat and grow less and may have a higher risk of dying than those exposed to lower temperatures (Ernst *et al.*, 1982). A high heat tolerance is therefore an important attribute. Hamdy *et al.* (1991b) reported that a 10% lower than normal incubation relative humidity did not negatively affect performance of chicks after heat exposure. Chicks hatching early had a higher risk of dying after exposure to heat (39°C for 48 h) than late-hatching ones. Chicks hatching late from eggs incubated at low relative humidity were most heat-tolerant.

Conclusions and Recommendations

1. High environmental temperatures have been shown to depress fertility and hatchability. Decreases in fertility are due to depressions in sperm production, while decreases in hatchability have been shown to be due to reduced feed consumption, which causes embryonic nutritional deficiencies and deterioration in egg quality.

2. The feeding of Leghorn-type breeders in hot climates is no different from that of the table egg layer, except for the extra fortification of the diet in vitamins and trace minerals known to be critical for optimum reproductive performance.

3. The feeding and feeding programmes of meat-type breeders are critical in hot climates because of the feed restriction used and differences in maintenance requirements as well as quality of feeds available in many hot regions. The use of a high-protein pre-breeder ration may be useful in those areas where breeder pullets are underweight at onset of production.

4. Certain studies in hot regions have shown that dietary vitamin A, vitamin E, vitamin C and riboflavin are important for maximizing fertility and hatchability in those areas.

5. Broiler breeder pullets should not be severely restricted early in their life cycle in hot climates since a moderate restriction at that age is less stressful. This allows for better uniformity and proper fleshing, both of which contribute to good hatching egg production.

6. Feed allowances for growing meat breeders should be determined on the basis of ME requirements per bird per day and these should be adjusted in relation to body weight and condition and uniformity of the pullets. More feed and feeder space is required when uniformity is low and when feed quality is poor.

7. The energy requirements of the meat breeder hen, especially early in the production cycle, are considered to be very critical and increase from about 300 kcal ME per bird per day at 20 weeks of age to about 400 kcal ME per bird per day at 28 weeks of age. Therefore, feed allowances during this period should exceed those requirements. The daily protein requirement of the meat breeder has been estimated to be about 20 g per bird per day and feed allowances providing less protein can reduce egg weight and body weight.

8. A feeding programme has been recommended which consists of an 18% protein chick starter with 2850 kcal/kg, a 15% protein grower with 2700 kcal/kg, a 16% protein female breeder with 2750 kcal/kg and a 12% protein male breeder with 2750 kcal/kg. When it is difficult to use a specific male

diet, the grower diet can be used for feeding males separately throughout the breeding period, provided that it is supplemented with the breeder vitamin and trace mineral premix rather than the grower premix.

9. In the majority of cases, poor hatching is the result of mismanagement at the breeder farm, i.e. poor male and female breeder management. Eggs failing to hatch after a normal incubation period should be examined on a regular basis (every 30 days) and compared with a known standard, which would indicate where the problem lies.

10. Many hatcheries are located in areas of convenience rather than areas best suited for a hatchery operation. Some critical factors in locating a hatchery are altitude (not exceeding 1000 m above sea level), proximity to poultry buildings, wind direction, clean and plentiful water supply, and design to avoid re-entry of used air.

11. Operation of a hygienic hatchery requires a number of key elements to be in place. Other than a clean location, a sound sanitation programme is needed, with a desire and knowledge on the part of management to maintain a hospital-like environment. The type of disinfectants to be used and how and when they are applied are important considerations. Disinfectants used should be effective against both Gram-positive and Gram-negative organisms. Formaldehyde fumigation (where allowed) should be applied at the appropriate times, i.e. eggs soon after lay, eggs just prior to setting and in the hatchers continuously.

12. The operation of hatcheries in tropical climates is not very different from that in a temperate climate. However, the cooling equipment, clean water and hygienic conditions are more critical factors in a hot climate.

13. Water supply in a hot climate is critical with regard to volume, temperature and purity. A substantial amount of water is required for cleaning and disinfection. This should be a separate supply from that used for cooling and humidification. For cooling, water-chilling equipment is needed, while for humidification a reverse-osmosis system is recommended.

14. During processing and delivery, chicks can be, and often are, damaged. Temperatures need to be monitored in and around the chick boxes to ensure that chicks are not overheated. In warm environments, horizontal fans should be used to move air through chick boxes while stationary in the hatchery. Thermo-King refrigeration is recommended for use in chick transportation. Chick boxes should be partially filled and not confined to one area during holding, in order to prevent overheating.

References

Arif, M., Joshi, K.L., Shah, P., Kumar, A. and Joshi, M.C. (1992) Hatchability of layer eggs stored and incubated in different positions at 5500 feet altitude. *Journal of Agricultural Science* 118, 133–134.

Askar, M. (1927) Egyptian methods of incubation. *Proceedings of the World's Poultry Congress,* Ottawa, Canada, pp. 151–156.

Babu, M., Mujeer, K.A., Prabakaran, R., Kalatharan, J. and Sudararasu, V. (1989) Effect of vitamin E and selenium supplementation on fertility and hatchability in White Leghorn breeders. *Cheiron* 18, 158–161.

Baker, R. (1987) Effect of storage on weight loss in eggs. *Poultry Digest* 46, 276–278.

Bennett, C.D. (1988) Feed and water intake patterns of broiler breeders. *Canada Poultryman*, February, 22–24.

Bennett, C.D. (1992) The influence of shell thickness on hatchability in commercial broiler breeder flocks. *Journal of Applied Poultry Research* 1, 61–65.

Boren, B. (1993) Basics of broiler breeder nutrition. *Zootechnica International*, December, 54–58.

Brake, J., Garlich, J.D. and Peebles, D. (1985) Effect of protein and energy intake by broiler breeders during the pre-breeder transition period on subsequent reproductive performance. *Poultry Science* 64, 2335–2340.

Brake, J., Walsh, T.J. and Scheideler, S.E. (1992) Nutritional influences in broiler breeders: frame size, feather drop syndrome and reproduction. *Poultry Science* 71 (Suppl.), 16 (Abstract 48).

Brown, H.B. and McCartney, M.G. (1986) Restricted feeding and reproductive performance of individually caged broiler breeder males. *Poultry Science* 65, 850–855.

Buckner, R.E. and Savage, T.F. (1986) The effects of feeding 5, 7 or 9% crude protein diets to caged broiler breeder males. *Nutrition Reports International* 34, 967–976.

Buckner, R.E., Renden, J.A. and Savage, T.F. (1986) The effect of feeding programmes on reproductive traits and selected blood chemistries of caged broiler breeder males. *Poultry Science* 65, 85–91.

Cave, N.A.G. (1984) Effect of a high protein diet fed prior to the onset of lay on performance of broiler breeder pullets. *Poultry Science* 63, 1823–1827.

Chaudhuri, D. and Lake, P.E. (1988) A new diluent and methods of holding fowl semen for up to 17 hours at high temperature. *Proceedings of the 18th World's Poultry Congress*, pp. 591–593.

Christensen, V.L. and Bagley, L.G. (1988) Improved hatchability of turkey eggs at high altitudes due to added oxygen and increased incubation temperature. *Poultry Science* 67, 956–960.

Cier, D., Rimsky, I., Rand, N., Polishuk, O., Gur, N., Benshashan, A., Frish, Y. and Ben-Moshe, A. (1992a) The effects of supplementing breeder feeds with ascorbic acid on the performance of their broiler offsprings. *Proceedings of the 19th World's Poultry Congress*, Vol. 1, p. 620.

Cier, D., Rimsky, I., Rand, N., Polishuk, O. and Frish, Y. (1992b) The effects of pellets, mash, high protein and antibiotics on the performance of broiler breeder hens in a hot climate. *Proceedings of the 19th World's Poultry Congress*, Vol.2, pp. 111–112.

Clark, C.E. and Sarakoon, K. (1967) Influence of ambient temperature on reproductive traits of male and female chickens. *Poultry Science* 46, 1093–1098.

Couto, H.P., Soares, P.R., Rostagno, H.S. and Fonseca, J.B. (1990) Nutritional protein requirements of broiler breeding hens. *Revista da Sociedade Brasileira de Zootecnia* 19, 132–139.

Damron, B.L., Wilson, H.R. and Harms, R.H. (1983) Sodium chloride for broiler breeders. *Poultry Science* 62, 480–482.

Dobrescu, O. (1987) Vitamin C addition to breeder diets. *Feedstuffs*, 2 March, 18.

Dudgeon, J. (1988) The management of broiler breeders in the laying period. *Proceedings of the 4th International Poultry Breeders Conference*, pp. 1–7.

El-Ibiary, H.M. (1946) The old Egyptian method of incubation. *World's Poultry Science Journal* 2, 92–98.

Ernst, R.A., Weathers, W.W. and Smith, J.M. (1982) Effect of heat stress on growth and feed conversion of broiler chicks. *Poultry Science* 61, 1460–1461.

Fasenko, G.M. (2007) Egg storage and the embryo. *Poultry Science* 86, 1020–1024.

Fasenko, G.M., Wilson, J.L., Robinson, F.E. and Hardin, R.T. (1992) Effects of nest holding time during periods of high environmental temperature on pre-incubation embryo development, hatchability and embryonic viability in broiler breeders. *Poultry Science* 71 (Suppl.), 125 (Abstract 373).

Flores-Garcia, W. and Scholtyssek, S. (1992) Effect of levels of riboflavin in the diet on the reproductivity of layer breeding stocks. *Proceedings of the 19th World's Poultry Congress*, Vol. 1, p. 622.

Frazier, M.N. (1990) Broiler breeder management. *Vineland Update*, No. 32.

Furuta, F., Shinzato, G., Higa, H. and Shinjo, A. (1992) Effects of shipping of broiler eggs from the temperate to the subtropical regions on their hatchability. *Proceedings of the 19th World's Poultry Congress*, Vol. 1, p. 682.

Ghany, M.A., Kheir-Eldin, M.A. and Rizk, W.W. (1967) The native Egyptian hatchery: structure and operation. *World's Poultry Science Journal* 23, 336–345.

Glatz, P.C. (1988) The effect of restricted feeding during lay on production performance of broiler breeder hens. *Proceedings of the 18th World's Poultry Congress*, pp. 959–960.

Gonzales, L.M., Elias, A., Valdivie, M., Berrio, I.I., Fraga, L.M. and Rodriquez, C. (1991) A note on the fertility and hatching rate in heavy hens and roosters fed saccharina. *Cuban Journal of Agricultural Science* 25, 191–193.

Gowe, R.S. (1965) On the hatchability of chicken eggs stored in plastic bags flushed with nitrogen gas. *Poultry Science* 44, 492–495.

Hagos, A., Devegowda, G. and Ramappa, B.S. (1988) Restricted feeding of broiler breeders during the growing period. *Proceedings of the 18th World's Poultry Congress*, pp. 966–967.

Hamdy, A.M.M., Van Der Hel, W., Henken, A.M., Galal, A.G. and Abid-Elmoty, A.K.I. (1991a) Effects of air humidity during incubation and age after hatch on heat tolerance of neonatal male and female chicks. *Poultry Science* 70, 1499–1506.

Hamdy, A.M.M., Henken, A.M., Van Der Hel, W., Galal, A.G. and Abid-Elmoty, A.K.I. (1991b) Effects of incubation humidity and hatching time on heat tolerance of neonatal chicks: growth performance after heat exposure. *Poultry Science* 70, 1507–1515.

Harms, R.H. (1987) Formulation of broiler and broiler breeder feed based on amino acid composition. *Monsanto Technical Symposium*, pp. 116–127.

Harms, R.H., Bootwalla, S.M. and Wilson, H.R. (1984) Performance of broiler breeder hens on wire and litter floors. *Poultry Science* 63, 1003–1007.

Hazan, A. and Yalcin, S. (1988) The effect of different feeding levels on egg production and hatchability of caged broiler breeders. *Proceedings of the 18th World's Poultry Congress*, pp. 1099–1101.

Hess, J.B., Wilson, J.L. and Wineland, M.J. (1994) Management influences early chick mortality. *International Hatchery Practice* 8, 27–29.

Heywang, B.W. (1944) Fertility and hatchability when the environment temperature of chickens is high. *Poultry Science* 23, 334–339.

Heywang, B.W. (1945) Gathering and storing hatching eggs in hot climates. *Poultry Science* 24, 434–437.

Hocking, P.M. (1994) Feeding broiler breeder males. *International Hatchery Practice* 8, 17–21.

Jayarajan, S. (1992) Seasonal variation in fertility and hatchability of chicken eggs. *Indian Journal of Poultry Science* 27, 36–39.

Kling, L.J., Howes, R.O., Gerry, R.W. and Halteman, W.A. (1985) Effects of early maturation of brown-egg type pullets, flock uniformity, layer protein level and cage design on egg production, egg size and egg quality. *Poultry Science* 64, 1050–1059.

Leeson, S. and Spratt, R.S. (1985) Nutrient requirements of the broiler breeder. *Proceedings of the Maryland Nutrition Conference for Feed Manufacturers*, pp. 75–80.

Leeson, S. and Summers, J.D. (1980) Effect of early light treatment and diet selection on laying performance. *Poultry Science* 59, 11–15.

Leeson, S. and Summers, J.D. (1985) Effect of cage versus floor rearing and skip a day versus everyday feed restriction on performance of dwarf broiler breeders and their offspring. *Poultry Science* 64, 1742–1749.

Leeson, S. and Summers, J.D. (1991) *Commercial Poultry Nutrition.* University Books, Guelph, Ontario, Canada, pp. 188–189.

Leeson, S., Walker, J.P. and Summers, J.D. (1977) Environmental temperature and the incidence of unabsorbed yolks in sexed broiler chicks. *Poultry Science* 56, 316–318.

Lilburn, M.S. and Myers-Miller, D.J. (1990) Effect of body weight, feed allowance and dietary protein intake during the prebreeder period on early reproductive performance of broiler breeder hens. *Poultry Science* 69, 1118–1125.

Lilburn, M.S., Ngiam-Rilling, K. and Smith, J.H. (1987) Relationship between dietary protein, dietary energy, rearing environment and nutrient utilization by broiler breeder pullets. *Poultry Science* 66, 1111–1118.

McDaniel, G.R. (1991) Management of breeder replacement stock. *Avian Farms Flock Report* 3, 1–5.

McDaniel, G.R., Balog, J.M., Freed, M., Elkin, R.G., Wellenreiter, R.H., Kuczek, T. and Hester, P.Y. (1992) Response of layer breeders to dietary ASA 3. Effects on fertility and hatchability of embryos exposed to control and elevated incubation temperatures. *Poultry Science* 72, 1100–1108.

McLeod, M.G. and Hocking, P.M. (1993) Thermoregulation at high ambient temperature in genetically fat and lean broiler hens fed ad libitum or on a controlled-feeding regime. *British Poultry Science* 34, 589–596.

Meijerhof, R., Noordhuizen, J.P.M. and Leenstra, F.R. (1994) Influence of pre-incubation treatment on hatching results of broiler breeder eggs produced at 37 and 59 weeks of age. *British Poultry Science* 35, 249–257.

Mikec, M., Bidin, Z., Valentic, A., Savic, V., Amsetzekwika, T., Raguz-Duric, R., Lukaenovak, I. and Balenovic, M. (2006) Influence of environmental and nutritional stressors on yolk sac utilization, development of chicken gastrointestinal system and its immune status. *World's Poultry Science Journal* 62, 31–40.

Miller, P.C. and Sunde, M.L. (1975) The effects of precise constant and cyclic environments on shell quality and other lay performance factors with Leghorn pullets. *Poultry Science* 54, 36–40.

Morrison, W.D., Braithwaite, L.A. and Leeson, S. (1988) Report of a survey of poultry heat stress during the summer of 1988. Unpublished report of Department of Animal and Poultry Science, University of Guelph, Ontario, Canada.

Mousi, A. and Onitchi, D.O. (1991) Effects of ascorbic acid supplementation on ejaculated semen characteristics of broiler breeder chickens under hot and humid tropical conditions. *Animal Feed Science and Technology* 34, 141–146.

Muiruri, H.K. and Harrison, P.C. (1991) Effect of roost temperature on performance of chickens in hot ambient environments. *Poultry Science* 70, 2253–2258.

Nasser, A., Wentworth, A. and Wentworth, B.C. (1992) Effect of heat stress on egg quality of broiler breeder hens. *Poultry Science* 71 (Suppl.), 51 (Abstract).

National Research Council (NRC) (1994) *Nutrient Requirements of Domestic Animals: Nutrient Requirements of Poultry*, 9th edn. National Academy Press, Washington, DC, pp. 32–34.

North, M.D. and Bell, D. (1990) *Commercial Chicken Production Manual*, 4th edn. Van Nostrand Reinhold, New York.

Obioha, F.C., Okorie, A.U. and Akpa, M.O. (1986) The effect of egg treatment, storage and duration on the hatchability of broiler eggs. *Archiv fur Geflugelkunde* 50, 213–218.

Oluyemi, J.A. and Roberts, F.A. (1979) *Poultry Production in Warm Wet Climates*. Macmillan Press, London.

Oyawoye, E.O. and Krueger, W.F. (1986) The potential of monensin for body weight control and *ad libitum* feeding of broiler breeders from day-old to sexual maturity. *Poultry Science* 65, 884–891.

Pankov, P.N. and Dogadayeva, I. (1988) Standards of feeding battery-caged broiler parent stock. *Proceedings of the 18th World's Poultry Congress*, pp. 973–975.

Peebles, E.D. and Brake, J. (1985) Relationship of eggshell porosity to stage of embryonic development in broiler breeders. *Poultry Science* 64, 2388–2391.

Peebles, E.D. and Brake, J. (1987) Egg shell quality and hatchability in broiler breeder eggs. *Poultry Science* 66, 596–604.

Pierre, P.E. (1989) Effects of feeding thiouracil on the performance of laying hens under heat stress. *Bulletin of Animal Health in Africa* 37, 379–383

Proudfoot, F.G. (1964) The effects of plastic packaging and other treatments on hatching eggs. *Canadian Journal of Animal Science* AA, 87–95.

Proudfoot, F.G. (1965) The effect of film permeability and concentration of nitrogen, oxygen and helium gases on hatching eggs stored in polyethylene and cryovac bags. *Poultry Science* 34, 636–644.

Proudfoot, F.G. and Stewart, D.K.R. (1970) Effect of preincubation fumigation with formaldehyde on the hatchability of chicken eggs. *Canadian Journal of Animal Science* 50, 453–465.

Renden, J.A. and Pierson, M.C. (1982) Production of hatching eggs by dwarf broiler breeders maintained in cages or in floor pens. *Poultry Science* 61, 991–993.

Robbins, K.R., Chin, S.F., McGhee, C.G. and Roberson, K.D. (1988) Effects of *ad libitum* versus restricted feeding on body composition and egg production of broiler breeders. *Poultry Science* 67, 1001–1007.

Robinson, F.E. (1993) What really happens when you over-feed broiler breeder hens? *Shaver Focus* 22, 2–5.

Rostagno, H.S. and Sakamoura, N.K. (1992) Environmental temperature effects on feed and ME intake of broiler breeder hens. *Proceedings of the 19th World's Poultry Congress*, Vol. 2, pp. 113–114.

Samara, M.H., Robbins, K.R. and Smith, M.D. (1996) Interaction of feeding time and temperature and their relationship to performance of the broiler breeder hen. *Poultry Science* 75, 34–41.

Spratt, R.S. and Leeson, S. (1987) Broiler breeder performance in response to diet protein and energy. *Poultry Science* 66, 683–693.

Spratt, R.S., Bayley, H.S., McBride, B.W. and Leeson, S. (1990a) Energy metabolism of broiler breeder hens. 1. The partition of dietary energy intake. *Poultry Science* 69, 1339–1347.

Spratt, R.S., McBride, B.W., Bayley, H.S. and Leeson, S. (1990b) Energy metabolism of broiler breeder hens. 2. Contribution of tissues to total heat production in fed and fasted hens. *Poultry Science* 69, 1348–1356.

Subiharta, P.T., Wiloeto, D. and Sabrani, M. (1985) The effect of nest construction on village chicken egg hatchability in rural areas. Research Report.

Surai, P.F. (1992) Vitamin E feeding of poultry males. *Proceedings of the 19th World's Poultry Congress*, Vol. 1, pp. 575–577.

Taylor, L.W. (1949) *Fertility and Hatchability of Chicken and Turkey Eggs*. Wiley, New York.

Thaxton, J.P. and Parkhurst, C.R. (1976) Growth, efficiency and livability of newly hatched broilers as influenced by hydration and intake of sucrose. *Poultry Science* 55, 2275–2279.

Tzschantke, B. (2007) Attainment of thermoregulation as affected by environmental factors. *Poultry Science* 86, 1025–1036.

Ubosi, C.D. and Azubogu, C.N. (1989) Evaluation of terramycin Q and fishmeal for combating heat stress in poultry production. *Bulletin of Animal Health in Africa* 37, 373–378.

Van Der Hel, W., Verstegen, M.W.A., Henken, A.M. and Brandsma, H.A. (1991) The upper critical ambient temperature in neonatal chicks. *Poultry Science* 70, 1882–1887.

Van Wambeke, F. (1992) The effect of a high protein pre-breeder ration on reproductive performance of broiler breeder hens. *Proceedings of the 19th World's Poultry Congress*, Vol. 2, pp. 260–261.

Waldroup, P.W., Hazen, K.R., Bussell, W.D. and Johnson, Z.B. (1976) Studies on the daily protein and amino acid needs of broiler breeder hens. *Poultry Science* 55, 2342–2347.

Warren, D.C. and Schnepel, R.L. (1940) The effect of air temperature on egg shell thickness in the fowl. *Poultry Science* 19, 67–72.

Weytjens, S., Meijrhof, R., Buyse, J. and Decuypere, E. (1999) Thermoregulation in chicks originating from breeder flocks of two different ages. *Journal of Applied Poultry Research* 8, 139–145.

Whitehead, C.C., Pearson, A.R. and Herron, K.M. (1985) Biotin requirement of broiler breeders fed diets of different protein content and effect of insufficient biotin on the viability of progeny. *British Poultry Science* 26, 73–82.

Wilson, J.L., McDaniel, G.R. and Sutton, CD. (1987) Dietary protein levels for breeder males. *Poultry Science* 66, 237–242.

Wilson, J.L., Lupicki, M.E. and Robinson, F.E. (1992) Reproductive performance and carcass characteristics of male broiler breeders fed *ad libitum* or feed restricted from 22 to 30 weeks of age. *Poultry Science* 71 (Suppl.), 50 (Abstract 150).

Yalcin, S. and Hazan, A. (1992) The effect of different strains and cage density on performance of caged broiler breeders. *Proceedings of the 19th World's Poultry Congress*, Vol. 2, p. 341.

Yalcin, S., Ozkan, S., Cabuk, M., Buyse, J., Decuypere, E. and Siegel, P.B. (2005) Pre- and postnatal conditioning induced thermotolerance on body weight, physiological responses and relative asymmetry of broilers originating from young and old breeder flocks. *Poultry Science* 84, 967–976.

Yang, N. and Shan, C. (1992) Housing broiler breeders in China: cage vs. floor. *Proceedings of the 19th World's Poultry Congress*, Vol. 2, p. 340.

Yu, M.E., Robinson, F.E., Charles, R.G. and Weingardt, R. (1992) Effect of feed allowance during rearing and breeding on female broiler breeders: 2. Ovarian morphology and production. *Poultry Science* 71, 1750–1761.

Yuan, T., Lien, R.J. and McDaniel, G.R. (1994) Effects of increased rearing period body weight and early photostimulation on broiler breeder egg production. *Poultry Science* 73, 792–800.

12 Waterfowl Production in Hot Climates

J.F. Huang[1], Y.H. Hu[1] and J.C. Hsu[2]

[1]Ilan Branch, Livestock Research Institute, Council of Agriculture, Executive Yuan, Taiwan; [2]Department of Animal Science, National Chung Hsing University, Taiwan

Introduction	331
Waterfowl genetics	333
Breeds	333
Traditional selection in ducks	335
Pedigree selection in ducks	335
Goose genetics	338
Biotechnology	339
Summary	340
Nutrition and management of ducks	340
Physiological response and productive performance under heat stress	340
Housing and management	341
Duck nutrition	347
Nutrition and management of geese	354
The feeds	354
Nutrient requirements of meat-type geese	355
Nutrient requirements of breeder geese	356
Feeding and management of goslings during 0–4 weeks of age	357
Feeding and management in the growing period	360
Feeding and management in the finishing period	360
Feeding and management of breeding geese	361
Selection of breeding geese	361
Ratio of male to female	363
Lighting management	363
Sexing	363
Forced moulting	364
Artificial insemination	365

© CAB International 2008. *Poultry Production in Hot Climates* 2nd edition (ed. N.J. Daghir)

		366
Sanitary management of waterfowl		366
Disease prevention		366
Other management practices		366
References		367

Introduction

Waterfowl production has become increasingly important in many countries in the past 35 years. Duck and goose meat accounted for only 3.3% and 1.5%, respectively, of poultry meat in 1970. Production increased to 4.2% and 2.9%, respectively, in 2005 (Windhorst, 2006). The production of ducks and geese has increased continuously around the world between 2001 and 2005. During this period, the average percentage of annual increase in duck production around the world was 3.7%. The lowest growth was 1.92%, observed between 2003 and 2004, due to avian influenza outbreaks (FAOSTAT, 2007).

The top ten duck-meat-producing countries in 2005 accounted for 90.75% of the world production (Table 12.1) (FAOSTAT, 2007). In 2005, China had 67.23% of the world duck production, followed by France (5.95%), Malaysia (3.00%), Vietnam (2.52%) and Thailand (2.43%) (FAOSTAT, 2007). Table 12.1 shows that the fastest growth in the production increase between 2004 and 2005 occurred in the USA (46.7%), followed by Taiwan (17.1%) and China (3.9%). The top five duck-egg-producing countries are China, Thailand, Indonesia, Philippines and Brazil (Table 12.2), which account for 96.73% of the duck eggs produced. Although the values obtained from FAOSTAT are

Table 12.1. Top ten duck-meat-production (1000 tonnes) countries in the world*.

Country	Year					% of the world (2005)
	2001	2002	2003	2004	2005	
1. China	1965.9	2087.7	2230.5	2262.3	2350.1	67.23
2. France	231.1	253.9	240.2	238.1	208.0	5.95
3. Malaysia	68.8	52.7	81.6	102.0	105.0	3.00
4. Vietnam	77.4	81.6	82.8	88.2	88.2	2.52
5. USA	56.3	52.9	50.8	58.0	85.1	2.43
6. Thailand	105.0	93.0	72.0	84.8	85.0	2.43
7. Taiwan	58.8	56.7	56.7	59.6	69.8	2.00
8. Hungary	45.5	66.8	64.7	65.0	68.0	1.95
9. India	57.2	59.8	62.4	65.0	65.0	1.86
10. Republic of Korea	45.0	56.0	46.0	46.0	48.0	1.37
World	3025.1	3181.9	3322.9	3386.8	3495.7	100.00
% change		+5.18	+4.43	+1.92	+3.22	

(FAOSTAT, 2007; Council of Agriculture, 2007)
* The ranking of countries is based on data from 2005.

eggs excluding hen eggs, these numbers in most of the Asian countries in Table 12.2 are believed to be derived mostly from duck eggs.

Based on data published by the Council of Agriculture (2007) and FAOSTAT (2007) goose-meat production changed to a positive trend after 2001, with the lowest growth of + 2.15% between 2003 and 2004, and the highest of + 6.81% between 2004 and 2005 (Table 12.3). In 2005, China accounted for the largest part (93.44%) in the world, followed by Hungary (1.94%), Egypt (1.82%), Taiwan (1.01%) and Madagascar (0.54%). Although the data in Table 12.3 includes both goose and guinea fowl meat, goose meat is believed to dominate (Evans, 2004; FAOSTAT, 2007).

The relative scarcity of available documentation in regard to waterfowl production in hot climates compared with chickens arises from the lower entrepreneurship in the waterfowl industry. Many duck producers are very small scale, with duck populations ranging from 20 to 2000 in different countries (Nind and Tu, 1998; Islam et al., 2002). Although the duck welfare issue has been recently discussed in the European Community (Rodenburg et al., 2005), the duck flocks are kept predominantly at subsistence level in many countries. The effect of heat stress on ducks and geese has received less attention in North America and Europe than in other areas. Birds in these two areas are exposed to high environmental temperature only a few days during the summer (Gonzalez-Esquerra and Leeson, 2006). Therefore, this chapter focuses mainly on duck and goose production in Asian countries.

Table 12.2. Top ten egg-production (excluding hen eggs) (1000 tonnes) countries in the world*.

Country	Year					% of the world (2005)
	2001	2002	2003	2004	2005	
1. China	3533.4	3721.4	3937.7	4111.2	4326.1	84.58
2. Thailand	297.5	304.0	304.0	305.0	310.0	6.06
3. Indonesia	157.6	169.7	185.0	173.2	180.3	3.52
4. Philippines	73.8	74.4	74.0	72.0	72.0	1.41
5. Brazil	65.0	59.0	59.5	59.5	59.5	1.16
6. Taiwan	31.4	30.7	31.0	27.1	31.7	0.62
7. Republic of Korea	23.5	25.0	28.0	26.0	28.0	0.55
8. Bangladesh	26.0	26.0	26.0	26.0	26.0	0.51
9. Myanmar	10.5	11.0	12.5	13.5	14.2	0.28
10. Romania	23.8	33.0	30.6	34.0	10.2	0.20
World	4295.3	4511.0	4745.1	4927.2	5115.0	100.00
% change		+5.02	+5.19	+3.84	+3.81	

(FAOSTAT, 2007; Council of Agriculture, 2007)
* The data for Taiwan is derived only from duck eggs.

Table 12.3. Top five goose and guinea fowl meat-production (1000 tons) countries in the world*.

Country	Year					% of the world (2005)
	2001	2002	2003	2004	2005	
1. China	1825.2	1894.7	1975.6	2026.2	2172.5	93.44
2. Hungary	38.1	43.6	47.9	43.6	45.0	1.94
3. Egypt	42.2	42.2	42.2	42.2	42.2	1.82
4. Taiwan	25.2	24.6	25.4	26.0	23.5	1.01
5. Madagascar	12.6	12.6	12.6	12.6	12.6	0.54
World	1970.3	2044.1	2131.1	2176.8	2325.0	100.00
% change		+3.74	+4.26	+2.15	+6.81	

(FAOSTAT, 2007; Council of Agriculture, 2007)
* The data for Taiwan is derived only from goose meat.

Waterfowl Genetics

One genus (*Anser*) of geese and two genera of ducks, Muscovy duck (*Cairina moschata*) and common duck (*Anas platyhynchos*), were domesticated for commercial production several hundred years ago (Crawford, 1990). The other prevalent commercial duck, mule duck, is a hybrid between Muscovy drake and common duck female. Although some reviews on waterfowl genetics and breeding have been reported (Rosinski *et al.*, 1996; Tai *et al.*, 1999; Wezyk, 1999; Cheng *et al.*, 2003), the information from most of the tropical countries is not included.

In this context, most waterfowl genetic studies are cited from literature available in China and Taiwan. The breeds in some other tropical, subtropical or desert countries are also described, such as Thailand, Malaysia, India, Indonesia, Vietnam, Cambodia, Bangladesh, Egypt and Iran. In these regions, the breeding methods applied by pure duck breeders include traditional selection and pedigree selection. Out-crossing is used to produce commercial products.

Breeds

China is one of the countries with abundant waterfowl breeds and varieties. There are 26 indigenous breeds of duck and 26 indigenous breeds of goose in China (Wang, G.Y. *et al.*, 2005). The main breeds of meat-type ducks are the Pekin, Muscovy, mule duck and some imported strains such as the Cherry Valley, Legard and Aobaixing duck. Laying ducks include the local breeds Shaoxing, Jinding, Youxian partridge, Sansui, Liancheng White, Putian Black, etc. (Qiu, 1988) and some imported breeds like Khaki Campbell and Cherry Valley (Wang, G.Y. *et al.*, 2005). In the goose industry, there are at least ten major breeds, including the Lion head goose, Wan Xi white goose,

Yan goose, Xu Pu goose, Zhe Dong white goose, Sichuan white goose, Tai Hu goose, Huo Yan goose, Wu Zong goose and Yi Li goose (Qiu, 1988; Qiu and Qiu, 2005).

In Taiwan, the major laying duck is the Brown Tsaiya, which is one of the best duck egg layers in the world (Table 12.4). The major meat-type duck breeds are the mule duck (75–80%), which is the hybrid product of male Muscovy and female Kaiya duck (hybrid of male Pekin duck and female white Tsaiya), the Pekin, and the Muscovy. The White Roman and Chinese goose are bred for meat.

In Thailand the laying duck breeds are mostly native breeds such as the Paknum, Nakorn Phatom and some native crossbreeds. The parent stocks of mule ducks, Cherry Valley and Pekin, have been imported since the 1980s (T. Sawat, Taiwan, 1990, personal communication). Before that date, mule ducks were produced from Thai native ducks, according to Sawat. The parent stocks of laying ducks are mainly imported, but the ducklings and eggs are usually produced for exportation (Thongwittaya, 1999).

In Malaysia, the meat-duck breeds are the Pekin duck and mule duck. The parental breeds of Pekin are imported. The progeny of these commercial breeds are used in several smaller breeding farms (C.T. Yimp, Taiwan, 1990, personal communication). The Muscovy was introduced in 1990. The laying ducks include the local Itik Java, Khaki Campbell, Taiwan Tsaiya and the crosses (Yimp, 1985).

In Vietnam, the duck breeds include exotic and local breeds, with the latter accounting for over 63% of the duck population in 1998. The most popular local common duck is the Co duck (also named Tau duck in the south). The second is the Bau duck. The exotic duck breeds include the Pekin duck, Cherry Valley, Khaki Campbell and Muscovy. A system of parent and commercial duckling operation has been established since 1991 for meat production. The local goose is the Co goose, also called the Sen goose. There are some exotic breeds such as the Lion head goose and the Rheinland goose (Hanh and Tieu, 1999).

Table 12.4. The major laying duck breeds in the world.

Breed	Adult body weight (kg)	Age at first egg (day)	Egg Production* (egg/yr)	Reference
Shaoxing	♂ 1.43 ♀ 1.27	110	280	Wang, G.Y. *et al.*, 2005
Jinding	♂ 1.76 ♀ 1.73	100–120	260–300	Wang, G.Y. *et al.*, 2005
Brown Tsaiya	♂ 1.27–1.35 ♀ 1.33–1.41	116–126	292–323*	Lee *et al.*, 1992
Khaki Campbell	♂2.4 ♀ 2.3	130	280	Wang, G.Y. *et al.*, 2005
Cherry Valley CV2000		140	285	Wang, G.Y. *et al.*, 2005

* Egg production (egg/year) is calculated from egg number at 72 weeks of age ÷ (72 × 7− age at first egg) × 365.

In Bangladesh, the major duck breed (about 95%) is Deshi duck, which is a local breed for egg production, with meat yield being only a by-product. Some of the Thai ducks from Thailand are raised for egg production. Recently, the Khaki Campbell and Indian Runner ducks have been introduced into this country. The meat-type ducks raised are the Pekin and Muscovy (Shafiuddin, 1985).

In India, the two important native breeds are Sythet mete and Nageswari. The Desi duck is one of the other local breeds. Some other varieties include Khaki Campbell, Indian Runner, Pekin and Muscovy ducks (Bulbule, 1985).

In desert areas, Egypt produces a great amount of goose meat with two native geese, whose plumage is white and grey (Kosba, 1999). Some 23 different local duck breeds in Iran have been identified, such as the Tenjeh, Anghut and Iranian Pekinese, etc. For production, two common egg-laying varieties, the Khaki Campbell and the Indian Runner, are used in modern duck-breeding farms (Nimruzi, 1998).

Traditional selection in ducks

The traditional breeding method is still used without pedigree and performance tests in China, where people use appearance judgement and random mating (Chen, 1990). In Taiwan, the farmers select breeders at the end of the production season using their own judgement. The sires and dams then mate randomly (Chou and Huang, 1970). In this way, it is difficult to always keep the ducks at the same production level. From time to time the farmers must introduce some birds from other farms to avoid the inbreeding problem. The average interval of introducing birds from other farms by the local meat-type duck farmers in Taiwan is about 3 years. Similar phenomena were observed in an experimental selection population (Hu *et al.*, 1999), where the changes of variability on averaged body weights and standard deviation were similar, as reported by Lerner (1937). The major reasons for changes were presumably due to the genetic segregation and inbreeding. In recent years, for improving the efficiency, the meat-type duck breeders started to adopt the performance test as the selection criterion, instead of appearance and phenotypic characteristics. The sire and dam are mated according to the family unit.

Pedigree selection in ducks

In recent years, pedigree selection has been well developed in experiment stations and in some commercial companies. In the breeding system, the 1-day-old ducklings are tagged or tattooed for identification. The bird performance criteria are tested throughout their life cycle (Hu, 2001). According to the breeding value or the index, a certain number of sires and dams are chosen based on a preset selection programme. Taiwan, being one of the highest duck producers per capita in the world (Pingel, 1999), was one of the earliest countries to improve the local breeders and to establish the modernized duck breeding industry in Asia. It began selecting the ducks by pedigree in 1966 (Tai, 1985). Certain pure lines have been created in this country. The genetic parameters are estimated by Restricted Maximum Likelihood in an animal model on laying performances in Brown Tsaiya for five generations, on growth performances in Muscovy duck for eight generations and on duration of fertility in female Brown Tsaiya for six generations, which were inseminated with pooled Muscovy semen to test the fertility on intergeneric crossing. The heritability values of some parameters are presented in Table 12.5. Although ducks are selected in a small population, favourable efficiency of selection is obtained when predicting it by additive genetic values in an animal model based on the parameters estimated (Hu et al., 1999, 2004; Chen et al., 2003; Cheng et al., 2005).

The preferred meat duck is the heavy ducks (Pekin, Muscovy and mule ducks), although some of the spent light laying ducks such as Tsaiya,

Table 12.5. The heritabilities of ducks estimated by REML in an animal model in Taiwan*.

Breed	Characteristic	h^2 M	h^2 F	Reference
WM	BW10	0.24 ± 0.03	0.31 ± 0.03	Hu et al., 1999
WM	BW18	0.36 ± 0.04	0.43 ± 0.04	Hu et al., 1999
WM	FL10	0.37 ± 0.04	0.14 ± 0.02	Hu et al., 1999
WM	AGE1EGG	–	0.20 ± 0.03	Hu et al., 2004
WM	NEGG52	–	0.27 ± 0.03	Hu et al., 2004
WM	NEGG22W	–	0.22 ± 0.03	Hu et al., 2004
BT	BW40	0.48 ± 0.08		Cheng et al., 1995
BT	NEGG40	–	0.15 ± 0.004	Cheng et al., 1995
BT	AGE1EGG	–	0.19 ± 0.004	Cheng et al., 1995
BT	ES30	–	0.12 ± 0.006	Cheng et al., 1995
BT	EW40	–	0.30 ± 0.004	Cheng et al., 1995
BT	NES	–	0.14 ± 0.04	Poivey et al., 2001
BT	NEF	–	0.30 ± 0.03	Poivey et al., 2001
BT	NED	–	0.06 ± 0.01	Poivey et al., 2001
BT	MD	–	0.28 ± 0.03	Poivey et al., 2001
BT	NEH	–	0.18 ± 0.02	Poivey et al., 2001

* REML: Restricted Maximum Likelihood; WM: White Muscovy; BT: Brown Tsaiya; BW10, BW18 and BW40: body weights at 10, 18 and 40 weeks of age; FL10: length of the 8th primary feather at 10 weeks of age; AGE1EGG: age at first egg; NEGG40 and NEGG52: number of eggs laid up to 40 and 52 weeks of age; NEGG22W: number of eggs laid during the first 22 weeks in the first laying cycle; ES30: eggshell strength at 30 weeks of age; EW40: egg weight at 40 weeks of age; NES: number of eggs set; NEF: number of fertile eggs at candling; NED: total number of dead embryos; MD: Maximum duration of fertility; NEH: number of hatched mule duckling.

Shaoxing duck, etc. are traditionally favoured by the local people for meat. Although the Pekin duck was bred in China 340 years ago, it was introduced into the USA, the UK, some European countries, Japan and Russia. Some Pekin duck lines have been created in these countries (Chen, 1990). The Muscovy was selected in France for about 30 years and in the UK in recent years. These countries with moderate climate are the major duck breeders' producers in the world. These duck breeders have been exported into Asian countries. However, the farmers in Asian countries began to create some lines with pedigree, and more and more research has been conducted in some countries, such as Taiwan and China.

Pekin duck is the major meat-duck breed in China. Some pedigree data have been established (Li, Z. et al., 2005). The heritability is 0.31 for body weight at 6 weeks of age by multi-trait Derivative-Free REML in an animal model (Li, Z. et al., 2005). In recent years, owing to the advantages for promoting the duck industry, China has paid more attention to the Muscovy duck, although this breed has been in southern China for at least 276 years (Wang, G.Y. et al., 2005). Certain paternal and maternal lines of White Muscovy have been created and some progress has been made by synthetic index selection in China (Wang et al., 2003).

The mule duck production has existed in Taiwan and south China since the early 18th century (Chou and Huang, 1970; Qiu, 1988). Owing to the demand for white mule ducks, one criterion system of six colour levels of mule ducklings' down had been created, first in Taiwan for selecting its parental lines (Huang, 1985), and then the improved criterion system of 15 colour levels has been developed since 1984. The selection of one of the parental lines, White Tsaiya, has reached 99.5–99.9% of mule duck, whose white plumage colour is accepted by farmers (Lee and Kang, 1997). In recent years, the mule duck production has also been more and more developed in China. The three-way-crossing mule duck (1.6–1.8 kg at 10 weeks of age) has been gradually replaced by the two-way-crossing mule duck (2.8 kg at 8 weeks of age) in China, and the pure breeder system was introduced to select White Muscovy (Tan, 2002). Some pedigree selection programmes have been studied in the parental female lines of mule ducks, in which selection is based on the progeny test with a criterion of six colour levels of mule ducklings' down (Chen et al., 2000). Although the mule duck production has been developing well for several decades with the technique of artificial insemination (AI) (Tai, 1985; Rouvier et al., 1987), some problems still exist. One of the problems is the short duration of fertility occurring in the intergeneric crossbreeding. To prolong the fertility duration, an experimental line of Brown Tsaiya tested with Muscovy duck has been created in Taiwan for selection on the number of fertile eggs since 1992. After 11 generations of selection, the results showed it is possible to improve the AI efficiency in the Brown Tsaiya without impairing the hatchability (Cheng et al., 2005). However, the results are not consistent with the heavy common ducks, and therefore further study of the selection model on the parental line is suggested (Brun et al., 2006).

Duck-egg production and consumption has existed for a long time in Asian countries, whereas laying ducks are still bred in a traditional way in

Table 12.6. The heritabilities of characteristics in the Tsaiya duck egg.

Breed*	Characteristic	Age (week)	h^2_s	h^2_d	h^2_{s+d}	Reference
WT	Albumen height	50	0.344	0.141	0.242	Tai et al., 1985
WT	Shell thickness	50	0.316	0.204	0.260	Tai et al., 1985
WT	Shape index	50	0.011	0.623	0.317	Tai et al., 1985
WT	Yolk weight	50	0.467	0.128	0.298	Tai et al., 1985
WT	Yolk index	50	0.150	0.358	0.254	Tai et al., 1985
BT	Green egg colour	52	0.38	0.98	0.68	Hu et al., 1993
BT	Green egg colour	33	0.15	0.27	0.21	Liu et al., 2001
WT	Yolk colour	40	0.018	0.55	0.28	Hu and Tai, 1993

* WT: White Tsaiya; BT: Brown Tsaiya.

most tropical countries. The caged system has been introduced for the Brown Tsaiya duck performance test since 1968 in Taiwan. One Brown Tsaiya duck population has been created using a pedigree system and selected on the multi-trait phenotypic index since 1984 (Tai et al., 1989; Lee et al., 1992) and on the genetic selection index since 1997 (Chen et al., 2003). The other experimental lines derived from this line were created in 1992 for selection on the number of fertile eggs (Cheng, 1995), in 1996 for selection on blue shell eggs (Liu et al., 2001) and in 1994 for divergent selection of eggshell strength in Taiwan (Lee et al., 1999; Huang, 2004). About 10 years ago, the caged performance test was also developed in China, and some pedigree selection programmes have been studied for the local laying ducks (Lu et al., 2002; Jiang, 2006). The heritabilities of some egg quality characteristics, such as shell thickness, albumen height, shell and yolk colours, are presented in Table 12.6.

Goose genetics

In tropical or subtropical areas, the three largest goose-producing countries are China, Egypt and Taiwan (Table 12.3). The breeding work started late in these countries. In the past two decades, the major areas working on goose breeding were Shangdong, Sichuan, Jiangshu and Jiling in China (Wang, L. et al., 2005). About 20 years ago, some lines of Wulong goose, which is called Huo Yan goose officially, were selected in the Laiyang Agriculture College in Shangdong. Annual egg production and body weight at 56 days of age is 90–120 eggs and 3 kg, respectively (Wang, L. et al., 2005). The Best Linear Unbiased Prediction (BLUP) program was used in the selection of Wulong for three generations (Wang, L. et al., 2002). About 10 years ago some lines of goose were selected in Sichuan, with annual egg production of 85–90 eggs and body weight of 4.2 kg at 10 weeks of age. In the university in Jiangshu, some lines were also created for about 10 years (Wang, L. et al., 2005). In the Jilin Agriculture University, one egg-production line from the Nong'an Zi goose was established based on the family selection for eight generations. Compared with the first generation, age and body weight at first egg in the eighth generation had decreased significantly by 11.7 days and 0.2 kg, respectively. Annual egg number per bird had significantly increased by 46.73% (Wu, W. *et al.*, 2005a).

In Taiwan, one pedigree breeding programme of white Roman goose has been conducted since 1992 (Yeh *et al.*, 1999). The age and body weight at first egg, and egg number at first and second laying period are 275 days, 5.33 kg, 35.1 eggs and 44.3 eggs, respectively. Their heritabilities are 0.26, 0.18, 0.06 and 0.24, respectively. In Egypt, the development of breeding work was difficult owing to shortage of good breeders, breeding knowledge and techniques (Kosba, 1999).

Biotechnology

Owing to the biological limits which may be reached in the near future and to the difficulty and cost of measuring some new traits, the application of genomics would be useful in the poultry industry (Burt, 2002). Some microsatellite loci or markers were studied in ducks in Europe (Burt *et al.*, 1999; Maak *et al.*, 2000; Genet *et al.*, 2003; Paulus and Tiedemann, 2003; Denk *et al.*, 2004), in Australia (Guay and Mulder, 2005), and in the USA (Stai and Hughes, 2003). Recently, the waterfowl-producing countries in Asia have begun to construct the genomic libraries enriched for $(CA)n$, $(CAG)n$, $(GCC)n$ and $(TTTC)n$ in Pekin duck in China (Huang, Y.H. *et al.*, 2005, 2006), enriched for $(CA)n$ in Tegal duck in Indonesia (Takahashi *et al.*, 2001), enriched for $(GATA)n$ in Brown Tsaiya duck in Taiwan (Hsiao *et al.*, 2006), and enriched for various repeat sequences in various Chinese geese in China (Chen *et al.*, 2005). A genetic linkage map for ducks was developed with 155 microsatellite markers in China (Huang, Y.H. *et al.*, 2006). A few chicken microsatellite markers found could be used in ducks (Liu *et al.*, 2005). The microsatellite markers were used to study the genetic diversity in the Pekin duck and Cherry Valley (Wu *et al.*, 2005b) and various ducks in China (Gong *et al.*, 2005; Li *et al.*, 2006). In addition to microsatellite markers, the amplified fragment length polymorphism (AFLP) markers are studied in Brown Tsaiya duck for the traits of duration of fertility. Two hundred and sixty AFLP markers are dispersed in 32 linkage groups. The linkage groups cover 1766 cM, with an average interval distance of 6.79 cM (Huang, C.W. *et al.*, 2006). AFLP markers have also been employed to study the green eggshell trait in Gaoyou duck (Li, H.F. *et al.*, 2005).

In terms of the genes in the duck, Chang *et al.* (2000) indicated that the polymorphism of the oestrogen receptor gene in Brown Tsaiya ducks was similar to that observed in pigs. To identify the homology of the oestrogen receptor gene between ducks and pigs, DNA sequencing needs to be done. Yen *et al.* (2005) have cloned cDNA fragments for the sterol regulatory element-binding protein 1 (SREBP1), SREBP2, fatty acid synthase, and HMG-CoA reductase and very low density apoVLDL-II from the liver of Tsaiya ducks. The gene expression associated with lipid metabolism in the laying duck has been studied. The results showed that the egg laying may affect particular aspects of lipid metabolism, especially biochemical pathways that involved apoVLDL-II and HMG-CoA reductase. In addition, the gene expression related to eggshell formation has also been investigated lately. The major protein (about 15 kDa) in Tsaiya duck eggshell cross-reacts with hen's anti-ovocleidin-17 antibody. A search of NCBI data reveals that the 15-kDa eggshell protein has a very high homology with the goose eggshell protein ansocalcin (Huang, J.F. et al., 2005, 2006).

One germ-line chimera between chicken and duck was developed in China for studying reproductive and development mechanisms. The germ-line chimerism was confirmed by hybridization with chicken DNA (Li et al., 2002). One germ-line chimera between Brown Tsaiya (*Anas platyrhynchos*) and Muscovy (*Cairina moschata*) was developed by transferring gonadal primordial germ cells to study the sterility between intergeneric crosses of ducks in Taiwan. The germ-line chimerism was confirmed by the reciprocal AI (Tai et al., 2004).

Summary

1. The genetic parameters of waterfowl economic traits should be continuously studied and breeding values should be applied to selection in the experimental and commercial population for improving the growth rate, meat production, higher egg production and feed efficiency.
2. In the future, research on breeding and genetics of waterfowl should focus on gene, marker and quantitative trait loci (QTL) mapping for use in the selection programs.
3. Due to the demand for more efficient waterfowl production it is necessary to pay more attention to preserve diversity of waterfowls.
4. In the future, study on the interaction of genotype with the environment in hot climate countries is needed.

Nutrition and Management of Ducks

Physiological response and productive performance under heat stress

The optimal room temperature for housed Pekin ducks ranges from 10 to 15°C (Hagen and Heath, 1976). Panting occurs when the ambient temperature is over 25°C (Bouverot et al., 1974). Surendranathan and Nair (1971) compared the physiological factors in laying ducks (Desi breed) with or without access to bathing water under diurnal atmospheric temperature variation at 24.3–37.2°C. The rectal temperature increases during the daytime and decreases from 42.1 to 40.9°C after ducks enter water for bathing for 1 h. Enlarged adrenal glands are observed in Pekin ducks exposed to heat stress (Hester et al., 1981). Body-weight gain decreases by 30% when the ambient temperature increases from 18.3 to 29°C (Bouverot *et al.*, 1974). Owing to the heat-alleviating effect on ducks with access to bathing water in Asian areas, there are relatively few documents related to measures for counteracting heat stress as compared with the broiler and laying hen industries. It has been reported that only 51.7% of hen housed egg production was observed when Khaki Campbell ducks were raised under a hot and humid climate for 196 days (Singh *et al.*, 1991).

Housing and management

Duck-house construction varies from region to region, depending on whether the management and husbandry practices are intensive, semi-intensive or extensive (Bird, 1985; Scott and Dean, 1991a). In some countries, cheap local materials are usually employed to build duck houses with an atap (or palm leaf) roof and narrow-timbered frame (Bird, 1985). Owing to high winds in some subtropical/tropical areas, duck houses in these areas are usually low, especially those near the seashore (Huang, 1973). About 70% of the world duck production is located in Asian areas. Most of these ducks are produced in extensive or semi-intensive practices near ponds, lakes, streams, canals or the seashore, because snails and cheap fish can be acquired more easily. Some environmentally controlled duck houses have been built in Taiwan. The major types of duck-raising practices are listed below.

Backyard duck raising

Backyard duck raising is characterized by little care and scarce supplementary feeds. Owing to low input, the production efficiency is low, e.g. only 60–90 eggs are laid annually per duck (indigenous) in Bangladesh. In this country, there are approximately 13 million ducks raised, mostly in this traditional system by women and children (Huque, 2006). The scavenging system in Indonesia, with its egg production ranging from 20 to 38%, is predominant in some areas. Flocks in this system have 4–20 ducks that are allowed to scavenge over part of the village area (Evans and Setioko, 1985; Ketaren, 1998). Owing to poor-quality feed and husbandry, these ducks are more susceptible to non-infectious (e.g. leg weakness and aflatoxicosis) and infectious diseases, such as duck virus hepatitis, duck virus enteritis, cholera, etc. (Aini, 2006).

Duck herding

Duck herding (Fig. 12.1) is practised in the rice fields after harvest (Huang, 1973; Edwards, 1985), or on the tidal seashore (Nho and Tieu, 2006). This type of practice accounted for about 50% in Taiwan in 1972, for both mule and Tsaiya duck. In this system, two men look after a flock of 500 ducks (Huang, 1973). This practice has disappeared in Taiwan, mainly owing to the application of chemical pesticides and a shortage of labour. In India, ducks are normally raised in low-lying areas. They are fed with broken rice, rice bran, coconut stem powder or similar products between hatching and 4 weeks of age. After that

Fig. 12.1. Duck herding in rice fields after harvest.

they are moved on to wet land, back water, irrigation water lines or paddy fields to find their own feed. After reaching the point of lay, at about 5.5 to 6 months, ducks are gathered in at night into 60–90-cm-high bamboo enclosures for laying. During daytime, ducks are taken to places with water or harvested paddy fields for foraging. Carbohydrates come from the remnants of fallen paddy grains, and proteins are provided by snails, small fish, earthworms and other insects. Annual egg production per bird is over 200 eggs under this system (Bulbule, 1985). When the Khaki Campbell is herded on the tidal seashore of the Red River delta in Vietnam, provided with paddy rice grain and aquatic creatures, egg production of 74.1% can be obtained, compared with 66.5% in the confined group (Nho and Tieu, 2006). Herding systems in Indonesia are divided into four categories: fully mobile, semi-mobile, home-based and opportunistic herding. Rice grain and aquatic molluscs (snails) represent more than 90% of the dry-weight crop content (Evans and Setioko, 1985; Setioko *et al.*, 1985a,b). The egg production of herded ducks is 26.9–41.3%, lower than the 55.6% in confined ducks (Setioko *et al.*, 1985b).

Duck–fish integrated system

The duck–fish integrated system (Fig. 12.2) is employed by some farmers, in which ducks have access to water for drinking and heat-stress alleviation (Bird, 1985; Edwards, 1985; Lee *et al.*, 1997). In Taiwan, the duck–fish integrated system is employed by some farmers for both meat ducks (mostly mule ducks) and laying Tsaiya ducks. The most common fish introduced into this system is tilapia (*Oreochromis nilotius* × *Oreochromis mossambicus*), harvested once a year with a mature body weight of 600 g. The annual average production on the farm is 27,000 ducks and 78,000 fishes. The profits from duck and fish are about equal, regardless of farm size (Lee *et al.*, 1997). To prevent a sediment flavour in fish meat, the pond sediment must be removed

Fig. 12.2. A duck–fish integrated system.

and cleaned properly at the end of each year. A daily volume of 130–190 m^3 ground water is pumped into a 1 ha pond to maintain the pond in good condition for both ducks and fish (Lee *et al.*, 1997). In China, the duck–fish integrated system usually has 450–900 meat ducks per hectare of fish pond, although 2250–3750 head of ducks per hectare of fish pond has also been suggested. Crucian carp, silver carp, grass carp and variegated carp are normally introduced (Liao and Luo, 2005). In southern India the multi-purpose communal tank is used for duck–fish integrated production. The farmers synchronize duck rearing, seasonal cropping and the rains of the south-west monsoon well. Ducklings are introduced into the tank at approximately 30 days of age (Rajasekaran, 2007).

Constructed swimming/drinking channel

In some countries, a wading/drinking channel is constructed for duck production (Fig. 12.3) using underground water, rain water, irrigation water or springs (Bird, 1985). Some ground is covered with pebbles. It is necessary to clean the ground when it is spread with excreta, for the prevention of disease dissemination and cleanliness of feathers. A pool of standing water left in swampy ground after abandoned tin mining is usually employed to raise ducks in Malaysia. Some unused mining pits are used to raise ducks in Malaysia (Yimp, 1985).

Duck–rice integrated system

The duck–rice integrated system has been practised in some countries, such as Taiwan and southern China (Fig. 12.4). Although this duck-raising method accounts for a relatively small part of the duck industry, it has attracted more attention owing to its connection to organic farming. The rice–duck system

Fig. 12.3. A wading/drinking channel constructed for duck production.

Fig. 12.4. A duck–rice integrated system.

provides a measure to benefit both rice paddy field and ducks. Insects, snails and weeds are the food sources for ducks, and the ducks' excreta becomes the fertilizer for the rice paddy. Water stirring caused by the ducks' activities inhibits the growth of weeds through photosynthesis reduction when the water becomes turbid. Their activities also enhance the rice root, stalk and leaf development, thereby accelerating rice growth. In addition, a reduced application of pesticides and fertilizers benefits the ecological system (FFTC, 2001; Furuno, 2001; Liao and Luo, 2005). In practice, an optimum population of 200–300 ducks/ha paddy field is recommended to obtain a good rice and duck harvest. This number is adjusted based on other feed sources (weeds, insect and snails) that are available in the paddy field (FFTC, 2001; W.H. Lin, unpublished results). Aigamo, with a mature body weight at 1.6 kg, and other indigenous small ducks can be raised (FFTC, 2001; Furuno, 2001). When mule ducks were raised at a density of 100, 300 and 500 ducks/ha paddy field, beginning at 2–3 weeks of age, the final body weight was similar to when fed a normal meat-type finisher feed for another 3 weeks before marketing (J.F. Huang, unpublished results). Other points that need to be taken into account in this practice are as follows:

1. Ducklings at 1–2 weeks of age are introduced into the field after transplanted seedlings become rooted, and before introduction they must be trained to get into the habit of flocking and oiling their feathers.
2. A protective fence (e.g. electric fence) is required to protect the ducks from predators such as dogs, cats and hawks and to prevent them from escaping.
3. Water must be kept at a level in which the ducks can both swim and walk (FFTC, 2001; Furuno, 2001).

Terrestrial duck raising

Although the duck is classified as a waterfowl, water for swimming is not absolutely necessary (Lee *et al.*, 1991; Huang *et al.*, 1993). Better egg production and feed-conversion efficiency is obtained when laying ducks are raised in cages equipped with nipples or cut-in-half pipes, as compared with being raised on a floor with a water area for swimming (Lee *et al.*, 1991; J.F. Huang, unpublished results). This is probably owing to less energy expenditure in the cage-feeding system. In another study in which laying ducks were reared on deep litter, it did not change their egg production, livability or feed consumption when compared with semi-intensive systems with a swimming area (Andrews, 1978). However, it is well known that feather growth and cleanliness of feathers and skin are better in ducks raised in an area with water provided for bathing (Bird, 1985; Huang *et al.*, 1993; Liao and Luo, 2005). Bathing in ponds is believed to promote metabolism in Pekin ducks, whereas the amount of swimming needs to be restricted to obtain a good growth performance (Yi and Zhou, 1980). It is recommended that the enclosed pond area be about 1–2 m^2/duck (Bird, 1985). In the caged system with water supply nipples, it is critically important to train the ducks to get water from the nipples right after they are hatched, with at least 1 month of acclimation to the nipples before they are caged. The cage design needs to minimize horizontal wires, to prevent the ducks' heads from being trapped. It is worth mentioning that the cage system efficiently decreases dioxin pollution for laying ducks in dioxin-contaminated areas (J.F. Huang, unpublished results).

For ducks raised on wired floors, the type of wire mesh needs to be considered, especially for sensitive breeds (e.g. Tsaiya duck) that are likely to panic when exposed to unexpected noise, animals or people (Huang *et al.*, 1993). It appears that meat-type Muscovy ducks acclimatize better to wire floors than mule ducks. The rapid movement of the mule ducks increases the chances of getting their webs stuck in the wires more easily (Huang *et al.*, 1993).

Breeder ducks with heavy body weights, like Pekin and Muscovy ducks, are more susceptible to bumble-foot when raised on wire for an extended period.

Pad cooling/tunnel air systems

Pad cooling/tunnel air systems (PCTA) have been well developed in the goose, chicken and pig industries (Hsiao and Cheng, 2000; Chiang and Hsia, 2005; Lee and Wu, 2006). However, this system has not been employed in the duck industry in Taiwan until recently (Fig. 12.5). There are two duck breeder farms that use this type of duck house. It is claimed to be profitable according to the results obtained. Several laying farms are expressing interest in introducing PCTA at this time in Taiwan. In a high-efficiency PCTA system, the temperature can be pulled down to 28°C, even when the outside environmental temperature reaches 35°C. The wind-chilling effect is able to make the animals inside feel like the temperature is under 23°C (Mabbett, 2006). Dirty shutters and belt slippage may cause a reduction in cooling efficiency. It is also very important to seal all air leaks around the house (Mabbett, 2006). A recent study was conducted to compare the environmental parameters and laying performance of caged Tsaiya ducks between the PCTA system and a traditional duck house during the hot season (from June to October). The PCTA system had lower room temperature and higher relative humidity (>93% versus < 88%). The results showed that the maximum wind velocity on the anterior (close to water pads), middle and posterior sides (close to extractor fans) in the PCTA duck house were 2.0, 1.9, and 1.7 m/s, respectively. Higher egg production and lower mortality was observed in the Tsaiya ducks raised in the PCTA system compared with the traditional duck house (Lin *et al.*, 2006). In addition, the impact from typhoons was less in the PCTA system compared with the traditional duck house. This is probably because the ducks in the PCTA systems were acclimatized to the noise from the

Fig. 12.5. A water-pad cooling/tunnel air system.

extractor fans. In the same study, it was also noticed that ducks in the PCTA had higher feed intake throughout the study and lower eggshell strength during the first half of the 20-week experimental period (Lin *et al.*, 2006). The lower eggshell strength is believed to be associated with higher egg production (Huang, 2004).

Duck nutrition

The feed

Owing to the unique anatomical structure of the duck bill, pelleted feed is strongly recommended for ducks to prevent waste. The pelleted diet is able to save up to 10% feed and improve feed efficiency by 10% compared with mash diets (Chen and Huang, 1993; Leeson and Summers, 1997). It is recommended that the pellet diameter in the starter diet (0–3 weeks of age) be 0.15 cm (Hu, 2001). The diameter and length of pellet in the finisher diet for meat-type ducks are 0.3 cm and 1 cm, respectively (Chen and Huang, 1993). In addition, ducks are one of the most sensitive domestic animals to aflatoxins (Leeson and Summers, 1997). Monitoring and reduction of aflatoxins in the feed is therefore critically important in hot climate areas, especially during the summer (see Chapter 8).

Nutrient requirements of laying ducks

Ducks are still raised in very traditional ways where agricultural by-products are used extensively as duck feed in many Asian countries (Yimp, 1985; Dong, 2005; Dong *et al.*, 2005). However, there are some countries in which nutrient requirements for ducks have been well established, and most duck farmers are using the complete ration, as in Taiwan (Shen, 1988, 2000, 2002). The recommended nutrient requirements for laying Brown Tsaiya ducks are listed in Table 12.7. The metabolizable energy (ME) and crude protein (CP) is 2730 kcal/kg and 18.7% in laying ducks, respectively, compared with an ME value of 2850–2900 kcal/kg and CP 16.0–17.0% in Leghorn hens (NRC, 1994a). Egg weight may be increased when the CP level is increased, whereas oversized eggs may increase the possibility of inflammation in the oviduct. Owing to the variations in breeds and body weights in ducks raised in different countries, the nutrient requirements and feeding stages may need to be adjusted (Ketaren, 1998). Special attention should be paid to niacin, owing to poor niacin conversion from the amino acid tryptophan in ducks (Scott and Dean, 1991c; Shen, 2001). The niacin requirement is 55–60 mg/kg in Tsaiya, mule duck, and Pekin ducklings at 0–3 weeks of age (Tables 12.7, 12.8; NRC, 1994b), compared with 26–27 mg/kg in chicks at 0–6 weeks of age (NRC, 1994a). Bowed legs with hard skeletal texture and severe crippling are the characteristic symptoms of niacin deficiency (Scott and Dean, 1991c; Shen, 2001). It is important to notice that the essential fatty acid linoleic acid needs to have a minimum level of 1.1% in the feed, which is similar to that for laying hens (Shen, 1988; NRC, 1994a).

Table 12.7. Recommended nutrient requirement for egg-laying Tsaiya ducks as percentage or unit per kg of diet (88% dry matter).

Nutrient	Growing stage			Laying stage
	0–4 wk	4–9 wk	9–14 wk	>14 wk
ME, kcal/kg	2890	2730	2600	2730
Crude protein, %	18.7	15.4	11.5	18.7
Amino acids, %				
Arginine	1.12	0.92	0.79	1.14
Histidine	0.43	0.35	0.32	0.45
Isoleucine	0.66	0.54	0.57	0.80
Leucine	1.31	1.08	1.09	1.55
Lysine	1.10	0.90	0.61	1.00
Methionine + cystine	0.69	0.57	0.52	0.74
Phenylalanine + tyrosine	1.44	1.19	1.04	1.47
Threonine	0.69	0.57	0.49	0.70
Tryptophan	0.24	0.20	0.16	0.22
Valine	0.80	0.66	0.61	0.86
Minerals				
Calcium, %	0.90	0.90	0.90	3.00
Total phosphorus, %	0.66	0.66	0.66	0.72
Available phosphorus, %	0.36	0.36	0.36	0.43
Sodium, %	0.16	0.15	0.15	0.28
Chloride, %	0.14	0.14	0.14	0.12
Potassium, %	0.40	0.40	0.40	0.30
Magnesium, mg	500	500	500	500
Manganese, mg	47	47	47	60
Zinc, mg	62	62	62	72
Iron, mg	96	96	96	72
Copper, mg	12	12	12	10
Iodine, mg	0.48	0.48	0.48	0.48
Selenium, mg	0.15	0.12	0.12	0.12
Vitamins				
Vitamin A, IU	8250	8250	8250	11,250
D, ICU	600	600	600	1200
E, IU	15	15	15	37.5
K, mg	3.0	3.0	3.0	3.0
Thiamin, mg	3.9	3.9	3.9	2.6
Riboflavin, mg	6.0	6.0	6.0	6.5
Pantothenic acid, mg	9.6	9.6	9.6	13.0
Niacin, mg	60	60	60	52
Pyridoxine, mg	2.9	2.9	2.9	2.9
Vitamin B_{12}, mg	0.020	0.020	0.020	0.013

(Continued)

Table 12.7. *continued.*

Nutrient	Growing stage			Laying stage
	0–4 wk	4–9 wk	9–14 wk	>14 wk
Choline, mg	1690	1430	1430	1300
Biotin, mg	0.1	0.1	0.1	0.1
Folic acid, mg	1.3	1.3	1.3	0.65

(Shen, 2000; Lin *et al.*, 2005a)

Nutrient requirements of meat-type ducks

MULE DUCKS. Approximately 80% of duck-meat consumption comes from mule ducks in Taiwan. This gave rise to a relatively thorough determination of the nutrient requirements for mule ducks in this country. The recommended nutrient requirements for mule ducks are listed in Table 12.8. Owing to dramatic differences in feed efficiency between 3 and 10 weeks of age, dividing the finishing period into two stages, as 3–7 and 7–10 weeks of age, is recommended, to save on feed cost (Lin *et al.*, 2004, 2005b). Mule ducks adapt well to ME levels between 2600 and 3050 kcal/kg, at both 0–3 and 3–10 weeks of age. They are able to adjust feed intake to obtain optimal body-weight gain. The ME-level decision is subject to the market price of feedstuffs. In addition, when ME was increased by 150 kcal/kg, the feed efficiency was improved by 4–6% (Shen, 1988, 2001). However, the fat content in the carcass increases with the increase of ME (Shen, 1988; Scott and Dean, 1991b). The recommended CP level at 0–3 and 3–10 weeks of age is 18.7 and 15.4%, respectively. In terms of amino acid requirements, special attention needs to be paid to lysine, methionine and tryptophan, because their values are on the margins of requirements in the maize–soybean-based diet. As mentioned above, niacin needs to be monitored to avoid deficiency.

PEKIN DUCKS. Not many studies on the nutrient requirements of Pekins have been done in tropical and subtropical areas, although Pekins are produced in many Asian countries. Owing to the higher percentage of fat deposition in the Pekin carcass than in mule and Muscovy ducks (Abdelsamie and Farrell, 1985; Leclercq and de Carville, 1985; Scott and Dean, 1991b), the nutrient requirements differ among these meat-type ducks. The nutrient requirements of Pekin ducks are reported by NRC (1994b).

MUSCOVY DUCKS. Most of the nutrient requirements for Muscovy ducks have been developed in France. The Muscovy has a body-weight dimorphism, which means that the body weights of male and female differ significantly. This dimorphism begins at 4–5 weeks of age (Scott and Dean, 1991b). When considering the nutrient requirements of Muscovy ducks, this dimorphism must be taken into consideration (Table 12.9). In practice, male and female Muscovy ducks are raised separately after 7 weeks of age owing to their different

Table 12.8. Recommended nutrient requirement for mule ducks as percentage or unit per kg of diet (88% dry matter).

Nutrient	0–3 weeks	3–10 weeks
ME, kcal/kg	2890	2890
Crude protein, %	18.7	15.4
Amino acids		
Arginine, %	1.12	0.92
Histidine, %	0.27	0.22
Isoleucine, %	0.66	0.54
Leucine, %	1.31	1.08
Lysine, %	1.10	0.90
Methionine + Cystine, %	0.69	0.57
Phenylalanine + Tyrosine, %	1.11	0.92
Glycine + Serine, %	1.22	0.71
Threonine, %	0.68	0.56
Tryptophan, %	0.24	0.20
Valine, %	0.80	0.68
Minerals		
Calcium, %	0.72	0.72
Nonphytate phosphorus, %	0.42	0.36
Sodium, %	0.21	0.21
Chloride, %	0.13	0.13
Potassium, %	0.49	0.49
Magnesium, mg	500	500
Manganese, mg	72	60
Zinc, mg	82	82
Iron, mg	96	96
Copper, mg	12	12
Iodine, mg	0.28	0.28
Selenium, mg	0.15	0.15
Vitamins		
Vitamin A, IU	8250	8250
D, ICU	600	600
E, IU	15	15
K, mg	3	3
Thiamin, mg	3.9	3.9
Riboflavin, mg	6	6
Pantothenic acid, mg	9.6	9.6
Niacin, mg	60	60
Pyridoxine, mg	2.9	2.9
Vitamin B_{12}, mg	0.02	0.02
Choline, mg	1690	1690
Biotin, mg	0.1	0.1
Folic acid, mg	1.3	1.3

(Shen, 2002)

Table 12.9. Nutrient requirements for Muscovy ducks.

Nutrient	0–3 weeks		3–7 weeks		7 weeks–marketing			
	Mixed		Mixed		Male		Female	
ME, kcal/kg	2800	3000	2600	2800	2800	3000	2800	3000
Crude protein, %	17.7	19.0	13.9	14.9	13.0	14.0	12.2	13.0
Amino acids, %								
Arginine	1.03	1.10	0.80	0.86	0.78	0.84	0.65	0.70
Isoleucine	0.80	0.85	0.58	0.62	0.57	0.61	0.47	0.51
Leucine	1.69	1.80	1.24	1.34	1.26	1.36	1.05	1.13
Lysine	0.90	0.96	0.66	0.71	0.65	0.70	0.54	0.58
Methionine	0.38	0.41	0.29	0.31	0.24	0.26	0.23	0.24
Methionine + cystine	0.75	0.80	0.57	0.61	0.50	0.54	0.46	0.50
Phenylalanine + tyrosine	1.57	1.67	1.15	1.23	1.15	1.24	0.96	1.03
Threonine	0.65	0.69	0.48	0.51	0.24	0.26	0.38	0.41
Tryptophan	0.19	0.20	0.14	0.15	0.13	0.14	0.11	0.12
Valine	0.87	0.93	0.64	0.69	0.64	0.69	0.53	0.57
Minerals:								
Calcium, %	0.85	0.90	0.70	0.75	0.65	0.70	0.65	0.70
Total phosphorus, %	0.63	0.65	0.55	0.58	0.49	0.51	0.49	0.51
Sodium, %	0.15	0.16	0.14	0.15	0.15	0.16	0.15	0.16
Chloride, %	0.13	0.14	0.12	0.13	0.13	0.14	0.13	0.14
Manganese, mg	70		60				60	
Zinc, mg	40		30				20	
Iron, mg	40		30				20	

(Continued)

Table 12.9. continued.

Nutrient	0–3 weeks Mixed	3–7 weeks Mixed	7 weeks–marketing Male
Vitamins:			
Vitamin A, IU	8000	8000	4000
D, IU	1000	1000	500
E, mg	20	15	–
K, mg	4	4	–
Thiamin, mg	1	–	–
Riboflavin, mg	4	4	2
Pantothenic acid, mg	5	5	–
Niacin, mg	25	25	–
Pyridoxine, mg	2	–	–
Vitamin B12, mg	0.03	0.01	–
Choline, mg	300	300	–
Biotin, mg	0.1	–	–

(Leclercq et al., 1987)

nutrient requirements (Table 12.9) (Leclercq *et al.*, 1987). Originating from the rainforest of South America, Muscovy ducks and the intergeneric crosses between Muscovies and common ducks are recognized to better adapt to hot environments compared with Pekins (Scott and Dean, 1991a; Chen and Huang, 1993). It can grow well in the completely terrestrial duck house without water for wading and bathing (Chen and Huang, 1993). A diet containing no more than ME of 3000 kcal/kg is recommended in hot climates (Chen and Huang, 1993). The dietary ME/CP ratio of the ducks between 0–3 and 3–7 weeks of age is 158 and 187, respectively. The ratio after 7 weeks of age is 214 and 230 for males and females, respectively (Chen and Huang, 1993). Imbalance of ME and CP may cause slipped wing in Muscovy ducks (Chen and Huang, 1993).

Nutritional measures to alleviate heat stress

Only a few papers discuss the measures to alleviate heat stress in ducks. The main reason is that the major duck-producing countries are located in Asia, where plenty of surface water is available. Ducks can dissipate remarkable amounts of heat in water through their feet and bills (Hagen and Heath, 1980; Scott and Dean, 1991a).

METABOLIZABLE ENERGY AND CRUDE PROTEIN. The hot climate causes a reduction in feed intake, which decreases productive performance. It has been reported that when breeders were raised in the UK, ducks consumed an average of 230 g/day per bird compared with only 170 g/day per bird when raised in tropical areas (Bird, 1985). In a 10-month period, their egg production and egg weight decreased from an excess of 210 eggs and 87 g to less than 160 eggs and 78 g, respectively (Bird, 1985). Reduction of ME from 2900 kcal/kg to 2700 kcal/kg and increases in CP from 18 to 21%, along with increases in micronutrients by 50%, improved egg production to 190 eggs and egg weight to 84 g (Bird, 1985). In Taiwan, duck farmers tend to add some fish meal or full-fat soybean meal, or both, to improve laying performance in duck layers during the summer. In general, increases in nutritional concentration of the feed may improve performance. Chin and Hutagalung (1984) obtained good results in Pekin ducks when highly concentrated diets (ME 3850 kcal, CP 24% at 0–6 weeks; ME 3850 kcal/kg, CP 22% at 7–10 weeks) were fed. Theoretically, diets containing high fat have low heat increment, which can alleviate heat stress on animals (Cronj'e, 2006).

ACID–BASE BALANCE. The dietary electrolyte balance (DEB) effects were investigated in laying Tsaiya ducks by Huang *et al.* (2002). In hot seasons, ducks fed a diet with a DEB of 228 meq/kg produced the highest egg-production rate and feed intake, and best eggshell quality. However, ducks fed diets with a DEB of 15 and 498 meq/kg had the worst performance. A positive correlation was observed between DEB and blood pH, HCO_3^- and base excess values. In cool seasons, birds fed diets with 324 and 403 meq/kg gave the best results. It has also been reported that the best DEB for meat-type ducks is around 200 meq/kg (Chen and Huang, 1993).

VITAMIN C.   It has been known for some time that ascorbic acid is able to improve the productive efficiency of poultry in hot climates. Lai *et al.* (2003) added graded levels of ascorbic acid between 50 and 300 p.p.m. into the diets of mule ducks in June and August. Body-weight gain response due to the addition of ascorbic acid differed between genders. Body-weight gain at 11 weeks of age was not significantly changed by the addition of ascorbic acid in females. However, body-weight gain of males was significantly lower in the 300 p.p.m. group than in the control group. Feed intake and feed efficiency were not changed by the supplementation of ascorbic acid (Lai *et al.*, 2003).

Nutrition and Management of Geese

Geese have been domesticated in many parts of the world for a longer period than any other fowl (about 3000 years or longer) (Sherow, 1975). Geese can consume a considerable portion of their nutrients from high-fibrous feeds and can live on grasses. Therefore, geese can be raised intensively or extensively. The purpose of goose-raising and the variety of breeds affect the choice of production systems (Romanov, 1997). Most goose farms rear White Roman geese under confinement in Taiwan (Yeh, 1995), but many countries still raise geese extensively. In addition, geese characteristically have considerable quantities of body fat, especially subcutaneous fat. In formulating diets for geese, the body fat deposition must be considered, such as the balance of protein and energy to minimize body fat deposition. Generally, geese are fed with a concentrated diet, in addition to small quantities of forage. The growth rate depends on the proportion of concentrate and forage. Because geese are raised under extensive conditions and provided with varying quantities of green forage in most tropical countries, there are few studies on goose nutrition. The information in this section is obtained mainly from recent works in Taiwan.

The feeds

Goose nutrient requirements are dependent upon the climatic conditions and forage supply. Because geese are good grazers, people often assume that they'll never have to feed them anything but grass. Actually, a supplement of grain and/or concentrated feed is recommended under most situations in tropical climate areas where high-quality forage is not available in sufficient quantity. Goose feed may be divided into three main classes by the different amounts of nutrients. Concentrates have the higher amount of digestible nutrients. There are cereal grains, milling by-products, oilseed meals, and by-products from processing of plants or animals. Roughages contain high crude fibre having low nutrient digestibility, such as soilages, hays, leaf vegetables, silages and sagebrush. Supplements are defined as small quantities of materials added in the feed to serve special physiological functions, such as vitamins, minerals, amino acids, antibiotics, enzymes and colouring agents.

Geese have neither teeth nor a crop. The feed is ground in the gizzard to disintegrate the plant fibres. Geese can consume a great deal of roughage but utilize only a part of the fibres. Fibre can accelerate stomach and intestinal peristalsis to speed up the rate of chyme passing through the digestive tract. The increased dietary fibre content shortens the period of chyme staying in the digestive tract, which lowers digestibility of crude fibre. In addition, fibres can reduce the digestibility of other nutrients in the diet, which lowers feed efficiency (Sue et al., 1995, 1996). Although geese can eat a large amount of pasture roughage, we should supply concentrates to meet the nutrient requirements for geese to increase the growth rate and raise efficiency, and also to shorten the raising period. However, it is not appropriate to have too little fibre in the goose diets because fibre can help stimulate digestive tract activity and clean wastes from the stomach and intestines to keep digestive functions normal. There must be an adequate amount of crude fibre in the diet to help geese to grow and develop normally. Studies indicated that the best dietary crude fibre content in meat-type geese was 6–9% (Hsu et al., 2000). In addition, it is also better to provide pelleted feed for geese, to reduce feed waste, and make a mixture of chopped grass and concentrates for geese, to decrease plant stalk waste.

Nutrient requirements of meat-type geese

Energy and protein

Meat-type goose diets are classified as starting diet (0–4 weeks) and growing–finishing diet (4 weeks to marketing). The latter diet may be divided into two stages: growing diet (4–8 weeks) and finishing diet (8 weeks to marketing). In Taiwan, Lu and Hsu (1994a) indicated that metabolizable energy (ME) and protein requirements at the starting stage for White Roman goose and White Chinese goose were 2900 kcal/kg and 19.9%, and 3050 kcal/kg and 21.6%, respectively. The requirements for these two nutrients in White Roman geese are similar to NRC (1994c) recommended ME and protein requirements at the starting stage: 2900 kcal/kg and 20%, respectively. Hsu (2002) recommended ME and protein requirements at the rearing and finishing stages of 2800 kcal/kg and 18.0%, and 2850 kcal/kg and 15%, respectively.

Amino acids

Research has shown that the requirement for methionine was 0.70% (Lu et al., 1992) and that of lysine 1.16% (Lu et al., 1996) in 0–4-week-old goslings. The tryptophan requirement was estimated at 0.23–0.28% when weight gain and feed efficiency were used as criteria, respectively. All of these nutrient requirements are higher than the NRC (1994c) recommendation and other published recommendations (Leeson and Summers, 1997). The requirements for lysine and sulfur amino acids during the rearing and finishing stages for meat-type goose were found to be 0.91 and 0.73%, and 0.72 and 0.58%, respectively (Hsu, 2002).

Minerals and vitamins

Reports about the mineral requirements by Lu and Hsu (1994b) indicated that the calcium and available phosphorus minimum requirements in 0–4-week-old goslings were 0.76% and 0.46%, respectively. These values are similar to those recommended by Leeson and Summers (1997) but higher than the NRC (1994c). It is recommended that the minimum manganese (Mn) requirement in diets of 0–4-week-old goslings is 85 p.p.m. (Hsu and Chen, 1998), which is higher than other reported values (Larbier and Leclercq, 1994; NRC, 1994c; Leeson and Summers, 1997). A study about the vitamin requirement by Lu *et al.* (1996) suggested that the requirement for niacin was 68–98 p.p.m., which was higher than the NRC (1994c) report.

Crude fibre

The marked difference between geese and other poultry is that geese can utilize a great deal of crude fibre in the diet. Crude fibre might affect the availability and metabolism of amino acids, dry matter and energy. A proper amount of crude fibre can stimulate the activities of digestive tract and clean wastes from the gastrointestinal (GI) tract to keep digestive functions normal and to reduce cannibalism. An excess of crude fibre in the diet might accelerate the rate of chyme passing through the digestive tract and damage the villi on the GI tract mucosa, affecting digestion and nutrient absorption. Studies showed that an optimal level of crude fibre in the diet of meat-type geese was 6–9% (Hsu *et al.*, 2000). However, the levels of crude fibre in most commercial diets are lower than 6%. The data suggested that the nutrient requirements for meat-type geese in tropical climates are higher than those in mild climates. Suggested meat-type goose rations are shown in Table 12.10.

Nutrient requirements of breeder geese

Breeder geese are characterized by seasonal egg production. The diet nutrient requirements are therefore divided into starting, rearing, holding, laying and non-laying diets, according to the different reproductive stages. The nutrient requirements for the starting and rearing stages are those used for meat-type geese. Geese are fed a holding diet after 12 weeks of age, until about 2 weeks before the onset of laying. The diet is then changed to the laying diet. The holding diet is supplied during the non-laying stages. Recent studies on the nutrient requirements of breeder geese are scarce and we refer to the recommendations made previously (Larbier and Leclercq, 1994; NRC, 1994c; Leeson and Summers, 1997). We should consider weather conditions in tropical climate areas, to adopt a higher level of nutrients as a standard. To avoid overweight breeder geese, an appropriate amount of crude fibre should be added to the diets. The recommended crude fibre level is 4.0–6.5%. Examples of holding and breeder goose commercial diets are shown in Table 12.11.

Table 12.10. Examples of commercial diet formulations for meat-type geese.

Ingredient	Starting, %	Growing, %	Finishing, %
Maize	51.60	54.55	61.50
Wheat bran	3.00	6.00	6.00
Soybean meal, 44%	29.10	20.20	15.70
Fish meal, 60%	3.00	3.00	–
Lucerne meal, 17%	7.00	11.00	11.00
Tallow	3.50	2.90	3.00
Dicalcium phosphate	1.32	1.02	1.32
Calcium carbonate, pulverized	0.50	0.35	0.43
Salt	0.40	0.40	0.40
L-lysine	–	–	0.10
DL-methionine	0.20	0.20	0.15
Choline chloride, 50%	0.08	0.08	0.10
Premix*	0.03	0.30	0.30
Total	100	100	100
Calculated value, %			
Crude protein	20.60	18.07	15.00
ME, kcal/kg	2886	2800	2854
Calcium	0.83	0.73	0.72
Non-phytate phosphorus	0.45	0.38	0.37

* Mixture of vitamins, minerals and other feed additives.

Feeding and management of goslings during 0–4 weeks of age

Preparation for brooding

The goose house, materials and heaters should be ready in the brooding areas before the arrival of a new batch of goslings. Fresh air should be available at all times, and brooding equipment must provide the required temperature. Also clean waterers should be filled with water and placed in the brooder area to warm the water to room temperature 4 h prior to goslings' arrival. To prevent the stress resulting from transporting and handling, vitamins and electrolytes are usually added to the drinking water.

Selection of goslings

Healthy goslings are the basis of successful breeding. Goslings should have the appropriate birth weight typical to each breed. The umbilicus of goslings after hatch should contract well, adhere without blood and not be wet. The down must grow tightly and shining, without adhering shell membranes or bloodstains. The abdomen should be soft, elastic and contract well when touched. The cloaca must be dried without adhering faeces around and contract well. Eyes should be bright, shining and without secretions around the spherical area. The shanks and webs should be bright, colourful with no oedema. The toes and webs should not be crooked.

Table 12.11. Examples of commercial diet formulations for breeder geese.

Ingredient	Holding, %	Laying, %
Maize	31.90	51.70
Sorghum	20.00	15.00
Wheat bran	15.90	3.00
Soybean meal, 44%	11.30	16.40
Defatted rice bran	14.00	–
Fish meal, 65%	–	2.00
Meat and bone meal	–	2.00
Lucerne meal, 17%	2.00	2.00
Molasses	1.00	1.00
Tallow	–	1.00
Dicalcium phosphate	1.70	1.21
Calcium carbonate, pulverized	1.40	3.80
Salt	0.40	0.40
DL-methionine	0.03	0.10
Choline chloride, 50%	0.07	0.09
Premix*	0.30	0.30
Total	100	100
Calculated value, %		
Crude protein	14.17	15.60
ME, kcal/kg	2600	2849
Calcium	1.13	2.30
Non-phytate phosphorus	0.42	0.40

*Mixture of vitamins, minerals and other feed additive

Brooding temperature

Artificial brooding is always used in industrial goose production. Artificial brooding of goslings includes two methods: floor-brooding and stilted or wire floor-brooding. In floor-brooding the litter materials must be spread about 5 cm on the floor. Litter materials can include fresh rice hulls, wood shavings or rice straw. Goslings should not be placed on a smooth floor, otherwise this will result in spraddled legs. Hover gas brooders are generally used for floor-brooding. A hover brooder with a 1.52-m diameter has a capacity to hold 100–125 goslings. Feeders and waterers could be set inside a circular-brooder area surrounded by adjustable fences. Each m^2 area has a capacity to keep 14–15, 7–8 or 2–3 goslings, during the first, second and third week, respectively. Metal or plastic nets are usually used for stilted floor-brooding, kept warm by hover brooders. This kind of brooding can increase the capacity of each area by 30%. The brooding temperature for goslings is lower than that for chicks. The edge of hover brooders is about 30–32°C, and then gradually decreases by 2.8°C per week. Goslings will spread under the brooder evenly, breathe gently, sleep peacefully and be comfortable if the temperature is appropriate. However, goslings will drink more water, keep away from the heaters, open their wings, lose down and fall asleep if it is too hot. If it is too chilly, the goslings will call disquietingly and weakly, crowd each other below the heaters and curl their bodies. Goslings are kept insufficiently warm if their feet and dorsum are contaminated by faeces at dawn. For this reason, managers should observe the response of the goslings at all times to modulate the temperature inside brooders. The period for brooding goslings is determined by the climatic conditions. In general, it is 1–2 weeks in summer and 2–3 weeks in winter. Brooding should be reduced gradually by eliminating it first during the day and then at night.

Brooding humidity

Wet litter may have adverse effects on goslings' growth and health. Furthermore, high humidity causes accumulation of ammonia, which makes birds more prone to respiratory diseases. High temperature and high humidity in tropical areas reduce the ease with which goslings can release body heat. Therefore brooders should be adequately ventilated, and relative humidity kept at 60–70%. Goslings like to play with water, so it is easy to wet the ground of a brooder with floor-brooding. In the rainy season, the litter materials should be replaced, or stilted or wire floor-brooding with adequate ventilation adopted, to resolve such problems.

Illuminations

Suitable illumination should be provided for goslings to have a sense of security, to prevent their being timid and frightened. It is appropriate to use a fluorescent lamp 20 lux or a light bulb 60 lux in each 3.3 m^2 area. It should not be over-bright, to prevent feather pecking, anus pecking and cannibalism on goslings.

Pasturing and swimming

Setting goslings to graze in a pasture can improve their appetite, metabolism and adaptation to a poor environment that they may be exposed to later in life. According to the climatic conditions, goslings can be put on pasture to let them graze grass freely on the sunny days when the goslings are 7 days old in summer or 10–20 days old in winter. Goslings can be driven to a water pool to swim at about 10–14 days of age. However, it is unsuitable for them to pasture or swim for a long time. In the first instance, goslings should be carried back to brooders after 1 h of pasturing or when the down is moist with water. The period in the pasture and in the water pool can be lengthened gradually for goslings, but it should be ceased in bad weather.

Water and feed

It is very important to confirm that each gosling can access drinking water. Goslings need enough drinking space, with at least two waterers of 4 l each allotted per 100 goslings. Regardless of the type of waterers, goslings should not be allowed to swim inside the waterer. The waterers should be washed every day to keep them clean. Goslings will increase their water intake after 1 week; therefore automatic drinkers should be provided for them. Waterers should be set on the slat to help water drain away easily and to prevent the litter materials becoming wet. Feed should be supplied for goslings after providing water for 2–3 h. In general, goslings should be given feed 24–36 h after hatching, and it would be better not to exceed 48 h. Goslings do not have a well-functioning digestive system just after hatching. Therefore they should be provided with a good-quality feed *ad libitum.* The green forage should be finely chopped. A feeding space of 3.5 cm and 5 cm is needed for each gosling at 2 and 4 weeks of age, respectively.

Bill trimming

Down pecking or feather pecking occurs frequently in goslings raised in a crowded area, fed unbalanced rations or provided with insufficient feeding or drinking space. In order to prevent these pecking phenomena, it is required to perform bill trimming in goslings at an early age.

Feeding and management in the growing period

Gosling feather will gradually grow, and after 3–4 weeks of brooding, the goslings can encounter the environmental change. They can be put outside the house with a wading/drinking channel and moved from stilted floor-feeding to floor-feeding or semi-stilted floor-feeding.

Farmers generally rear geese in pens with concentrates and forages. The amount of feed for geese is regulated by the size of the breeds and the fluctuation of the market price. In general, each goose is fed 200–250 g concentrate per day and fed forage *ad libitum*. If the goose is to be raised as a breeder, the feed should be restricted to about 70% of full feeding, and exercise is needed for the goose. In the large-scale goose farms, meat-type geese have few chances to graze in pastures. Therefore, it is better to have concentrates containing some ingredients with high fibres. This will increase the capacity of the digestive tract to consume more feed during the finishing period and to prevent pica syndrome if the level of crude fibre in the diet exceeds 6%.

Feeding and management in the finishing period

In the large commercial goose farms, meat-type geese are usually raised to 12–13 weeks of age for sale. Geese should be supplied with a high-energy diet, to improve the quality of carcass, about 3–4 weeks before being sold to market. Geese should be divided into small groups according to their body size. In addition, they should be administered anthelmintic to enhance the finishing effect. Geese can be reared in pens and given increasing amounts of concentrate. About 5–6 geese are raised in each 3.3 m² of space. At the same time, they should be reared with low lighting intensity and kept quiet to reduce activities. The finishing method includes *ad libitum* feeding and force feeding. Geese are usually fed finishing diet *ad libitum* in the large commercial farms. During the finishing period, note should be taken of the variations in the faeces from the geese. When the faeces look black, thread-like and hard, it means that the digestive tract of the goose is starting to deposit fat, in particular around the intestinal tract. At this time, the quantity of roughage should be reduced and the amount of concentrates increased, to accelerate the fattening rate and to shorten the finishing period, which will get economic benefits. Geese are marketed when they have a full abdomen and pectorals, thick neck, square bodies and shiny feathers.

Feeding and management of breeding geese

Feeds and feeding

When the goslings are hatched, they have almost no feathers and are covered with down. However, the goslings experience first moulting of down at 3–4 weeks and grow feathers. They are fully feathered by the time the birds are 8–9 weeks of age. The second moulting occurs when birds reach sexual maturity, by the time they are about 7–8 months of age, according to nutritional conditions. Ganders begin to exhibit sexual behaviour after the second moulting. To make ganders thoroughly mature they should be raised apart from females. The most important thing in raising breeding geese is to make their framework well developed but not overweight. When the geese reach 70–80 days old, they are chosen as the breeders. In general, female geese moult earlier than ganders. Geese need more nutrients during the moulting period, so the diet formula and feeding method should be appropriately regulated according to the moulting rate and health condition of the geese. Geese are switched to the maintenance diet, which contains 12–14% CP and 2400–2600 kcal/kg, at about 80 g, after 12 weeks of age. When the second moulting is complete, geese are fed with concentrates at about 120–150 g for each goose daily and given enough roughage, such as pangola grass, napier grass, lucerne and clover, to prevent them becoming overweight. Thus, female geese will delay the laying time to correct the disadvantage of early egg production and excessively small eggs. Furthermore, the onset of the laying period of the geese will be more identical. Geese are switched feed to the breeding diet, which contains 14–16% CP and 2650–2850 kcal/kg ME, and given more concentrates at 1 month before the reproductive season to store nutrients for laying. At this time, geese are fed about 160–200 g concentrates and enough roughage according to its body fitness. At onset of laying, the bird is given 200–250 g concentrates and has free access to roughage.

Selection of breeding geese

Selection of breeding pullets

Not all geese are suitable for breeding, so poor geese should be culled out to maintain the vitality of the breeding stock. Selected breeder geese should look healthy, with perfect and strong legs. For selecting thoroughbred geese, they should have breed characteristics, excellent appearance and graceful postures. Goose breeders should normally be updated every 3–5 years to keep up their reproductive performance. The best laying performance of a female goose is about 2–4 years of age, and will decrease gradually after 5 years of age. The best reproductive performance for ganders is reached between 2 and 3 years of age, and will decrease gradually after 4 years.

Geese with the undesirable characteristics listed below should not be selected as breeders:

1. Drooping shoulders: drooping shoulders occur when geese are raised with imbalanced feeds, improper management and a muddy housing environment. Geese in this condition will have low vitality, poor egg production and hatchability.

2. Abnormal sex organs: the oviduct of female geese may easily droop from expanded abdomens. The sex organ of the gander will be atrophied or drooped if the penis cannot be contracted normally. Feeding mouldy diet to the gander causes its penis to develop unhealthily. To ensure excellent hatchability in goose eggs, ganders should be inspected to determine the healthy condition of the penises before the breeding season.

3. Weak legs: weak legs occur when geese consume an imbalanced diet mainly deficient in Ca, P and vitamin D_3 or because of an inherent genetic disorder.

4. Crooked toes: one or more crooked toes may be inherent or arising from injury.

5. Slipped wing: the wing twists from the elbow and the primary feathers stick out because of lack of nutrients in the diet or a genetic defect.

6. Wry tail.

7. Kinked and bowed necks.

8. Crossed bill.

9. Blindness.

Selection of breeding geese

Breeding geese should be selected 20–30 days before the onset of laying.

1. Breeding gander: the gander should possess breeding characteristics, moderate physiques and strong development, and should have wide and deep chests, full, long body and sharp eyes. In addition, ganders should have long and powerful legs, a wide distance between the two legs, loud calls and bright feathers. Ganders with a well-developed penis should be selected. If AI is performed, the superior-quality semen from ganders should be used for breeding. Breeding ganders should be bigger in size than females, but not overweight. Ganders are unsuitable for breeding if they are overweight.

2. Breeding female goose: the female goose should possess breeding characteristics, moderate body size, meticulous head and neck, and bright and watchful eyes. In addition, they should have moderate weight, because poor productive capacity may arise from overweight.

Ratio of male to female

The ratio of male to female is very important. For lightweight breeds such as Chinese and White Roman goose, each gander can mate with 4–5 females; for heavyweight breeds, such as Toulouse and Embden goose, each gander can mate with 2–3 females. Each Canadian gander always mates with only one female. The fertilization of female geese will be affected if there are too few head of ganders in a group of female geese. However, if there are too many ganders, the feed consumption and raising cost will be increased. In addition, ganders would fight for pecking order, so there should be a proper ratio of male to female geese.

Lighting management

Many research reports indicate that reproductive performances are affected by extreme lighting, but the results are usually dissimilar because of different researchers, breeds and environmental conditions. In Taiwan, long-day lighting (16 h of lighting) will increase the egg production of White Roman female geese temporarily, but also may promote moulting at the same time. There is a tendency for higher egg production when geese are exposed to 12 h of lighting per day (Hsu *et al.*, 1990). Some evidence shows that geese will not start to lay eggs if they are exposed to continued lighting for 16 h before the laying season. However, by gradually decreasing lighting to 8 h per day, geese will start to lay 3 months after the beginning of the decrease in lighting length. Because geese respond to a short photoperiod, a short-day lighting programme can be used to induce the onset of laying in the non-breeding season. A windowless goose house has been developed with a water-pad cooling/tunnel air system (Fig. 12.6), to regulate the laying period of female geese by making use of different photoperiods. In order to get maximum response in the laying period from the lighting programme, 12–13-h or 8–10-h light with 40 lux light intensity is recommended for geese raised in open-sided houses or in lightproof houses, respectively.

Sexing

Sexing by feather colours

Geese are different from chickens because of their unobvious secondary sex characteristics. Only a few breeds, such as Pilgrim goose, can be sexed by feather colours in goslings. Newly hatched Pilgrim males have silver-grey down and light-coloured bills and feet, while females have dark grey coats, bills and feet. At maturity, ganders are white, often with some grey on the

Fig. 12.6. A windowless goose house with a water-pad cooling/tunnel air system.

wings, rump and thigh coverts, while females are grey with some white on the head and neck. Embden and White Roman goslings can also be sexed by down colour in the first week of life, males having lighter grey and wider neck marking than females.

Sexing by voice

The sexually mature gander has a bigger body size and possesses a higher-pitched voice than the female.

Vent sexing

Waterfowls have obvious penises, which are the basis for sexing. Geese can be sexed by squeezing or pinching the sides of the vent to observe the penis as goslings and as adults. In the male gosling (1 day old), a pinkish and screwed penis about 3 mm long should be visible, while there is a flat and light-coloured genital eminence in the female gosling. The penis of an adult gander is screwed, about 3–5 cm inside the cloaca. It should be carefully identified that there is a tiny protuberance in the front of the cloaca, which is the gander penis.

Forced moulting

The goose laying period can be regulated by forced moulting. For instance, geese can be induced to cease egg-laying by forced moulting in the last third of December. After that, geese can be restored to laying in April of the next year. Methods of forced moulting are described below.

Removal of the primary and secondary feathers

Geese are deprived of feed in December for 10 days to reduce egg production, and then the primary and secondary feathers are pulled by manpower or machine to cease egg production. After that, the feed should be gradually increased. Then, the other feathers will fall out gradually. After 2 months, new feathers will grow and the goose will start laying eggs in April of the following year. This method is not usually used because of animal welfare restrictions.

Inappropriate amounts of feed and water deprivation

This method is similar to that used in chickens. First, stop giving geese feed for 1 day, and then stop drinking water for 1–3 days, thereafter, limit feed to 30% of full feed for 10–20 days. The limited feed duration is determined by the fitness of the geese. The geese are then gradually restored to full feed over a 14-day period and then the geese will begin to moult. At this time, it is suggested that the geese should receive a 20 L:4 D lighting programme.

Artificial insemination

In a large breeding farm, geese are usually mated naturally to save manpower and the cost of equipment. In such a mating method, we can get neither individual data nor the data of their ancestors for breeding geese, so we cannot establish the pedigree. Furthermore, sometimes it is difficult for the gander and female to mate, so some females may not mate with ganders, resulting in a decrease in their reproductive efficiency. To solve such problems, AI can be used for geese. The AI technology has been developed in Taiwan. The breeding goose is cage fed and the gander is massaged around the cloaca and the semen collected. Two or three persons should cooperate to collect the semen of larger ganders, which should be trained for 1–2 weeks for semen collection by massage. The semen volume of the gander differs with different frequencies of semen collection, usually about 0.1–1.2 ml are collected. Semen collection is performed about once every 4–5 days. After collecting, semen is injected into the reproductive tract of the female. The amount of semen injected is different depending on differences in concentration and quality. For instance, the amount of superior semen injected immediately after collection is about 0.05 ml. It should be diluted with saline of an equal volume to prevent agglutination, and the amount should be raised to 0.1 ml. However, the amount of inferior semen should be increased if the semen has been collected for over half an hour.

Sanitary Management of Waterfowl

Pathogens spread easily in hot climates owing to high temperature and humidity. Preventing disease is more effective than trying to cure disease. Ignorance of biosecurity could cause pecuniary losses, drug resistance and residual drugs in poultry products. Hence, strict adherence to disease prevention measures and sanitary management practices is crucial.

Disease prevention

The gate into the farm must always be locked, with only qualified visitors allowed inside. Disinfecting tanks must be placed at the entrance and exit. Facilities, equipment and transportation vehicles should be disinfected before moving on to the farm. Clean clothing, caps, masks and disinfected rubber footwear should be worn before qualified people enter the farm. Workers must minimize any visits to other farms. Special attention must be paid to selecting the proper disinfectants and applying them in the appropriate way. Vaccination programmes must be conducted based on the diseases that occur in that area, in line with current government regulations. Professional guidance in vaccination programmes should be obtained from official veterinary services. Furthermore, serological inspections of infectious diseases should be adopted on a regular time schedule. For disease prevention, it is very important to prevent contact between waterfowl and other animals, including pets, pests, other species of poultry and wild birds. Sick and moribund ducks must be transported to disease control centres for diagnosis. Dead birds must be properly buried or incinerated.

Other management practices

In practical operations, the 'all-in, all-out' principle must be practised. It is important to disinfect the house and leave it empty for at least 7–10 days before waterfowl are moved in. Healthy goslings or ducklings must be purchased from sanitary hatcheries free of pathogenic organisms. Waterfowl should be raised in raising fields instead of fields established along streams. The soil and origin of the water should be tested for contamination. Waterfowl from different sources, lines and ages should be raised individually at the proper density. Special attention must be paid to avoid scaring waterfowl with loud noises and undue activity. For waterfowl health, balanced rations are necessary and old feed must be avoided. To maintain productive performance, feed should not be suddenly changed. Antibiotics and feed additives must be used correctly, in accordance with standard practices regarding the safety of waterfowl products. Keeping up-to-date records on disinfectants, antibiotics, drugs and feed additive use is of great importance. Good ventilation and drainage must be maintained in waterfowl farms.

References

Abdelsamie, R.E. and Farrell, D.J. (1985) Carcass composition and carcass characteristics of ducks. In: Farrell, D.J. and Stapleton, P. (eds) *Duck Production: Science and World Practice.* University of New England, Amidale, Australia, pp. 83–101.

Aini, I. (2006) Diseases in family duck farming in south-east Asia. The First INFPD/ FAO Electronic Conference on Family Poultry. http://www.fao.org/ag/aga/AGAP/LPA/Fampol/contents.htm (accessed 4 June 2006).

Andrews, C.V. (1978) Evaluation of production performance of desi dusks reared in confinement. Thesis. Karala Agricultural University, Vellannikkara.

Bird, R.S. (1985) The future of modern duck production, breeds, and husbandry in south-east Asia. In: Farrell, D.J. and Stapleton, P. (eds) *Duck Production and World Practice.* University of New England, Amidale, Australia, pp. 229–237.

Bouverot, P., Hildwein, B. and LeGoff, D. (1974) Evaporative water loss, respiratory pattern, gas exchange and acid-base balance during thermal panting in Pekin ducks exposed to moderate heat. *Respiration Physiology* 21, 255–269.

Brun, J.M., Dubos, F., Richard, M.M., Sellier, N. and Brillard, J.P. (2006) Results of the INRA research concerning duration of fertility of the common duck (*Anas platyrhynchos*) In: *Proceedings of Symposium Scientific Cooperation in Agriculture between Council of Agriculture (Taiwan, R.O.C.) and Institut National de la Recherche Agronomique (France).* Council of Agriculture (Taiwan, R.O.C.) and Institut National de la Recherche Agronomique (France), Taiwan, pp. 91–94.

Bulbule, V.D. (1985) Duck production in India. In: Farrell, D.J. and Stapleton, P. (eds) *Duck Production: Science and World Practice.* University of New England, Armidale, Australia, pp. 351–363.

Burt, D.W. (2002) Applications of biotechnology in the poultry industry. *World's Poultry Science Journal* 58, 5–13.

Burt, D.W., Paton, I.R., Martin, D. and Wilson, B. (1999) The isolation and characterization of microsatellite genetic markers from Pekin duck. In: *Proceedings of the 1st World Waterfowl Conference.* Taiwan and European Branches of World's Poultry Science Association, National Chung-Hsing University, and Taiwan Livestock Research Institute, Taiwan, pp. 70–75.

Chang, T., Hung, T.S., Chang, L.Y., Hsieh, M.H., Cheng, Y.S. and Tai, C. (2000) The duration of fertility in ducks: studies on the polymorphism of the oestrogen receptor gene. In: *Proc 9th CAAAAPS and 23rd Biennial Conference of ASAP*, Australia, p. 125.

Chen, Y.X. (1990) *The Chinese Waterfowl.* Agricultural Publishing, Beijing, China.

Chen, B.J. and Huang, H.H. (1993) Poultry feeds. In: *Animal Husbandry Encyclopedia: Feed*, 2nd edn. Chinese Society of Animal Science, Taipei, Taiwan. pp. 313–412.

Chen, D.T., Lee, S.R., Hu, Y.H., Huang, C.C., Cheng, Y.S., Tai, C., Poivey, J.P. and Rouvier, R. (2003) Genetic trends for laying traits in the Brown Tsaiya (*Anas platyrhynchos*) selected with restricted genetic selection index. *Asian-Australasian Journal of Animal Sciences* 16(12), 1705–1710.

Chen, H., Tan, J.Z., Liu, Y.T. and Song, J.J. (2000) The selection of mule duck's white plumage and its response analysis. *Acta Veterinaria et Zootechnica Sinica* 31(5), 406–410. *Asian-Australasian Journal of Animal Sciences* 16(12), 1705–1710.

Chen, K.W., Tu, Y.J., Tang, Q.P., Zhang, S.J., Gao, Y.S. and Li, H.F. (2005) Dynamic genetic analysis of Chinese indigenous geese breeds. In: *Proceedings of the 3rd*

World Waterfowl Conference. China Branch of World's Poultry Science Association and South China Agricultural University, Guangzhou, China, pp. 59–64.

Cheng, Y.S. (1995) Selection de la cane Tsaiya Brune sur la ponte et la duree de la fertilite croisement avec le canard de Barbarie. PhD Thése, Institut National Polytechnique de Toulouse, France.

Cheng, Y.S., Rouvier, R., Poivey, J.P. and Tai, C. (1995) Genetic parameters of body weight, egg production and shell quality traits in the Brown Tsaiya laying duck. *Genetics Selection Evolution* 27, 459–472.

Cheng, Y.S., Rouvier, R., Hu, Y.H., Tai, J.J.L. and Tai, C. (2003) Breeding and genetics of waterfowl. *World's Poultry Science Journal* 59(4), 509–519.

Cheng, Y.S., Rouvier, R., Poivey, J.P., Huang, H.C., Liu, H.L. and Tai, C. (2005) Selection responses in duration of fertility and its consequences on hatchability in the intergeneric crossbreeding of ducks. *British Poultry Science* 46, 565–571.

Chiang, S.H. and Hsia, L.C. (2005) The effect of wet pad and forced ventilation house on the reproductive performance of boar. *Asian-Australasian Journal of Animal Science* 18(1), 96–101.

Chin, D.T.F. and Hutagalung, R.I. (1984) Energy and protein requirements of Pekin broiler ducks in a tropical environment. In: *Proceedings of the 8th Annual Conference of Malaysian Society of Animal Production*, Malaysia, pp. 60–66.

Chou, K.Y. and Huang, H.H. (1970) *The duck industry in Taiwan. Animal Industry Series No. 8*. Chinese-American Joint Commission on Rural Reconstruction, Taipei, Taiwan.

Council of Agriculture (2007) *AG. Statistics Yearbook*. Taipei, Taiwan.

Crawford, R.D. (1990) Origin and history of poultry species. In: Crawford, R.D. (ed.) *Poultry Breeding and Genetics*. Elsevier, Amsterdam, Netherlands, pp. 1–41.

Cronj'e, P. (2006) Fighting heat stress: diet, gut integrity, and gut health. *Feed nternational* May–June, 11–19.

Denk, A.G., Gautschi, B., Carter, K. and Kempenaers, B. (2004) Seven polymorphic microsatellite loci for paternity assessment in the mallard (*Anas platyrhynchos*). *Molecular Ecology Notes* 4, 506–508.

Dong, N.T.K. (2005) Evaluation of agro-industrial by-product as protein sources for duck production in the Mekong Delta of Vietnam. PhD Thesis, Swedish University of Agricultural Sciences, Uppsala, Sweden.

Dong, N.T.K., Elwingerm K., Lindberg, J.E. and Ogle, R.B. (2005) Effect of replacing soyabean meal with soya waste and fish meal with ensiled shrimp waste on the performance of growing crossbred ducks. *Asian-Australasian Journal of Animal Science* 18(6), 825–834.

Edwards, P. (1985) Duck/fish integrated farming system. In: Farrell, D.J. and Stapleton, P. (eds) *Duck Production: Science and World Practice*. University of New England, Amidale, Australia, pp. 267–291.

Evans, A.J. and Setioko, A.R. (1985) Traditional system of layer flock management in Indonesia. In: Farrell, D.J. and Stapleton, P. (eds) *Duck Production: Science and World Practice*. University of New England, Amidale, Australia, pp. 306–322.

Evans, T. (2004) Significant growth in duck and goose production over the last decade. *Poultry International* 43(11), 38–40.

FAOSTAT (2007) http://faostat.fao.org/site/569/default.aspx (accessed on Jan. 14, 2007).

FFTC (Food and Fertilizer Technology Center for the Asian and Pacific region) (2001) Symbiotic culture of rice plants with aigamo, a cross-breed of wild and domestic ducks. Leaflet for Agriculture No. 2001–1.

Furuno, T. (2001) *The Power of Duck*. Tagari Publications, Tasmania, Australia.

Genet, C., Vignal, A. and Larzul, C. (2003) Isolation and characterization of microsatellite genetic markers from Peking and Muscovy ducks. *British Poultry Science* 44, 794–795.

Gong, D.Q., Zhang, H., Zhang, J., Zhao, X.T., Duan, X.J., Yang, T.G., Chen, Z.Y. and Gao, Y.S. (2005) The genetic relationship analysis among 11 duck populations using microsatellite markers. *Acta Veterinaria et Zootechnica Sinica* 36(12), 1256–1260.

Gonzalez-Esquerra, R. and Leeson, S. (2006) Physiological and metabolic responses of broilers to heat stress – implications for protein and amino acid nutrition. *World's Poultry Science Journal* 62 (2), 282–295.

Guay, P.J. and Mulder, R.A. (2005) Isolation and characterization of microsatellite markers in musk duck (*Biziura lobata*: Aves), and their application to other waterfowl species. *Molecular Ecology Notes* 5, 249–252.

Hagen, A.A. and Heath, J.E. (1976) Metabolic responses of white Pekin duck to ambient temperature. *Poultry Science* 55, 1899–1906.

Hagen, A.A. and Heath, J.E. (1980) Regulation of heat loss in the duck by vasomotion in the bill. *Journal of Thermal Biology* 5, 95–101.

Hanh, D.T. and Tieu, H.V. (1999) Waterfowl production in Vietnam. In: *Proceedings of the 1st World Waterfowl Conference*, Taiwan and European Branches of World's Poultry Science Association, National Chung-Hsing University, and Taiwan Livestock Research Institute, Taiwan, pp. 429–437.

Hester, P.Y., Smith, S.G., Wilson, E.K. and Pierson, F.W. (1981) The effect of prolonged heat stress on adrenal weight, cholesterol and corticosterone in white Pekin ducks. *Poultry Science* 60, 1583–1586.

Hsiao, M.C., Hsu, Y.C., Hu, Y.H., Li, S.H., Lee, S.R. and Liu, S.C. (2006) Isolation and characterization of microsatellite markers in Tsaiya duck *(Anas platyrhynchos)*. In: *Proceeding of Symposium Scientific Cooperation in Agriculture between Council of Agriculture (Taiwan, R.O.C.) and Institut National de la Recherche Agronomique (France)*. Council of Agriculture (Taiwan, R.O.C.) and Institut National de la Recherche Agronomique (France), Taiwan, pp. 221–226.

Hsiao, T.H. and Cheng, M.P. (2000) Efficiency of temperature control in water pad poultry house. *Journal of the Chinese Society of Animal Science* 29 (suppl.), 120.

Hsu, J.C. (2002) *Nutrition and Feed*. San Min Book Shop Company Press, Taipei, Taiwan, pp. 82–83.

Hsu, J.C. and Chen, J.C. (1998) Effect of dietary crude fiber and manganese levels on the growth performance of goslings during 0–4 weeks of age. In: *Proceedings of the 10th European Poultry Conference*, Jerusalem, Israel, p. 103. (Abstract)

Hsu, J.C., Chen, Y.H. and Peh, H.C. (1990) Effects of lighting program on the laying performance of geese. II. Effects of the length of photoperiod on the laying performance of geese. *Journal of Agriculture and Forestry* 39, 27–36.

Hsu, J.C., Chen, L.I. and Yu, B. (2000) Effects of levels of crude fiber on growth performances and intestinal carbohydrates of domestic goslings. *Asian-Australasian Journal of Animal Science* 13(10), 1450–1454.

Hu, Y.H. (2001) Feeding and management of ducks. In: *Animal Husbandry Encyclopedia: Poultry*, 2nd edn. Chinese Society of Animal Science, Taipei, Taiwan, pp. 357–392.

Hu, Y.H. and Tai, C. (1993) Estimation of genetic parameters of yolk color in I-Lan Tsaiya. *Journal of Taiwan Livestock Research* 26, 131–138.

Hu, Y.H., Liao, Y.W., Lee, S.R., Pan, C.M. and Wang, C.T. (1993) Estimates of genetic parameters of green eggs of the Brown Tsaiya. *Journal of the Chinese Society of Animal Science* 22 (suppl.), 39.

Hu, Y.H., Poivey, J.P., Rouvier, R., Wang, C.T. and Tai, C. (1999) Heritabilities and genetic correlations of body weights and feather length in growing Muscovy selected in Taiwan. *British Poultry Science* 40, 605–612.

Hu, Y.H., Poivey, J.P., Rouvier, R., Liu, S.C. and Tai, C. (2004) Heritabilities and genetic correlations of laying performances in Muscovy selected in Taiwan. *British Poultry Science* 45, 180–185.

Huang, C.W., Huang, M.C., Liu, H.L., Cheng, Y.S., Hu, Y.H. and Rouvier, R. (2006) Quantitative trait locus (qtl) detection for duration of fertility in common duck (*Anas platyrhynchos*) bred for mule ducks. In: *Proceeding of Symposium Scientific Cooperation in Agriculture between Council of Agriculture (Taiwan, R.O.C.) and Institut National de la Recherche Agronomique (France)*. Council of Agriculture (Taiwan, R.O.C.) and Institut National de la Recherche Agronomique (France), Taiwan, pp. 227–232.

Huang, H.H. (1973) Management and feeding. In: *The Duck Industry of Taiwan*. Chinese-American Joint Commission on Rural Reconstruction. Animal Industry Series No. 8. Taipei, Taiwan, pp. 22–29.

Huang, H.H. (1985) Selection of white mule ducks. *Journal of the Chinese Society of Animal Science* 14 (3, 4), 111–122.

Huang, J.F. (2004) Comparisons of blood, uterine fluid and shell gland muscosa traits between Tsaiya ducks with high and low eggshell strength. PhD thesis, National Taiwan University, Taipei, Taiwan.

Huang, J.F., Lee, S.R., Lin, T.T., Chen, B.J. and Wang, C.T. (1993) The effect of different environment on growth performance and carcass traits of 3-way mule ducks. *Journal of Taiwan Livestock Research* 26(3), 203–211.

Huang, J.F., Hsiao, M.C., Wang, S.H., Tsao, P.H., Geng, N.T., Hu, Y.H. and Shen, T.F. (2005) Sequencing and expression of the Tsaiya duck eggshell major protein (15 kDa). *Journal of the Chinese Society of Animal Science* 34 (suppl.), 215.

Huang, J.F., Hsiao, M.C., Lin, C.C., Nys, Y., Jiang, Y.N., Gautron, J. and Shen, T.F. (2006) Electrophoresis of Tsaiya duck eggshell proteins and their cross-reaction with hen's anti-ovocleidin 17 antibody. In: *Proceedings of Symposium Scientific Cooperation in Agriculture between Council of Agriculture (Taiwan, R.O.C.) and Institut National de la Recherche Agronomique (France)*. Council of Agriculture (Taiwan, R.O.C.) and Institut National de la Recherche Agronomique (France), Taiwan, pp. 237–240.

Huang, S.C., Shen, T.F. and Chen, B.J. (2002) The effects of dietary electrolyte balance on the blood parameters and laying performance of laying brown Tsaiya ducks. *Journal of the Chinese Society of Animal Science* 31(3), 189–200.

Huang, Y.H., Tu, J.F., Cheng, X.B., Tang, B., Hu, X.X., Liu, Z.L., Feng, J.D., Lou, Y.K., Lin, L., Xu, K., Zhao, Y.L. and Li, N. (2005) Characterization of 35 novel microsatellite DNA markers from the duck (*Anas platyrhynchos*) genome and cross-amplification in other birds. *Genetics Selection Evolution* 37, 455–472.

Huang, Y.H., Zhao, Y.H., Haley, C.S., Hu, S.Q., Hao, J.P., Wu, C.X. and Li, N. (2006) A genetic and cytogenetic map for the duck (*Anas platyrhynchos*). *Genetics* 173, 287–296.

Huque, Q.M.E. (2006) Family poultry production and utilization pattern in Bangladesh. The First INFPD/FAO Electronic Conference on Family Poultry. http://www.fao.org/ag/aga/AGAP/LPA/Fampol/freeco12.htm (accessed 4 June 2006).

Islam, R., Mahanta, J.D., Barua, N. and Zaman, G. (2002) Duck farming in north-eastern India. *World's Poultry Science Journal* 58, 567–572.

Jiang, X.B. (2006) Effects of breeding modes on performance of layer duck. *Animal Production* 42(9), 43–45.

Ketaren, P.P. (1998) Feed and feeding of duck in Indonesia. *Indonesian Agricultural Research and Development Journal* 20(3), 51–57.

Kosba, M.A. (1999) The situation of geese production in Egypt. In: *Proceedings of the 1st World Waterfowl Conference*, Taiwan and European Branches of World's Poultry Science Association, National Chung-Hsing University, Taiwan Livestock Research Institute, Taiwan, pp. 501–502.

Lai, M.K., Huang, J.F., Lin, C.Y. and Lin, R.H. (2003) Effects of ascorbic acid supplementation on growth performance and carcass traits of mule ducks in summer season. *Journal of Taiwan Livestock Research* 36(4), 283–290.

Larbier, M. and Leclercq, B. (1994) *Nutrition and Feeding of Poultry*. Nottingham University Press, England.

Leclercq, B. and de Carville, H. (1985) Growth and body composition of Muscovy ducks. In: Farrell, D.J. and Stapleton, P. (eds) *Duck Production: Science and World Practice*. University of New England, Amidale, Australia, pp. 102–110.

Leclercq, B., Blum, J.C., Sauveur, B. and Stevens, P. (1987) Nutrition of Ducks. In: *Feeding of Non-ruminant Livestock*, Butterworths, London, pp. 102–109. (Translation of INRA (1984) L'Alimentation des animaux monogastriques by Julian Wiseman.)

Lee, S.R. and Kang, C.L. (1997) Improvement on percentage of white plumage in mule ducks. *Journal of Taiwan Livestock Research* 30(3), 293–299.

Lee, S.R. and Wu, K.G. (2006) Utilization of environment-controlled goose house to regulate reproductive season of goose. *Livestock Newsletter* 55, 20–21.

Lee, S.R., Pan, S.T., Shyu, S.T. and Chen, B.J. (1991) Study on the cage-feeding system for laying Tsaiya duck (*Anas platyrhynchos* var. *domestica*). *Journal of Taiwan Livestock Research* 24(2), 177–185.

Lee, S.R., Huang, J.F., Sheu, N.S., Chen, S.Y., Chen, B.J., Jiang, Y.N., Tai, L.J.J. and Tai, C. (1992) Study on the performance of Brown Tsaiya duck (*Anas platyrhynchos* var. *domestica*). *Journal of Taiwan Livestock Research* 25, 35–48.

Lee, S.R., Huang, A.J., Kang, C.L., Wang, C.T. and Tai, C. (1997) Integrated duck and fish production in Taiwan. In: *Proceedings of the 11th European Symposium on Waterfowl*, Nantes, France, pp. 499–503.

Lee, S.R., Shen, T.F. and Jiang, Y.N. (1999) Comparison on blood parameters of Tsaiya duck after eggshell strength selection for one generation. *Journal of the Chinese Society of Animal Science* 28, 19–32.

Leeson, S. and Summers, J.D. (1997) Feeding programs for waterfowl. In: Leeson, S. and Summers, J.D. (eds) *Commercial Poultry Nutrition*, 2nd edn. University Books, Guelph, Ontario, Canada, pp. 324–340.

Lerner, I.M. (1937) Relative growth and hereditary size limitation in the domestic fowl. *Hilgardia* 10, 511–560.

Li, H., Yang, N., Chen, K., Chen, G., Tang, Q., Tu, Y., Yu, Y. and Ma, Y. (2006) Study on molecular genetic diversity of native duck breeds in China. *World's Poultry Science Journal* 62, 603–611.

Li, H.F., Chen, K.W., Zhao, Z.G. and Wu, G.Y. (2005) Screening molecular markers of green eggshell trait of Gaoyou duck using fluorescent AFLP. In: *Proceedings of the 3rd World Waterfowl Conference*. China Branch of World's Poultry Science Association and South China Agricultural University, Guangzhou, China, pp. 183–186.

Li, Z.D., Liu, C.H., Huang, J.S., Sha, J., Wei, H., Shun, M.J., Zhao, C., Jiang, S.H. and Kagami, H. (2002) Preparation of interspecies chicken (*Gallus domesticus*)/duck (*Anas domesticus*) chimeras. *Acta Zoologica Sinica* 48(4), 543–548.

Liao, X.D. and Luo, S.M. (2005) Duck production systems and their safeties in southern China. In: *Proceedings of the 3rd World Waterfowl Conference*, China Branch of

World's Poultry Science Association and South China Agricultural University, China, pp. 27–29.

Lin, C.Y., Huang, J.F., Hu, Y.H., Lin, J.H., Chen, M.Y. and Lee, S.R. (2006) The comparison of laying performance of Tsaiya ducks raised in the duck house with water pad and forced ventilation and traditional duck house in the hot season. *Journal of Taiwan Livestock Research* 39(3), 175–182.

Lin, Y.H., Huang, A.J.F., Lin, Y.A. and Lin, C.Y. (2004) Determination of crude protein and metabolizable energy requirements in growing mule ducks derived from large Kaiya ducks. *Journal of the Chinese Society of Animal Science* 33 (suppl.), 237.

Lin, Y.H., Huang, J.F., Lin, C.Y., Hu, Y.H., Lin, Y.A. and Lin, C.Y. (2005a) Effects of dietary protein and metabolizable energy levels of growth period on the laying performance of Brown Tsaiya ducks. *Journal of the Chinese Society of Animal Science* 34 (Suppl.), 255.

Lin, Y.H., Huang, J.F., Lin, C.Y., Lin, Y.A., Hu, Y.H. and Lin, C.Y. (2005b) Determination of crude protein and metabolizable energy requirements in finishing mule ducks derived from large Kaiya ducks. *Journal of the Chinese Society of Animal Science* 34 (Suppl.), 254.

Liu, S.C., Chen, D.T., Huang, J.F. and Hu, Y.H. (2001) Selection for blue shell egg from Brown Tsaiya I. Establishment of foundation stock. *Journal of Taiwan Livestock Research* 34(3), 265–270.

Liu, S.C., Hsiao, M.C. and Hu, Y.H. (2005) The application of chicken microsatellite markers in Tsaiya ducks. *Journal of the Chinese Society of Animal Science* 34(3), 225–229.

Lu, J.J. and Hsu, A. (1994a) Energy and protein requirements of goslings. *Taiwan Journal of Veterinary Medicine and Animal Husbandry* 64, 13–22.

Lu, J.J. and Hsu, A. (1994b) Calcium and available phosphorus requirements of goslings. *Journal of the Chinese Society of Animal Science* 23, 139–150.

Lu, J.J., Hsu, A., Chen, Y.H. and Chen, C.C. (1992) The choline and methionine requirements of goslings. *Journal of the Chinese Society of Animal Science* 21, 157–168.

Lu, J.J., Chou, S.R., Hong, Y.M., Wu, K.C., Yeh, L.T. and Chen, C.C. (1996) Lysine requirements of White Roman goslings. *Journal of the Chinese Society of Animal Science* 25, 139–148.

Lu, L.Z., Tao, Z.R., Wang, Y.L., Zhang, C., Zhao, A.Z., Shen, J.D. and Xu, J. (2002) Generation selection on Shaoxing green shell line duck. *Journal of ZheJiang Agricultural Science* 4, 200–201.

Maak, S., Neumann, K., von Lengerken, G. and Gattermann, R. (2000) First seven microsatellites developed for the Pekin duck (*Anas platyrhynchos*). *Animal Genetics* 31, 233.

Mabbett, T. (2006) Keep your cool: evaporative cooling alleviates heat stress. *Poultry International* 45(3), 14–16.

Nho, L.T. and Tieu, H.V. (2006) Egg production and economic efficiency of Khaki Campbell ducks reared on locally available feedstuffs in the coastal stretch of the Red River delta. http://www.cipav.org.co/lrrd/lrrd9/1/nho91a.htm. (accessed on 4 June 2006).

Nimruzi, R. (1998) Iran has room for expansion in duck breeding. *World Poultry* 14(4), 24–25.

Nind, L. and Tu, T.D. (1998) Traditional systems of duck farming and duck egg incubation in south Vietnam. *World's Poultry Science Journal* 54, 375–384.

NRC (National Research Council) (1994a) Nutrient requirement of chickens. In: *Nutrient Requirements of Poultry,* 9th edn. National Academy Press, Washington, D.C., pp. 19–34.

NRC (National Research Council) (1994b) Nutrient requirement of ducks. In: *Nutrient Requirements of Poultry,* 9th edn. National Academy Press, Washington, D.C., pp. 42–43.

NRC. (1994c) Nutrient requirement of geese. In: *Nutrient Requirements of Poultry,* 9th edn. National Academy Press, Washington, D.C., pp. 40–41.

Paulus, K.B. and Tiedemann, R. (2003) Ten polymorphic autosomal microsatellite loci for the Eider duck *Somateria mollissima* and their cross-species applicability among waterfowl species (Anatidae). *Molecular Ecology Notes* 3, 250–252.

Pingel, H. (1999) Influence of breeding and management on the efficiency of duck production. *Lohmann Information* 22, 7–13.

Poivey, J.P., Cheng, Y.S., Rouvier, R., Tai, C., Wang, C.T. and Liu, H.L. (2001) Genetic parameters of reproductive traits in Brown Tsaiya ducks artificially inseminated with semen from Muscovy drakes. *Poultry Science* 80, 703–709.

Qiu, X.P. (1988) *Poultry Breeds in China.* Shanghai Scientific and Technical Publishers, Shanghai, China.

Qiu, X.P. and Qiu, M.H. (2005) Goose breeds in China. In: *Proceedings of the 3rd World Waterfowl Conference.* China Branch of World's Poultry Science Association and South China Agricultural University, Guangzhou, China, pp. 153–155.

Rajasekaran, B. (2007) An indigenous duck–fish production system in South India: impact on food and nutritional security. Saginaw, Michigan: Consortium for International Earth Science Information Network (CIESIN). Draft. http://www.ciesin.org/docs/004–200/004–200.html (accessed on 2 February 2007).

Rodenburg, T.B., Bracke, M.B.M., Berk, J., Cooper, J., Faure, J.M., Guemene, D., Guy, G., Harlander, A., Jones, T., Knierim, U., Kuhnt, K., Pingel, H., Reiter, K., Serviere, J. and Ruis, M.A.W. (2005) Welfare of ducks in European duck husbandry systems. *World's Poultry Science Journal* 61, 633–646.

Romanov, M.N. (1997) Impact of genotype, nutrition and management systems on goose production efficiency. *Proceedings of 11th European Symposium on Waterfowl,* Nantes (France), pp. 33–42.

Rosinski, A., Rouvier, R., Guy, G., Rousselot-Pailley, D. and Bielinska, H. (1996) Possibilities of increasing reproductive performance and meat production in geese. *Proceedings of the 20th World's Poultry Congress.* India Branch of World's Poultry Science Association, New Delhi, India, pp. 725–734.

Rouvier, R., Babilé, R., Salzmann, F., Auvergne, A. and Poujardieu, B. (1987) Repetabilite de la fertilite des canes Rouen et Pekin (*Anas platyrynchos*) en croisement interspecifique avec le Barbarie (*Cairina moschata*) par insemination artificielle. *Genetics Selection Evolution* 19(1), 103–112.

Scott, M.L. and Dean, W.F. (1991a) Housing and management. In: *Nutrition and Management of Ducks.* M. L. Scott of Ithaca, New York, pp. 12–34.

Scott, M.L. and Dean, W.F. (1991b) Energy, protein, and amino acid requirements of ducks. In: *Nutrition and Management of Ducks.* M. L. Scott of Ithaca, New York, pp. 55–88.

Scott, M.L. and Dean, W.F. (1991c) Vitamin allowances for ducks. In: *Nutrition and Management of Ducks.* M. L. Scott of Ithaca, New York, pp. 89–107.

Setioko, A.R., Hetzel, D.J.S. and Evans, A.J. (1985a) Duck production in Indonesia. In: Farrell, D.J. and Stapleton, P. (edn) *Duck Production: Science and World Practice.* University of New England, Amidale, Australia, pp. 418–427.

Setioko, A.R., Evans, A.J. and Raharjo, Y.C. (1985b) Productivity of herded ducks in West Java. *Agricultural System* 16, 1–5.

Shafiuddin, A. (1985) Duck production in Bangladesh. In: Farrell, D.J. and Stapleton, P. (eds) *Duck Production: Science and World Practice*. University of New England, Armidale, Australia, pp. 342–350.

Shen, T.F. (1988) *Manual of Nutrient Requirement of Ducks*. Department of Animal Science, National Taiwan University, Taipei, Taiwan.

Shen, T.F. (2000) Nutrient requirements of egg-laying ducks. *Asian-Australasian Journal of Animal Sciences* 13 (special issue), 113–120.

Shen, T.F. (2001) Nutrient requirements of poultry. In: *Animal Husbandry: Poultry*, 2nd edn. Chinese Society of Animal Science, Taipei, Taiwan. pp. 165–196.

Shen, T.F. (2002) Review on nutrient requirements of mule and Tsaiya ducks. *Scientific Agriculture* 50(1, 2), 129–134.

Sherow, D. (1975) *Successful Duck and Goose Raising*. Stromberg Publishing Company, Pine River, Minnesota.

Singh, N.J., Sahoo, G., Kanungo, H.K. and Nayak, J.B. (1991) Performance of Khaki Campbell ducks in hot and humid climate. *Indian Journal of Animal Production and Management* 7(4), 230–232.

Stai, S.M. and Hughes, C.R. (2003) Characterization of microsatellite loci in wild and domestic Muscovy ducks (*Cairia moschata*). *Animal Genetics* 34, 384–389.

Sue, C.J., Hsu, J.C. and Yu, B. (1995) Effects of dietary fiber levels on nutrient utilization of diet in goslings. I. Utilization of amino acids. *Journal of the Chinese Society of Animal Science* 24, 19–30.

Sue, C.J., Hsu, J.C. and Yu, B. (1996) Effects of dietary fiber levels on nutrient utilization of diet in goslings. II. Utilization of dry matter, crude fat, gross energy, neutral detergent fiber and acid detergent fiber. *Journal of the Chinese Society of Animal Science* 25, 129–138.

Surendranathan, K.P. and Nair, S.G. (1971) Environmental influences on certain physiological factors in ducks (*Anas platyrhynchos domesticus*). *The Indian Veterinary Journal* 48(6), 587–592.

Tai, C. (1985) Duck breeding and artificial insemination in Taiwan. In: Farrell, D.J. and Stapleton, P. (eds) *Duck Production: Science and World Practice*. University of New England, Armidale, Australia, pp. 193–203.

Tai, C., Tai, L.J.J. and Huang, H.H. (1985) Estimation of genetic parameters for egg quality characters in laying ducks. *Journal of the Chinese Society of Animal Science* 14, 105–110.

Tai, C., Rouvier, R. and Poivey, J.P. (1989) Genetic parameters of some growth and egg production traits in laying Brown Tsaiya (*Anas platyrynchos*). *Genetics Selection Evolution* 21, 377–384.

Tai, C., Wang, C.T. and Huang, C.C. (1999) Production systems and economic characters in waterfowl. In: *Proceedings of the 1st World Waterfowl Conference*. Taiwan and European Branches of World's Poultry Science Association, National Chung-Hsing University, Taiwan Livestock Research Institute, Taiwan, pp. 19–31.

Tai, J.J.L., Liu, J.F., Chen, L.R., Chang, M.C., Huang, M.C. and Tai, C. (2004) Studies on the transgenic poultry. In: *Proceedings of 2004 Asian Poultry Science Symposium*, Taipei, Taiwan, pp. 35–45.

Takahashi, H., Satoh, M., Minezawa, M., Purwadaria, T. and Prasetyo, H. (2001) Characterization of duck microsatellite repeat sequences. *Japan Agricultural Research Quarterly* 35(4), 217–219.

Tan, J.Z. (2002) Practicing of the selection of the parental stock of mule duck. *China Poultry* 24(18), 6–8.

Thongwittaya, N. (1999) Waterfowl production in Thailand. In: *Proceedings of the 1st World Waterfowl Conference*. Taiwan and European Branches of World's Poultry

Science Association, National Chung-Hsing University, Taiwan Livestock Research Institute, Taiwan, pp. 438–443.

Wang, B.W., Pan, Q.J., Zhu, X.C., Zhang, T.R., Ge, W.H. and Lian, A.L. (2002) Summary on breeding and selecting of Wulong Goose. *China Poultry* 24, 28–32.

Wang, C.K., Li, A., Wang, G.Y. and Zheng, N.Z. (2003) Application of synthetic index selection in Muscovy duck breeding. *Journal of Fujian Agriculture and Forestry University (Natural Science Edition)* 32(2), 234–236.

Wang, G.Y., Li, A., Wang, C.K. and Zhang, D.Y. (2005) Improved duck and related production techniques. In: *Proceedings of the 3rd World Waterfowl Conference*. China Branch of World's Poultry Science Association and South China Agricultural University, Guangzhou, China, pp. 78–82.

Wang, L., Wang, B.W. and Yang, Z.G. (2005) Talk about new ways of goose breeding. *Chinese Poultry Science* 6, 39–42.

Wezyk, S. (1999) Current problems of waterfowl genetics and breeding. In: *Proceedings of the 1st World Waterfowl Conference*. Taiwan and European Branches of World's Poultry Science Association, National Chung-Hsing University, Taiwan Livestock Research Institute, Taiwan, pp. 50–62.

Windhorst, H.W. (2006) Changes in poultry production and trade worldwide. *World's Poultry Science Journal* 62, 585–602.

Wu, W., Xu, R., Gao, G. and Sun, Y. (2005) Breeding of high egg production line of Jilin white goose. In: *Proceedings of the 3rd World Waterfowl Conference*. China Branch of World's Poultry Science Association and South China Agricultural University, Guangzhou, China, pp. 192–194.

Wu, Y., Hou, S.S., Liu, X.L. and Huang, W. (2005) Analysis of genetic diversity of microsatellite on Peking duck and Yingtaogu duck. In: *Proceedings of the 3rd World Waterfowl Conference*. China Branch of World's Poultry Science Association and South China Agricultural University, Guangzhou, China, pp. 230–234.

Yeh, L.T. (1995) The current status of goose breeding in Taiwan. *Taiwan Agriculture* 31, 73–77.

Yeh, L.T, Chiou, T.S., Wang, S.D., Wu, K.C., Chang, H.L. and Cheng, Y.H. (1999) Goose breeding: progress and prospect for improvement of egg production in White Roman geese. In: *Proceedings of Symposium Scientific Cooperation in Agriculture between Institut National de la Recherche Agronomique (France) and Council of Agriculture (Taiwan, R.O.C.)*. Institut National de la Recherche Agronomique (France) and Council of Agriculture (Taiwan, R.O.C.), Toulouse, France, pp. 117–122.

Yen, C.F., Jiang, Y.N., Shen, T.F., Wong, I.M., Chen, C.C., Chen, K.C., Chang, W.C., Tsao, Y.K. and Ding, S.T. (2005) Cloning and expression of the genes associated with lipid metabolism in Tsaiya duck. *Poultry Science* 84, 67–74.

Yi, J. and Zhou, Y.P. (1980) The Pekin duck in China. *World Animal Review* 34, 11–14.

Yimp, C.T. (1985) Duck production in peninsular Malaysia. In: Farrell, D.J. and Stapleton, P. (eds) *Duck Production: Science and World Practice*. University of New England, Amidale, Australia, pp. 399–417.

Index

acclimatization 53, 249–251, 267, 281
acetylsalicylic acid 150–151, 304
acid-base balance 49, 153, 231, 282, 353
ACTH (adrenocorticotrophic hormone) 60–62
adrenal glands: hypothalamic-pituitary-adrenal axis 60–62
aflatoxins *see under* mycotoxins
Africa 6–7
air inlets
    characteristics 04–106
    types 06–108
Algeria 6, 7, 178
alkalosis, respiratory 55–56, 231
allylisothiocyanate 76–177
altitude 321, 324
amino acids 35, 137(fig), 138–142, 230–231, 355
    effect of temperature on carcass content 236–237
ammonia 171, 215, 244, 276, 285, 359
antibiotics 150, 303–304
anticoccidial drugs 151, 241
antifungal agents 215
antioxidants 234, 252, 266(tab), 307(tab)
arginine 139–140
arginine vasotocin (AVT) 58–59
aprinocid 151
ascorbic acid *see* vitamins: vitamin C
Asia 7–9
aspirin 150–151, 304

attic inlet ventilation 107, 108(fig)
Australorp 15

bacteria: for dietary protein 177
bambara groundnut meal 183
Bangladesh 8
barley 161–164
    improvement of nutritional value 162–164
beaks: trimming 243, 268
Bedouin (Sinai) breed 15
behaviour 51–53
biosecurity 315–316
biotin 302
blood
    electrolytes 56–57
    flow rate and pressure 57–58
    lipids 59
    pH 55–56
body weight 35, 41, 85, 139, 151, 346, 349
    gain 84(tab), 148(tab), 228
bones 147
bran, rice 164–166
Brazil 7
breadfruit meal 180
breast blister 244
breeders
    disease control and prevention 313–316

breeders *continued*
  feeding *see under* feedstuffs and feeding
  geese *see under* geese
  housing *see under* housing
broilers
  benefits of tunnel ventilation 83–86
  breeding
    featherless birds 37, 43
    see also under breeders
  chick viability 296
  dewinging 247
  energy metabolism 229–230
  feather coverage and heat tolerance
    discover of scaleless *(Sc)* gene 37
    effects of naked neck *(Na)* gene 36–37, 38
    featherless *vs.* feathered 38–39
    featherless *vs.* naked-neck 37–38
    physiological consequences 36
  feeding *see under* feedstuffs and feeding
  growth rate
    effects of heat stress 31–32, 228–229
    and feed intake 35–36
    genetic association with heat tolerance 32–34
    selection in hot conditions 34–35
  management
    beak trimming 243
    coccidiostats 151
    dietary aspirin 151
    drug administration 241–242
    housing *see under* housing
    litter 244
    vaccination 242
  meat quality and yield
    effects of heat stress 31
    superiority of featherless birds 39–41
  optimum temperature range 133–135
  pullets *see* pullets, broiler
  seasonal effects on performance 235–236
  temperature and body composition 236–237
  water consumption 247–249
bruising 201
Brunei 9
buffalo gourd meal 182
bursal disease, infectious 205

caging 283–284, 312–313
calcium 145–147, 275, 302–303, 356
  binding by sesame meal 175–176
Cambodia 9, 333
cannibalism 16, 88, 243, 356
capital: constraint to development of poultry industry 10
carbon dioxide 55, 145, 244, 252
carbonic anhydrase 279
cardiovascular system
  bruising 201
  heat exchange mechanisms 54
  responses to heat stress 57–58
cassava root meal 177–178
cathecholamines 61, 62
ceilings, dropped 94
chaperons, molecular 66
China 1, 7, 333–334
chlorination 248
chlorogenic acid 174
choline 167, 234
circulation fans 98–100
citrinin 207–208
climate: constraint to development of poultry industry 11
coccidiostats 151
coconut meal 169–170
complement: effect of mycotoxins 204
composition, body 236–237
conditioning, early: to heat 250
conduction 82
consumption, feed *see under* feeds
consumption: of eggs and poultry meat
  Africa 6–7
  Asia 7
    Indian subcontinent 8
  Latin America 7
  of meat by region 10(tab)
  Middle East 7–8
  selected countries 2005 4(tab)
contamination: nests and eggs 310, 311
convection 82
cooling, evaporative 51–52, 82, 83
  part of ventilation system *see under* ventilation
copra 169–170
corticosterone 60–62

corticotrophin-releasing factor (CRF) 60
cottonseed meal 170–172
cross-ventilation 106
Cucurbita foetidisima 182
cucurbitacins 182
curtains, side-wall 87–88
cyanide 178
cyclopiazonic acid 206–207
cyclopropene fatty acids 171

dates and by-products 178–179
deflectors, air 110, 111(fig)
deoxynivalenol 212–213
dewinging 247
digestive tract 252
diseases
    enhanced by mycotoxins 205
    importance of diagnosis and control 11
    prevention and control 313–316
disinfectants 246–247
domestication, poultry: history 1–2
drinking systems 85, 269, 279–281
ducks
    biotechnology 339–340
    breeds 333, 334–335
    housing and management
      backyard 341
      constructed swimming/drinking channel 343
      duck-fish integrated system 342–343
      duck-rice integrated system 343–345
      herding 341–342
      terrestrial duck raising 345–346
    meat production 331–332
    nutrition 347–354
      alleviation of heat stress 353–354
    performance under heat stress 340–341
    selection 335–338
dw (dwarf) gene 18–19, 21

education: constraint to development of poultry industry 11
eggs
    aflatoxin residues 202–203

consumption *see under* consumption, eggs and poultry meat
contamination 310, 311
discoloration by feedstuffs 171, 172, 174
effect of protein intake on quantity and mass 138
effect of temperature on quality 281–282
elevated n-3 fatty acids with dietary linseed 176
floor- and slat-laid 309–310
gathering 308–309
hatchability 295–297, 309
non-hen egg production 332(tab)
prewarming 320
quality improvement measures 145–148, 276
shells
    adverse effects of saline water 278–279
    formation 55
storage 316–317
Egypt, Ancient 1, 295
electrolytes
    dietary 139–140, 149–150, 231–233
    effect on water consumption 248
    plasma 56–57
energy
    and body temperature 49–51
    intake and loss 49, 229–230
    requirements 135–137, 271–275, 299–301
epinephrine 62
Ethacal 276
ethoxyquin 166
evaporative cooling *see under* cooling, pad systems, ventilation
exhaust fans *see under* fans

F (frizzle) gene 17–18, 19, 20, 21
fans
    circulation 98–100
    exhaust
      calculation of capacity 101–104
      performance factors 117–119, 121
      positioning 106–107
      safety guards 121
      types 117, 119(fig), 120(fig)

fans *continued*
  tunnel
    capacity 109–110
    placement 113, 115
fats 135–137, 230, 274
  protect against aflatoxicosis
    200–201
fatty acids 171, 176
feathers
  and heat tolerance 36–39, 54–55
feedstuffs and feeding
  breeders: meat-type *vs.* egg-type
    297–298
  breeders: nutritional experiments
    298
  broiler breeder hens 300–304,
    306–308
    energy and protein requirements
      299–300
    feed restriction 298–299, 311–312
  broiler breeder males 304–306,
    307–308
  broilers
    amino acid balance in feeds
      139–141, 230–231
    cassava 178
    cooked bread fruit meal 180
    dietary fat 135–137, 230
    dietary protein 137–138, 230–231
    dual feeding 240
    early feed restriction 240–241
    effects of triticale in diet 168–169
    featherless broilers 41
    feed withdrawal 240
    feeding programmes 238–240
    minerals 148(tab), 149, 231–235
    optimum dietary electrolyte
      balance 150
    palm kernel meal 179
    sunflower seeds 174
    vitamin supplementation
      142–143, 144–145, 233–235
    wet *vs.* dry mashes 245–246
  constraint to development of
    poultry industry 10
  ducks
    laying ducks 347–349
    meat-type ducks 349–353
  efficiency 136(fig), 138
  geese 354–355
    breeder birds 356, 361
    growing and finishing periods
      360–361
    meat-type birds 355–356, 357(tab)
  intake 32, 35–36, 53–54
    effect of heat 134–135
  laying hens
    ascorbic acid supplementation
      143
    calcium 275
    cottonseed meal 172
    dietary aspirin 151
    dietary fat 135, 274
    dietary protein 138
    dried poultry waste in diets 181
    effects of sorghum tannins 168
    effects of triticale 169
    fats 274
    ipil ipil leaf meal 181
    minerals and vitamins 275–276
    palm kernel meal 179
    proteins 271–272
    recommendations 275, 276–278
    sunflower seeds 174–175
    vitamin A supplementation
      143–144
  non-nutrient additives 150–152
  nutrient content *see under* nutrition
  physical properties 245–246, 303
  pullets, replacement
    effect of temperature on
      requirements 263–266
    protein and amino acids 264
    recommendations 265–266
  recommendations 185(tab),
    238(tab), 265–266, 275, 276–278,
    305–308
  salmonellosis control 315
  types
    bambara groundnut meal 183
    barley 161–164
    bread fruit meal 180
    buffalo gourd meal 182
    cassava root meal 177–178
    coconut meal (copra) 169–170
    cottonseed meal 170–172
    dates and by-products 178–179
    dried poultry waste 181–182
    groundnut meal 172–173
    guar meal 182–183
    ipil-ipil leaf meal 180–181
    jojoba meal 183–184

linseed meal 176
millet 164
mungbean 180
mustard seed meal 176–177
palm kernel meal 179
rice by-products 164–166
safflower meal 175
salseed 181
sesame meal 175–176
single-cell protein 177
sorghum 166–168
sunflower seeds 173–175
triticale 168–169
fermentation, microbial 177
fertility 151, 202, 210, 212, 295, 296, 297, 301, 302, 305(tab), 313, 323, 337
fibre 356
flooring 52, 269, 311–312
flunixin 152
flutter, gular 55
fogging systems 122–124
folic acid 145
formaldehyde 311, 324
free fatty acids 59
fumonisins 200, 208–209

gases, blood 55–56
geese 332, 333(tab)
  breeder birds
    artificial insemination 365
    forced moulting 364–365
    lighting 363
    male to female ratio 363
    selection 361–363
    sexing 363–364
  breeds 333–334
  feeds and feeding 354–355
    breeder birds 356–357, 361
    meat-type birds 355–356, 357(tab)
  genetics 338–339
  goslings 357–360
  growing and finishing periods 360–361
  housing
    preparation for brooding 357
    temperature, humidity, lighting 358–359, 363
genes
  dwarf *(dw)* 18–19, 21
  frizzle *(F)* 17–18, 19, 20, 21

genetic association between heat tolerance and growth rate 32–34
heat shock proteins 67–68
interactions 19–20
naked neck *(Na)* 16–17, 19–20, 36–37, 38
scaleless *(Sc)* 37
slow feathering *(K)* 19, 20
genestein 152
β-glucans 162
glucomannan 216
gossypol 170–171
Greece
  history of poultry domestication 1
groundnut meal 172–173
growth hormone 60
growth rate: in broilers see *under* broilers
guar meal 182–183
guinea fowl 333(tab)
gular flutter 55

hatchability 295–297, 309
  and altitude 321
hatcheries
  chick processing and delivery 322–323
  design 317
  hygiene 317–318
  incubation time 320
  incubator conditions 318–319
  incubator problems 319
  native Egyptian 295
  prewarming of eggs 320
  recommendations 323–324
  storage of hatching eggs 316–317
  temperature requirements 295, 296
  water supply 320–321
hatching
  effects of heat 295–297
heart see cardiovascular system
heat, latent see latent heat
heat, sensible see sensible heat
heat exchange mechanisms 54
heat increment 138
heat shock proteins
  gene transcription 66–67
  high degree of genetic conservation 64
  intracellular interactions 65–66

heat shock proteins *continued*
   role in tolerance 65, 67, 68–69
   studies in chickens 67–69
heat stress
   alleviation in ducks 353–354
   behavioural responses 51–53
   and body temperature 49–51
   effects on hatching 295–297
   effects on performance 31–32, 35–36, 133–134
   heat shock proteins *see* heat shock proteins
   hormonal responses
     cathecholamines 61, 62
     growth hormone 60
     hypothalamic-pituitary-adrenal axis 60–62
     melatonin 62–63
     neurohypophyseal hormones 58–60
     reproductive hormones 63
     thyroid 59, 63–64
   and nutrition (reviews) 133
   physiological responses
     acclimatization *see* acclimatization cardiovascular system 57–58
     consumption of feed and water 53–54
     feather cover 54–55
     heat exchange mechanisms 54
     plasma electrolytes 56–57
     respiration and blood pH 55–56
   resistance *see* tolerance: to heat
hepatotoxicity 201
hormones
   cathecholamines 61, 62
   growth hormone 60
   hypothalamic-pituitary-adrenal axis 60–62
   melatonin 62–63
   neurohypophyseal (posterior pituitary) 58–60
   reproductive 63
   thyroid 59, 63–64
housing
   breeders
     cage *vs*, floor 312–313
     egg gathering 308–310
     egg hygiene 311
     nests 310–311
     water-cooled perches 313
     water supply 311–312
     see also roofs; ventilation
   broiler density 243–244
   cage space and shape 283–284
   cooling devices 244–245
   disinfection 246–247
   drinking systems 85, 247–249, 269, 279–281, 311–312
   ducks *see under* ducks
   flooring 52, 269
   geese
     preparation for brooding 357
     temperature, humidity, lighting 358–359, 363
   hatcheries *see* hatcheries
   lighting programmes 251–252, 268, 283, 363
   open-sided 269
   operant control in determination of optimum environment 52–53
   superiority of featherless birds at high stocking densities 40–41
   waterfowl
     pad cooling/tunnel air systems 346–347
humidity 83, 235
   and cooling air velocity 85–86
   effects on egg composition 145–146
hydrocortisone 61
hydrogen sulfide 244
hydrogen peroxide 311
hyperthermia 50
hypothalamic-pituitary-adrenal axis 60–62
hypothermia 50

immune system
   benefits of feed restriction 241
   effect of mycotoxins 203–205, 210, 212
   effects of corticosteroids and ACTH 61–62
   effects of vitamin C 142–143
increment, heat 138
incubators *see* hatcheries
India 1, 9
Indonesia 9, 164, 165, 180, 198, 311, 331, 341, 342

infectious bursal disease 142, 203(tab), 204, 205, 241
inlets, air *see* air inlets
inlets, tunnel 115–117
insulation
    roof 94–97
    wall 97
International Association of Poultry Instructors and Investigators 2
ipil-ipil leaf meal 180–181
Iraq 175, 178, 283

Japanese quail 64
jojoba meal 183–184
jungle fowl 1, 68

K (slow feathering) gene 19, 20
kani (broken rice grains) 166
kidney toxicity 208, 209, 210–211
koa haole 180–181

Laos 9
Lasalocid 242
latent heat 82
Latin America
    status of poultry industry 7
laying hens
    acclimatization 281
    effect of temperature on feed consumption 269–270
    energy and protein requirements 271–275
    feeding *see under* feedstuffs and feeding
    management
      cage space and shape 283–284
      lighting 283
      optimum temperature range 134
      parasites 284–285
      prevention of bone breakage 147
      water quantity and quality 278–281
      wet droppings 285, 286(fig)
      recommendations 287
Lebanon 8, 198
light
    for goslings 359

lighting programmes 251–252, 268, 283
    photostimulation of broiler pullets 300
    sunlight 91–92
linamarin 178
linseed meal 176
lipids, blood 59
litter: management 244, 312
liver toxicity *see* hepatotoxicity
Luecaena leucocephala 180–181
luteinizing hormone 63
lysine 135, 137(fig), 138–140, 355

Maitland, R.T 2
maize 151, 161, 164, 166, 169, 198–199, 208, 217, 239
Malaysia 9, 18, 24, 331, 333, 334
malvalic acid 171
manganese 356
Manual and Standards Book for the Poultry Amateur (R.T. Maitland) 2
meat
    aflatoxin residues 202–203
    consumption *see under* consumption, eggs and poultry meat
    effects of heat stress on quality and composition 236–237
    elevated n-3 fatty acids with dietary linseed 176
    see also production
melatonin 62–63
mesotocin 59–60
metabolism: zone of minimal metabolism (ZMM) 50–51
methionine 139, 167, 355
Mexico 7
microbes
    for dietary protein 177
Middle East 7–8
millet 164
mimosine 180–181
minerals 231, 234, 248, 265, 275, 321, 356
monensin 151, 304
Morocco 7
moulting, forced 364–365
Mucuna spp. 184
mungbean 180
muscle, skeletal 250

mustard seed meal 176–177
mycotoxins
    aflatoxins
        in broilers 200–202
        correlation with feed zinc content 206
        in cottonseed meal 171–172
        definition and occurrence 199–200
        effect on resistance and immunity 203–205
        extraction and detoxification 215–216
        in groundnut meal 173
        interactions with vitamins 205–206
        in layers and breeders 202
        residues in eggs and meat 202–203
    citrinin 207–208
    control 214–217
    cyclopiazonic acid 206–207
    detection 213–214
    feed contamination levels 198–199
    fumonisins 200, 208–209
    ochratoxins 204, 209–210
    oosporein 210–211
    sterigmatocystin 207
    T-2 toxin 211–212
    vomitoxin 212–213
    zearalenone 213

Na (naked neck) gene 16–17, 19–20, 36–37, 38
nephrotoxicity 208, 209, 210–211
nests 310–311
neurohypophyseal hormones 58–60
Newcastle disease 204
niacin 356
nicarbazine 151, 241–242
Nigeria 6–7
norepinephrine 62
normothermia 49–50
nutrition
    dietary electrolytes 139–140, 149–150
    fats 135–137, 230
    under heat stress (reviews) 133
    minerals 145–149, 232
    protein and amino acids *see* amino acids; proteins
    vitamins *see* vitamins
    see also feedstuffs and feeding

obesity 297, 307
ochratoxins 204, 209–210
oil, rice 165–166
oosporein 210–211
openings, side-wall 87–88
operant control: for selection of optimal environment 52–53
oxytocin 59
oyster shell 145, 146(tab), 153, 275
ozone 216–217

pad systems (for cooling)
    incoming air temperature 126(fig)
    installation 125–128
    size 124–125
Pakistan 8
palm kernel meal 179
palm oil 143
panting 55–56, 83
parasites 284–285
pathogenicity: increased by aflatoxins 203(tab)
peanut meal *see* groundnut meal
pearl millet 164
perches, water-cooled 313
pH, blood 55–56
phase feeding 276–277
Philippines 9, 180, 331, 332(tab)
phosphorus 147–148, 303, 356
photostimulation 300
phytic acid 175
pigeons 56, 64
pineal gland 62–63
pituitary
    hypothalamic-pituitary-adrenal axis 60–62
    posterior pituitary hormones 58–60
polishings, rice 164–166
potassium 148–149, 231–232
pressure, static *see* static pressure
production
    effects of heat stress and heat tolerance *see under* heat stress; tolerance: to heat

featherless broilers: advantages *see under* broilers
future development and constraints 9–11
  global 2(tab)
  by region 3(tab), 5(tab)
    Africa 6–7
    Asia 7
    developed *vs.* developing countries 3(tab), 4(tab)
    Indian subcontinent 8–9
    Latin America 7
    Middle East 7–8
    South-east Asia 9
progesterone 63
propionic acid 214–215
proteins
  dietary 33, 271–272
    for broilers 41, 238–239, 299–300, 301
    for geese 355, 357(tab), 358(tab)
    and heat increment 138
    from microbial fermentation 177
    for replacement pullets 264
    effect of temperature on carcass content 237
    heat shock *see* heat shock proteins
pullets, broiler
  body weight and photostimulation 300
  feeding *see under* feedstuffs and feeding
pullets, replacement
  acclimatization 267
  body weight 262–263
  feeding *see under* feedstuffs and feeding
  management 267–269
  recommendations 285–287
  water consumption 266, 267(tab)
pyridoxine 145

quail, Japanese 64, 151, 152
quality, meat
  effects of heat stress 31
  superiority of featherless birds 39–41

radiation 82
red jungle fowl 1

reflective insulation 97
reserpine 151–152
respiration 55–56
reteopthalmicum 54
riboflavin 302
rice
  bran and polishings 164–166
  kani (broken grains) 166
rigid board insulation 94
roofs 246
  insulation 94–97
  overhang 92, 93(fig)
  reflective coatings 97–98
  slope 93–94

safflower meal 175
salmonellosis 314–315
salseed 181
Saudi Arabia 7–8
Sc (scaleless) gene 37
selection
  for growth rate in hot conditions 34–35
  for heat tolerance 25
    experiments 21–23
    feasibility of developing commercial stocks 23–24
semen 83, 297, 302, 303(tab), 304, 305, 336, 362, 365
sensible heat
  definition 81–82
  loss 50(fig), 51, 54–55
    effect of wind speed 84(fig)
sesame meal 175–176
shells, egg *see under* eggs
Shorea robusta 181
side-wall inlet ventilation 106–107
side-wall openings 87–88
Singapore 9, 19, 231, 251
single-cell protein 177
skeletal muscle 250
slow feathering *(K)* gene 19, 20
sodium bicarbonate 140, 151, 232, 282
sodium chloride 139, 232, 279, 297
sorghum 166–168
South Africa 6
South-east Asia 9
soyabean meal 175, 214
sperm 297, 302, 303(tab), 304–305
spray polyurethane insulation 97

sprinklers
    for evaporative cooling 124
    roof-based 98
Sri Lanka 8, 180
static pressure 82
sterculic acid 171
sterigmatocystin 207
stress: caused by heat *see* heat stress
sunflower seeds 173–175
sunlight 91–92

T-2 toxin 211–212
tannins 167–168, 181
temperature, body
    effect of heat stress 49–51, 81
    upper and lower lethal temperature 83
temperature, thermoneutral
    definition 82–83
    range 81
terramycin Q 303–304
Thailand 4(tab), 9, 180, 331, 334
thermoregulatory effort, zone of least (ZLTE) 50
thiamine 144–145
thyroid hormones 59, 63–64
tissues: net energy 49
tolerance: to heat
    in the context of other traits 14
    in dewinged broilers 247
    enhanced by early feed restriction 240–241
    feasibility of developing commercial stocks 23–24
    of featherless broilers 37–39
    genes involved
      dwarf *(dw)* 18–19, 21
      frizzle *(F)* 17–18, 19, 20, 21
      interactions 19–20
      naked neck *(Na)* 16–17, 19–20, 36–37, 38
      other potential genes 19
      scaleless *(Sc)* 37
      slow feathering *(K)* 19, 20
      use in development of resistant strains 20–21
    genetic association with growth rate 32–34
    heritability 25
    importance in production 25
    population differences 15–16
    role of heat shock proteins 65, 67, 68–69
    selection experiments 21–23
    use of indigenous species 26
training
    as constraint to development of poultry industry 11
transcription
    of heat shock protein genes 66–67
trichothecenes 211, 212
triglycerides 59
trimming, beak 243, 360
tryptophan 175, 231, 347, 349, 355
tunnel ventilation *see under* ventilation
turkeys 200

vaccination 242, 313
    effect of mycotoxins 204, 205
velvet beans 184
ventilation
    evaporative cooling
      benefits 121–122
      fogging systems 122–124
      pad systems 124–128
      sprinkling systems 124
    natural
      circulation fans 98–100
      house dimensions 86–87
      house orientation 91–92
      house spacing 88–89
      ridge openings 89, 91
      roof coatings 97–98
      roof insulation 94–97
      roof overhang 92, 93(fig)
      roof slope 93–94
      roof sprinkling 98
      side-wall openings 87–88
      surrounding vegetation 89, 90(fig)
      wall insulation 97
    power
      air exchange 101–104
      air inlet characteristics 104–106
      air inlet types 106–108
      air movement with inlet ventilation 108–109
      exhaust fans 117–121
      house construction 101
      positive- *vs.* negative-pressure systems 100–101

tunnel
   air flow pattern 113–114
   air velocity distribution 110–113
   air velocity recommendations 109(tab)
   benefits 83–86
   fan capacity 109–110
   fan placement 115
   inlets 115–117
vessels, blood *see* cardiovascular system
vitamins
   effects of mycotoxins 205–206
   niacin 356
   stability in premixes and feeds 233–235
   vitamin A 143–144, 302
   vitamin B6 145
   vitamin C 142–143, 275–276, 302
   vitamin D3 144
   vitamin E 144, 302
Voandizeia subterrenea 183
vomitoxin 212–213

waste, dried poultry (feedstuff) 181–182
water
   carbonated water prevents eggshell thinning and bone breakage 146–147
   consumption 53–54, 247–249, 266, 267(tab), 311–312
   and medication 313
   deprivation 312
   effect on hormonal responses 58, 60
   effect on serum electrolytes 57
   for goslings 359–360
   hatcheries 320–321
   for laying hens 278–281
waterfowl
   biotechnology 339–340
   breeds 333–338
   egg production 332(tab)
   genetics 332
   goose meat production 332
   see also ducks; geese
weight: replacement pullets 262–263
White Leghorn 15
whitewash 246
wind speed
   effect on sensible heat loss 84(fig)
   impact on broiler performance 84–86
World's Poultry Science Association 2

yeasts
   for dietary protein 177
yield, meat: superiority of featherless birds 39–41

zearalenone 213
zeolites 215, 216
Zimbabwe 6
zinc 149
   correlation with feed aflatoxin content 206
ZLTE (zone of least thermoregulatory effort) 50
ZMM (zone of miminal metabolism) 50–51